ELECTRONICS
FOR TODAY AND TOMORROW

Tom Duncan

JOHN MURRAY

Also by Tom Duncan

Physics for Today and Tomorrow
Exploring Physics, Books 1 to 5
Electronics and Nuclear Physics
Advanced Physics: Materials and Mechanics
Advanced Physics: Fields, Waves and Atoms
Physics: A Textbook for Advanced Level Students
Adventures with Physics
Adventures with Electronics
Adventures with Microelectronics
Adventures with Digital Electronics
Success in Electronics
Science for Today and Tomorrow
(with M A Atherton and D G Mackean)

First published 1985
by John Murray (Publishers) Ltd
50 Albemarle Street London W1X 4BD

Typeset by Fakenham Photosetting Ltd,
Fakenham, Norfolk
Printed and bound in Great Britain by
The Alden Press, Oxford

British Library Cataloguing in Publication Data

Duncan, Tom
Electronics for today and tomorrow.
1. Electronic apparatus and appliances
I. Title
621.381 TK7870

ISBN 0-7195-4183-2

Preface

One- or two-year courses leading to an AO or O level examination in *Electricity and Electronics* are increasing in popularity for A level physics students and others. While syllabuses differ in content, this book meets most of the requirements of those GCE boards who offer the subject.

The organization and treatment of the 100 or so topics is such that the order in which they are studied may be varied, within limits, according to individual situations and preferences.

Thanks are due to the examination bodies listed below for permission to use questions from recent papers (answers given being the sole responsibility of the author), to my wife for typing the manuscript and to my daughter, Dr Heather Kennett, for help in various ways.

Oxford and Cambridge Schools Examination Board (*O. and C.*)
The Associated Examining Board (*A.E.B.*)
The University of London University Entrance and School Examinations Council (*L.*)
University of Cambridge Local Examinations Syndicate (*C.*)
University of Oxford Delegacy of Local Examinations (*O.L.E.*)

T.D.

Acknowledgements

Thanks are due to the following for permission to reproduce copyright photographs:
Figs. 9.1a(i) (ii) (iv), 9.3, 9.4, 10.2, 10.5b, 14.4a, b, 15.2, 15.3, 15.4, 17.4, 17.6, 18.1, 19.1, 19.4, 28.7, 30.1, 36.7, 39.3, 39.5, 40.6, 42.5, 45.1, 45.3, 83.2, A3, A5a, A6, RS Components Ltd; 9.1a(iii), A4, Gaye Allen; 10.5a, 14.4c, 18.3, Maplin Electronics Ltd; 21.3, by courtesy of Sinclair Research; 34.1, Ferranti plc; 34.4, 34.5a, Mullard Ltd; 34.5b, Department of Industry; 34.6, IBM United Kingdom Ltd; 46.1, 47.1, 102.5, A2, Unilab Ltd; 82.6, EMI Records Ltd; 82.9, Hitachi (UK) Ltd; 84.1, Sony (UK) Ltd; 90.2, 90.3, 93.3, 94.2, 94.3, 94.4, 94.10, 95.3, 95.6, 95.7, 95.9, 96.1, 96.2, 98.1a, c, British Telecommunications plc; 94.9, 95.4, STC Components Ltd; 95.1, BICC plc; 98.1b, Austin Rover Group Ltd; 98.1d, The Singer Company (UK) Ltd; 99.1, 3M United Kingdom plc; 102.1, Texas Instruments Incorporated; A1, A M Lock Ltd.
Thanks are also due to Pitman Publishing Ltd, London, for permission to base Fig. 93.1 on an illustration in P J Povey, *The Telephone and the Exchange* (Pitman, 1979).

Contents

Analogue electronics

Digital electronics

Information and electronics

Audio systems

Radio and television

Telephone system

Computers and microprocessors

Basic electricity

1 Electric current

Atoms and electric charges

An atom consists of a nucleus containing *protons*, which have a positive (+) electric charge, surrounded by an equal number of *electrons* with a negative (−) charge. The charge on a proton is the same size as that on an electron, making the whole atom electrically neutral. The nucleus also contains uncharged *neutrons*. Fig. 1.1 is a simplified 'picture' of an atom.

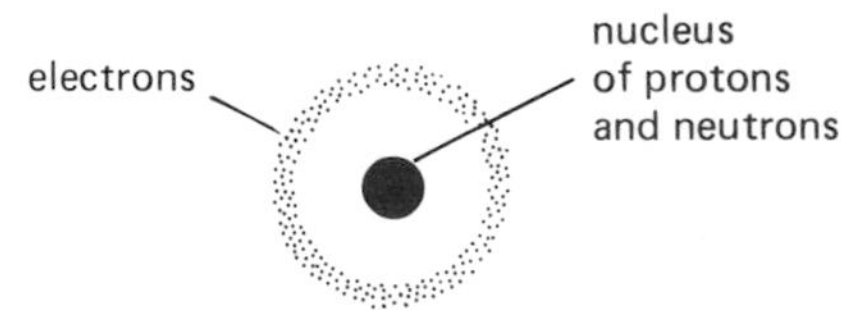

Fig. 1.1

An atom can lose one or more electrons. If it does it becomes positively charged, because it then has more protons than electrons, and is called a *positive ion*. If it gains one or more electrons it becomes a *negative ion*, Fig. 1.2.

Fig. 1.2

Electric charges exert forces on one another and

like charges repel, opposite charges attract.

Electric current: what is it?

An electric current is produced when electric charges (electrons or ions) move in a definite direction.

In *metals*, the outer electrons are held loosely by their atoms and are free to move around the fixed positive metal ions making up the 'body' of the metal. This free electron motion is normally haphazard with as many electrons moving in one direction as in the opposite direction at any time, Fig. 1.3a. Overall there is no net flow of charge in any direction, i.e. no current.

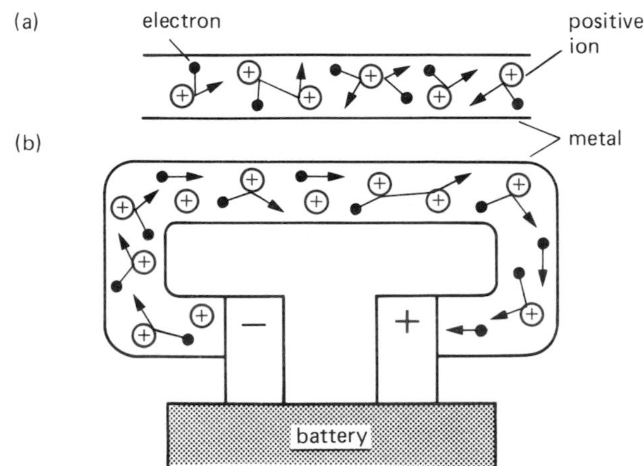

Fig. 1.3

When the metal is part of a circuit connected to a battery, the battery acts as an electron pump. It forces the free electrons to drift through the metal, in the direction from its negative (−) terminal towards its positive (+) terminal and then, in effect, through the battery itself, Fig. 1.3b. This flow of electrons in one direction is an electric current which can reveal itself by making the metal warmer and by deflecting a nearby magnetic compass.

The drift speed of the free electrons through the metal may, surprisingly, be less than 1 mm per second due to their frequent collisions with the fixed positive metal ions. But they *all* start drifting in the same direction as soon as the battery is connected, just as all the links in a bicycle chain do at the instant the pedals are pushed. If the circuit is broken the current stops but the haphazard motion of the free electrons goes on.

In some *liquids* and *gases*, under certain conditions, there are free positive and negative ions which can move when a battery is connected. The current produced consists of positive ions moving through the liquid or gas towards the − terminal of the battery and negative ions moving towards the + terminal, as shown in Fig. 1.4.

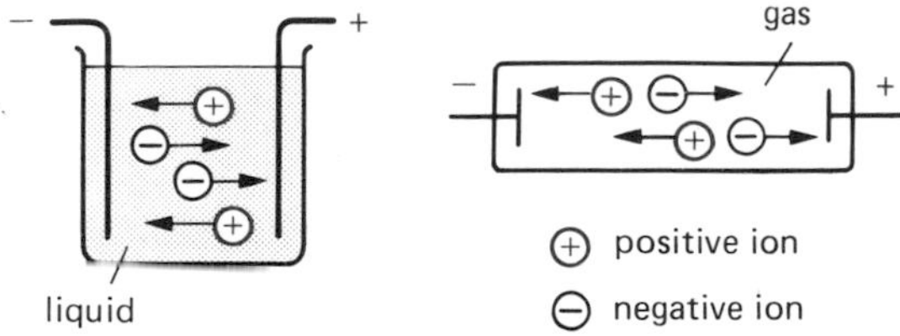

Fig. 1.4

Conductors and insulators

Electrical conductors are materials containing electric charges that are free to move. They allow currents to pass through them easily. The best conductors are the metals silver, copper and gold.

In electrical insulators such as polythene, Perspex and PVC (polyvinyl chloride) all electrons are bound firmly to their atoms, making charge flow difficult. They can however be charged by rubbing. For example, when polythene is rubbed with a cloth, electrons are transferred from the cloth to the polythene. The cloth is left with a positive charge and the polythene becomes negatively charged, Fig. 1.5. The charges produced cannot move on the insulator, i.e. they are static electric charges.

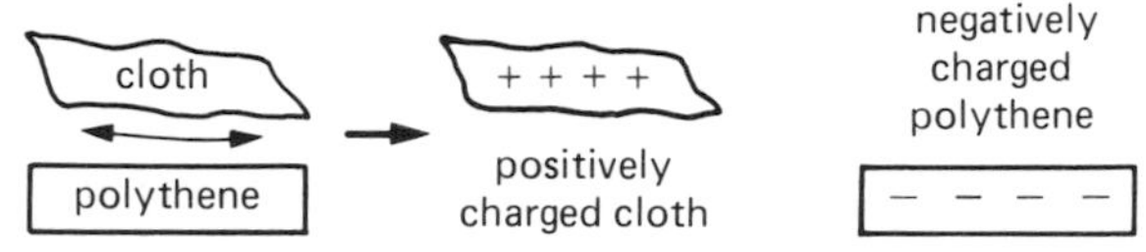

Fig. 1.5

Direction of current

Before the electron was discovered it was thought that a current consisted of positive charges moving from the battery's + terminal round the circuit to its − terminal. We now know that this is partly true for conduction by liquids (electrolytes) and gases; it is not for metals.

The original choice has been kept because it does not matter which direction is chosen provided we always keep to the same one. Also, the laws of electricity were drawn up assuming it was true.

The conventional direction of current, called the *conventional current*, is the direction in which positive charges would flow. It is opposite to the direction of electron flow, Fig. 1.6.

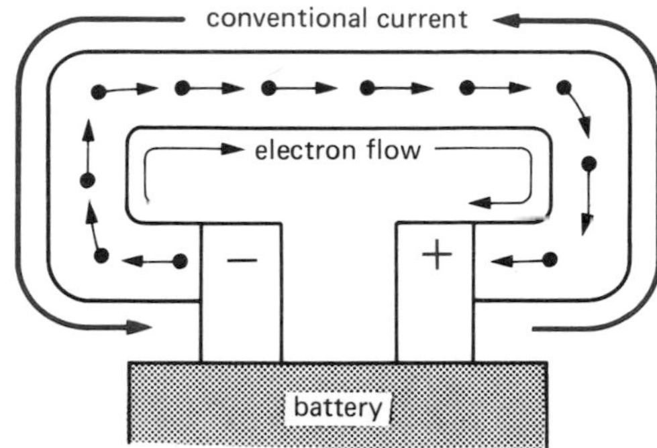

Fig. 1.6

Units of charge and current

A huge number of charges drift past each point in a circuit per second. The quantity of charge carried by about six million million million (6×10^{18}) electrons is called 1 *coulomb* (C).

When 1 coulomb passes each point in a circuit every second, the current is 1 *ampere* (A). That is,

1 ampere = 1 coulomb per second (1 A = 1 C/s)

If 2 C pass in 1 s, the current is 2 C/s = 2 A. If 6 C pass in 2 s, the current is 6 C/2s = 3 A. In general if Q coulomb pass in t seconds, the current I in amperes is given by

$$I = \frac{Q}{t} \text{ or } Q = It$$

Two smaller units of current are:

1 milliampere (mA) = 1/1000 A = $1/10^3$ A = 10^{-3} A
1 microampere (μA) = 1/1000 000 A = $1/10^6$ A = 10^{-6} A

Electric current is measured by an *ammeter*.

Circuits and diagrams

Currents require complete conducting paths (circuits). Wires of copper are used to connect batteries, lamps, switches, etc. in a circuit. If the wires are covered with insulation, e.g. a plastic such as PVC, the insulation must be removed from the ends before connecting up. Symbols (signs) used in circuit diagrams are shown in Fig. 1.7.

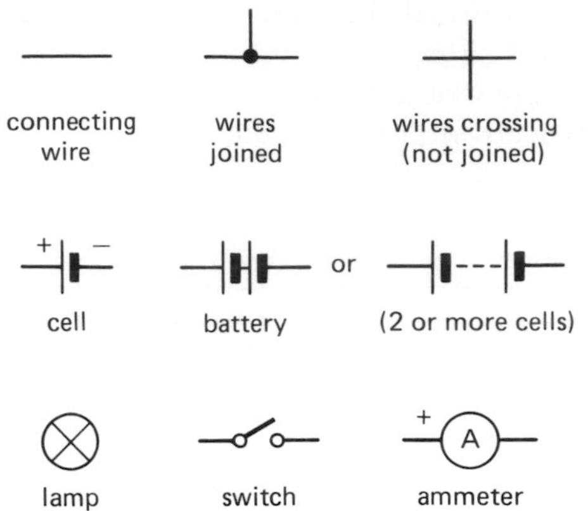

Fig. 1.7

Current is not used up in a circuit. In Fig. 1.8a the two lamps are in series and the readings on ammeters A_1, A_2, A_3 will all be the same. In Fig. 1.8b the lamps are in parallel and the total current equals the sum of the branch currents, i.e. $I = I_1 + I_2$.

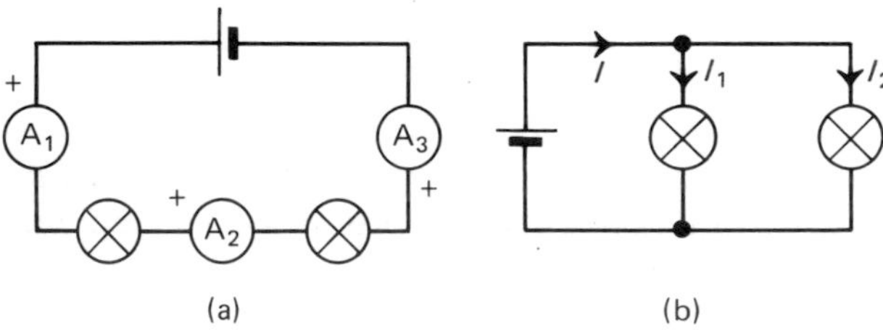

Fig. 1.8

Questions

1. Why are (a) metals, (b) some liquids and gases, conductors?

2. If the current through a floodlamp is 5 A, what charge passes in (i) 1 s, (ii) 10 s, (iii) 5 minutes?

3. What is the current in a circuit if the charge passing each point is (a) 10 C in 2 s, (b) 20 C in 40 s, (c) 240 C in 2 minutes?

4. (a) Express the following in mA: (i) 1 A, (ii) 0.5 A, (iii) 0.02 A.
(b) Express the following in μA: (i) 2 mA, (ii) 0.4 mA, (iii) 0.005 mA.

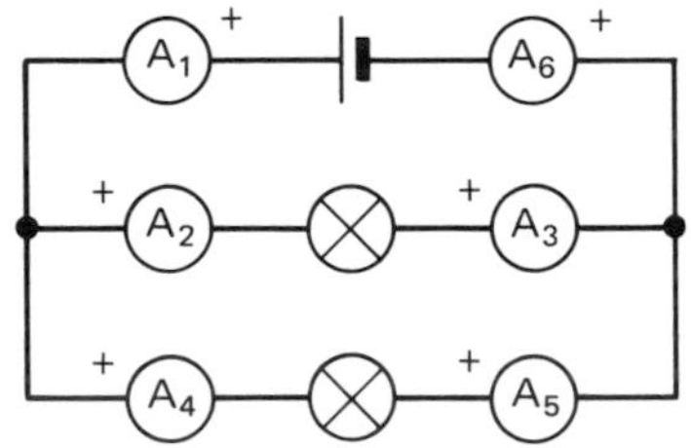

Fig. 1.9

5. All lamps in Fig. 1.9 are the same. If A_1 reads 1 A, what do A_2, A_3, A_4, A_5 and A_6 read?

2 E.M.F., P.D. and voltage

Electromotive force of a battery

The battery in Fig. 2.1 supplies the electrical force and energy which drives the free electrons round the circuit as a current. As they drift along they give up most of their energy as heat and light in the lamp. It is useful to *imagine* that as each coulomb leaves the battery, it receives a fixed amount of electrical energy which depends on the battery.

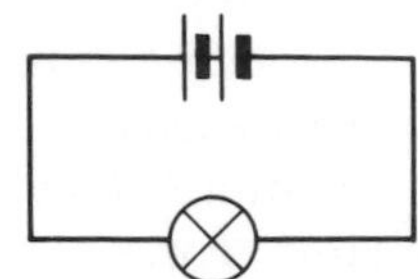

Fig. 2.1

The *electromotive force* (e.m.f.) E of a battery is defined to be 1 *volt* if it *gives* 1 joule of electrical energy to each coulomb of charge passing through it. That is

1 volt = 1 joule per coulomb (1 V = 1 J/C)

A 6 V battery gives 6 J of energy to each coulomb passing through it.

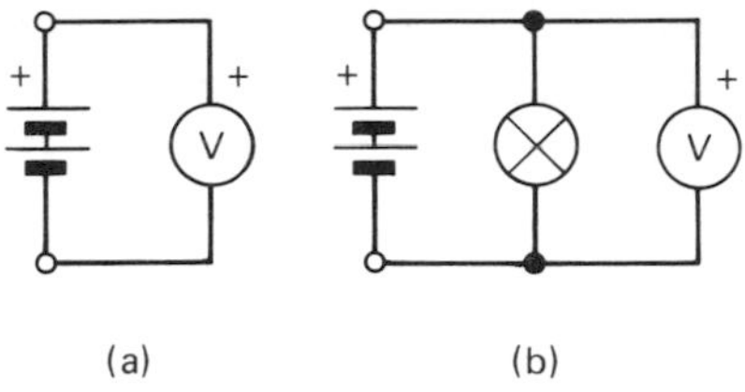

Fig. 2.2

The e.m.f. of a battery can be measured by connecting a *voltmeter* across it, Fig. 2.2a.

Potential difference across a device

The lamp in the circuit of Fig. 2.2b changes most of the electrical energy carried by the free electrons into heat and light. A negligible amount is lost as heat in the copper connecting wires.

The *potential difference* (p.d.) across a device (e.g. a lamp) in a circuit is 1 volt if it *changes* 1 joule of electrical energy into other forms of energy (e.g. heat and light) when 1 coulomb passes through it.

The p.d. across a device is 2 V if it changes 2 J when 1 C passes through it. In general, if W joule is changed when Q coulomb pass, the p.d. V in volts across the device is

$$V = \frac{W}{Q}$$

A voltmeter is also used to measure p.d., being connected *across* the device as in Fig. 2.2b.

E.M.Fs and p.ds are usually called *voltages*, since both are measured in volts.

The flow of electric charge in a circuit can be compared with the flow of water in a pipe. A pressure difference is required to make water flow. To move electric charge we consider that a p.d. is needed and whenever there is current between two points in a circuit there must be a p.d. across them.

Terminal p.d. of a battery

When a battery drives current round a circuit some of the electrical energy carried by the charge is needed to get the current through the battery itself. This energy is changed to heat in the battery. There is therefore less energy available to drive each coulomb round the rest of the circuit.

The *terminal p.d.* V of a battery when it is driving current, i.e. on closed circuit, is less than the e.m.f. E of the battery because some energy and volts v are 'lost'. On open circuit when the battery is not driving current, the terminal p.d. equals the e.m.f. We can say for each coulomb:

energy supplied by battery = energy changed in devices + energy 'lost' inside battery

or, e.m.f. = terminal p.d. + 'lost' volts

i.e. $E = V + v$

A voltmeter connected across a battery on open circuit only measures the e.m.f. if it does not require the battery to drive current through it.

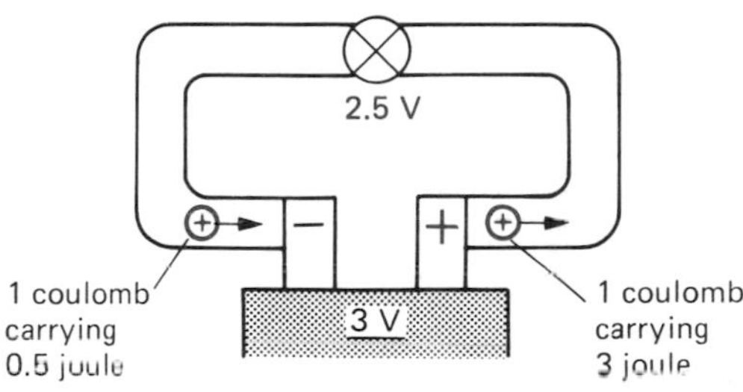

Fig. 2.3

In Fig. 2.3 if the e.m.f. of the battery is 3 V and considering conventional current, we can think of each coulomb as

(i) leaving the + terminal of the battery with 3 J of electrical energy,

(ii) changing 2.5 J of electrical energy into heat and light in the lamp,

(iii) arriving at the − terminal of the battery with 0.5 J of electrical energy which it requires to get through the battery, and

(iv) picking up another 3 J of electrical energy as it leaves the + terminal.

In this example, the useful (terminal) p.d. available for driving current through the lamp is 2.5 V and the 'lost' volts is 0.5 V, making a total of 3 V, which is the e.m.f. of the battery.

In general

e.m.f. = sum of p.ds across devices + 'lost' volts

Cells and batteries

Greater e.m.fs are obtained when cells are joined in series, i.e. + of one to − of next, to give a battery. In Fig. 2.4a the two 1.5 V cells have a total e.m.f. of 3 V.

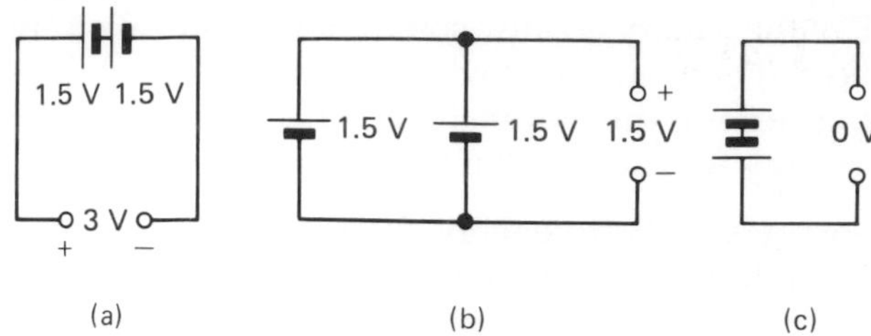

Fig. 2.4

If two 1.5 V cells are connected in parallel, Fig. 2.4b, the e.m.f. is still 1.5 V but they behave like one larger cell with more energy and last longer.

The two 1.5 V cells in Fig. 2.4c are in opposition and their combined e.m.f. is zero.

Potential at a point

Although it is usually the p.d. between two points in a circuit we have to consider, there are times when it is a help to deal with the *potential at a point*. To do this we choose one point in the circuit as having zero potential, i.e. 0 V. The potentials of all other points are stated with reference to it; the potential at any point is then the p.d. between the point and 0 V.

For example, if we take point C in Fig. 2.5a (in effect the − terminal of the battery) as our zero, the potential at B is +2.5 V and at A (the + terminal of the battery) it is +6 V. Alternatively, taking A as being at 0 V, then B is at −3.5 V and C at −6 V.

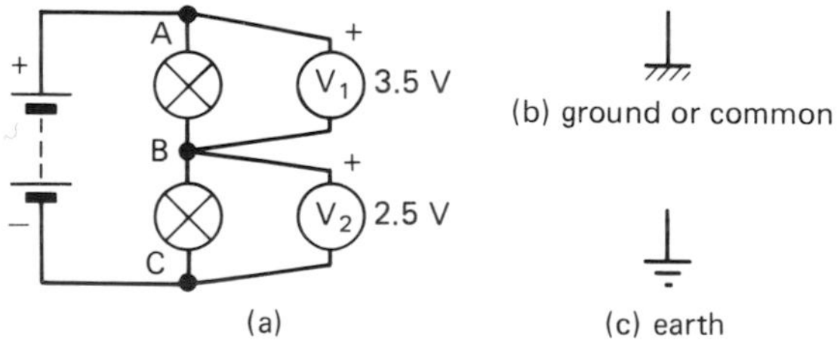

(b) ground or common

(c) earth

Fig. 2.5

In electronic circuits the point chosen as being at 0 V is called *ground* or *common*, or *'earth'* if it is connected to earth (e.g. via the earthing pin on a 3-pin plug). The signs used are given in Fig. 2.5b,c.

Questions

1. If the e.m.f. of a battery is 12 V, how many joules of electrical energy are given to a charge passing through of (a) 1 C, (b) 3 C?

2. The p.d. across a lamp is 6 V. What does this mean?

3. In the circuits of Fig. 2.6, the voltmeter V requires almost zero current to take a reading. What is (a) the e.m.f. of the cell, (b) its terminal p.d. when driving current through the lamp L and (c) the 'lost' volts?

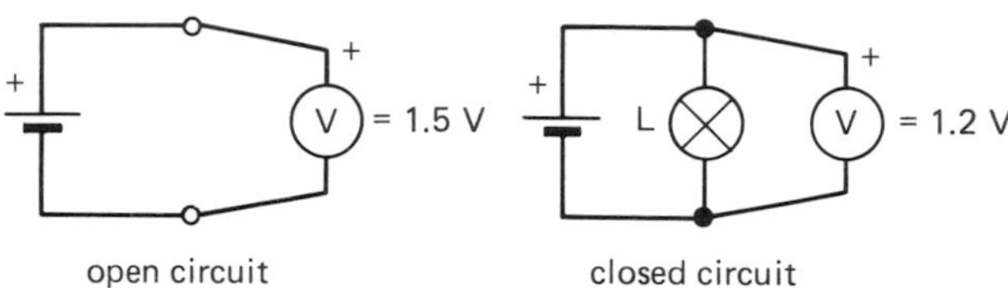

Fig. 2.6

4. What are the e.m.fs of the batteries of 1.5 V cells connected as in Fig. 2.7a,b?

Fig. 2.7

5. Three voltmeters V, V_1 and V_2 are connected as in Fig. 2.8. Copy and complete the table of voltmeter readings shown below which were obtained with three different batteries.

V	V_1	V_2
	12	6
6	4	
12		4

Fig. 2.8

6. In Fig. 2.9, L_1 and L_2 are identical lamps. Assuming the 'lost' volts in the 6 V battery is zero, what are the potentials at points A, B and C if 'ground' (0 V) is taken as (i) C, (ii) A, (iii) B?

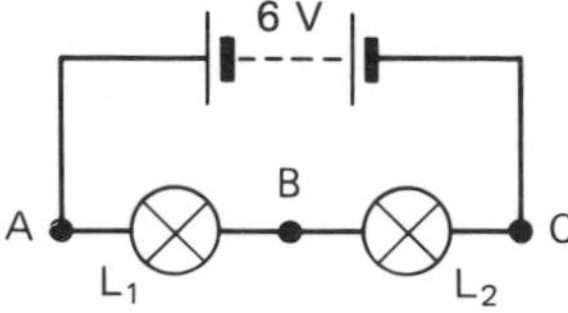

Fig. 2.9

3 Resistance and Ohm's law

Resistance

Conductors oppose current to a certain extent, some more than others. The *resistance* R of a conductor in *ohms* (Ω: pronounced 'omega') is defined by

$$R = \frac{V}{I}$$

where V is the p.d. across the conductor in *volts* and I is the resulting current in *amperes*. When $V = 6\,\text{V}$ and $I = 2\,\text{A}$, $R = 6/2 = 3\,\Omega$: but if $I = 1\,\text{A}$ then $R = 6/1 = 6\,\Omega$.

Larger units of resistance are the *kilohm* ($1\,\text{k}\Omega = 1000\,\Omega = 10^3\,\Omega$) and the *megohm* ($1\,\text{M}\Omega = 1000\,\text{k}\Omega = 10^6\,\Omega$). In electronics I is often in mA (or μA) and V in volts, this gives R in kΩ (or MΩ).

Conductors especially made to have resistance are called *resistors* (symbol ─▭─).

The equation $R = V/I$ can also be written $V = IR$ and $I = V/R$. All are used in calculations and the triangle in Fig. 3.1 is a memory aid. If you cover the quantity you want with your finger, e.g. I, it equals what you still see, i.e. V/R.

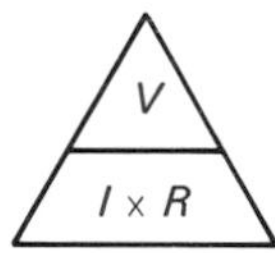

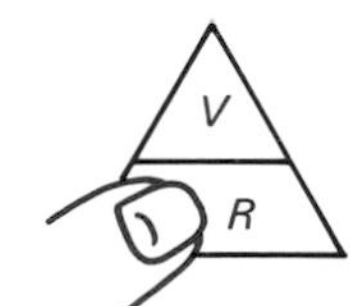

Fig. 3.1

If there is current through a resistor, the *potential at one end must be greater than at the other*, i.e. a p.d. exists across it. In Fig. 3.2 the potential at X is 9 V if S is open and 0 V if it is closed. This fact is the basis of electronic switching circuits (p. 66).

Fig. 3.2

Ohm's law

For metals, carbon and some alloys, V/I is constant whatever the value of V, if their temperature does not change. Since $R = V/I$ it follows that the resistance of such conductors is constant for different p.ds. We can write

$$\frac{V}{I} = \text{constant or } I \propto V$$

This is Ohm's law and states that *the current through a conductor is directly proportional to the p.d. across it if the temperature is constant*. Doubling or trebling V, therefore doubles or trebles I.

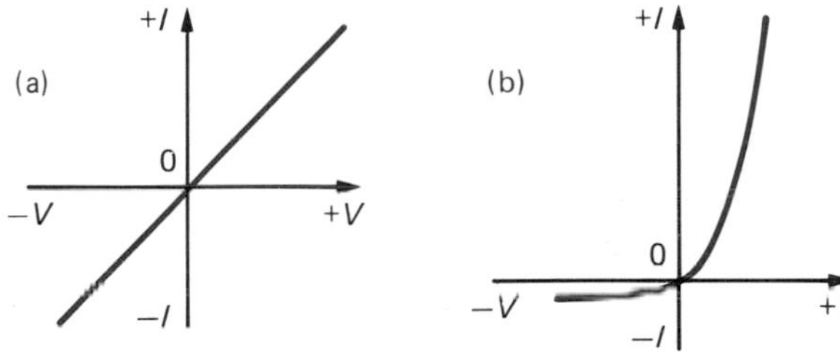

Fig. 3.3

Ohmic or *linear* conductors obey Ohm's law and a graph of I against V, called a *characteristic curve*, is a straight line through the origin, Fig. 3.3a. The resistance of a non-ohmic or non-linear conductor varies with the p.d. and its I–V graph is curved. The one in Fig. 3.3b is for a semiconductor diode (p. 58).

Resistor networks

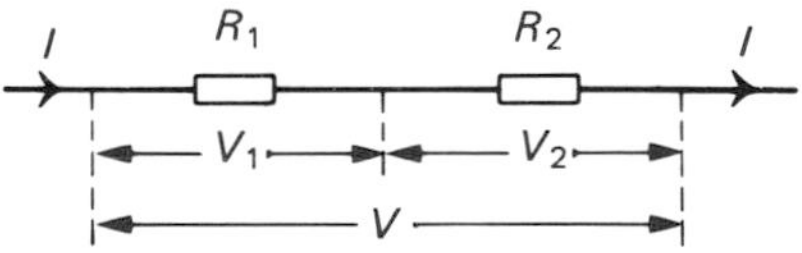

Fig. 3.4

(a) Series, Fig. 3.4. Note two points:

(i) I is the same in R_1 and R_2, and

(ii) $V = V_1 + V_2$ where $V_1 = IR_1$ and $V_2 = IR_2$, i.e. $V_1/V_2 = R_1/R_2$.

It can be shown that the combined resistance R is

$$R = R_1 + R_2$$

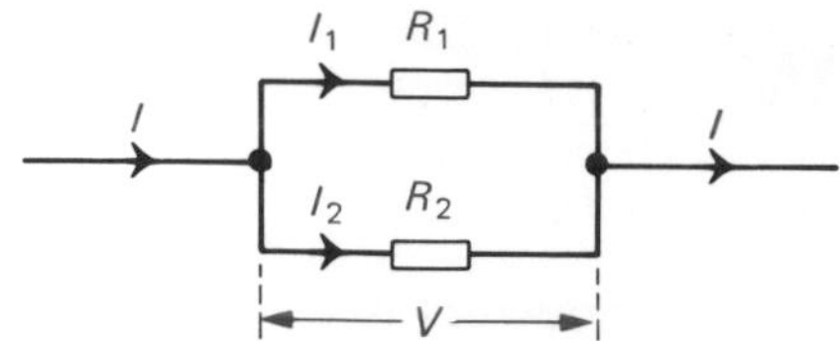

Fig. 3.5

(b) Parallel, Fig. 3.5. Note two points:

(i) $I = I_1 + I_2$ and $I_1 > I_2$ if $R_1 < R_2$ and vice versa, and

(ii) V is the same across R_1 and R_2, i.e. $V = I_1R_1 = I_2R_2$.

It can be shown that if the combined resistance is R then

$$\frac{1}{R} = \frac{1}{R_1} + \frac{1}{R_2} \quad \text{or} \quad R = \frac{R_1 \times R_2}{R_1 + R_2}$$

If $R_1 = R_2$ then $R = R_1^2/2R_1 = R_1/2$.

(c) Example. For the network in Fig. 3.6 calculate **(i)** the combined resistance R_p of 3Ω and 6Ω in parallel, **(ii)** the total resistance R of the network, **(iii)** I, **(iv)** V_p and V_3, and **(v)** I_1 and I_2.

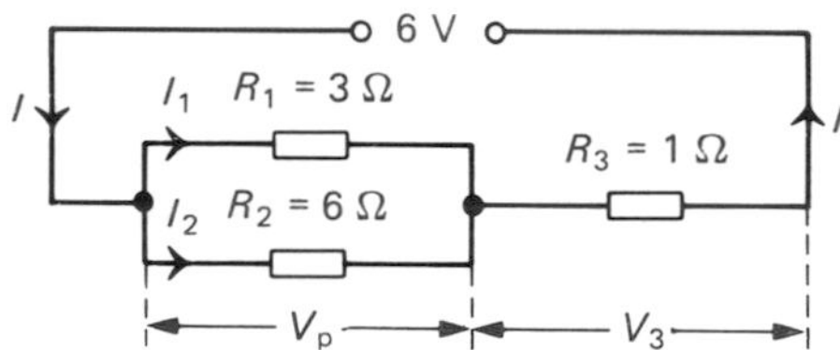

Fig. 3.6

(i) $\frac{1}{R_p} = \frac{1}{3} + \frac{1}{6} = \frac{2}{6} + \frac{1}{6} = \frac{3}{6} \therefore R_p = \frac{6}{3} = 2\Omega$

Note that R_p is less than either resistance.

(ii) $R = R_p + R_3 = 2 + 1 = 3\Omega$

(iii) The p.d. V across the network is 6 V

$\therefore I = \frac{V}{R} = \frac{6}{3} = 2\,\text{A}$

(iv) $V_p = IR_p = 2 \times 2 = 4\,\text{V}$

But $V = V_p + V_3 \quad \therefore V_3 = V - V_p = 6 - 4 = 2\,\text{V}$

(v) $I_1 = \frac{V_p}{R_1} = \frac{4}{3}\,\text{A}$

But $I = I_1 + I_2, \quad \therefore I_2 = I - I_1 = 2 - \frac{4}{3} = \frac{2}{3}\,\text{A}$

Resistance and temperature

If the temperature of a conductor rises, due for example to the heat produced by the current in it, its resistance increases if it is a metal but decreases if it is a semiconductor or carbon.

The change is given by the *temperature coefficient of resistance* α, defined by

$$\alpha = \frac{R_\theta - R_o}{R_o \times \theta}$$

where R_θ and R_o are the resistances at θ and 0°C. For metals α is positive, for semiconductors and carbon it is negative.

To find α for say copper, the resistance of a coil of thin copper wire is measured at different temperatures on a metre bridge (p. 15). A graph of resistance against temperature is plotted. Between 0 and 100°C it is a straight line, Fig. 3.7, and α is calculated from it.

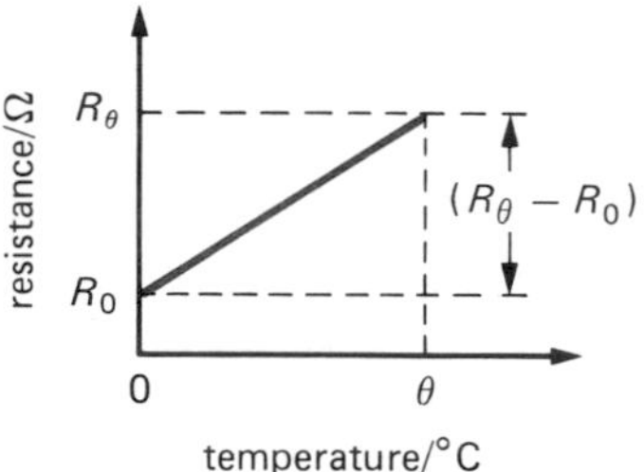

Fig. 3.7

The different behaviour of metals and semiconductors is due to the fact that in metals the positive ions vibrate more vigorously if the temperature rises and make it more difficult for the free electrons to flow. In semiconductors this also happens but more electrons are 'freed' as well, thereby, in effect, decreasing the resistance.

Resistivity

The resistance R of a conductor is directly proportional to its length l, i.e. $R \propto l$, and inversely proportional to its cross-section area A, i.e. $R \propto 1/A$. Combining these two statements we get $R \propto l/A$ or

$$R = \rho \frac{l}{A}$$

where ρ (rho) is a constant called the *resistivity* of the material. If $l = 1$ m and $A = 1\,\text{m}^2$ then $\rho = R$. That is, the resistivity equals the resistance of a 1 m length of cross-section area 1 m². Knowing ρ (in ohm metre) of a material, the resistance of different-sized samples can be calculated.

Variable resistors

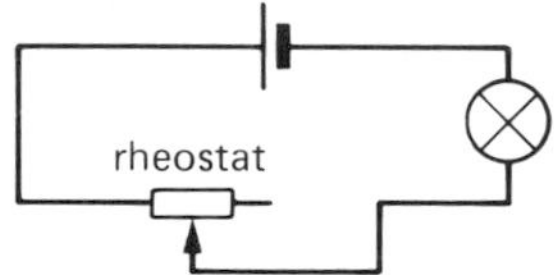

Fig. 3.8

A variable resistor has a connection at each end and one which can slide along it. When used as a *rheostat* to control the current in a circuit, connections are made to one end only and to the sliding contact, Fig. 3.8.

Questions

1. (a) If $V = 9$ V and $I = 5$ mA, find R.
(b) Find V when $I = 0.5$ mA and $R = 10$ kΩ.
(c) What is R if $I = 3\mu$A when $V = 6$ V?

2. For the network in Fig. 3.9 calculate the effective resistance, in ohms, between (a) A and B, (b) B and C, (c) A and C. *(A.E.B. 1982 Control Tech.)*

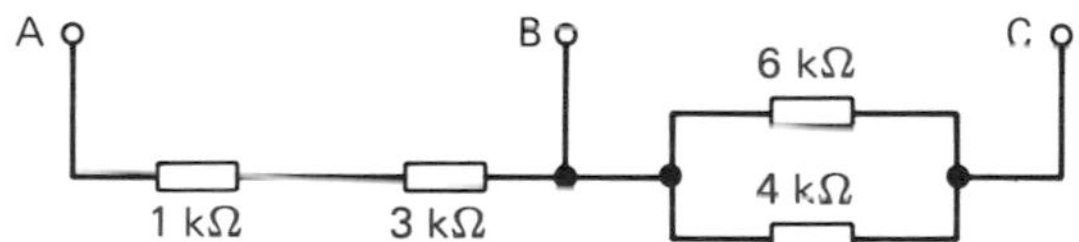

Fig. 3.9

3. (a) Calculate the total effective resistance between P and Q in the network of Fig. 3.10.
(b) 6 V is now applied between P and Q. Calculate the voltage between R and S. *(A.E.B.)*

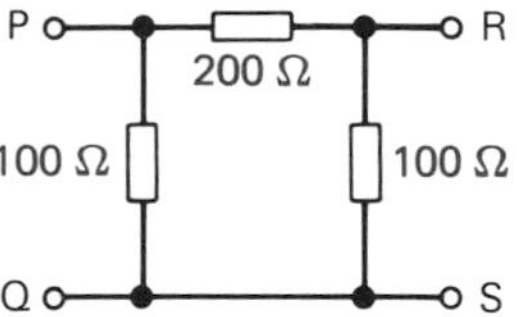

Fig. 3.10

4. Calculate I, I_1 and I_2 in the network of Fig. 3.11.

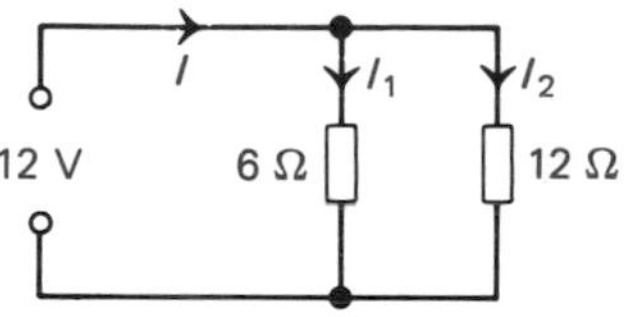

Fig. 3.11

5. In Fig. 3.12 what are the potentials at A and B when S is (a) open, (b) closed?

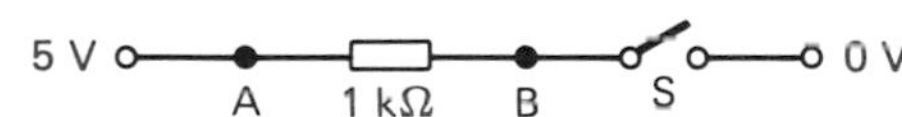

Fig. 3.12

4 Meters and measurement

Moving coil meter

Most ammeters and voltmeters are moving coil microammeters (galvanometers) adapted to measure a range of currents and voltages.

A moving coil meter consists of a coil of wire pivoted so that it can turn between the poles of a U-shaped magnet, Fig. 4.1. The coil has a pointer fixed to it which moves over a scale. When current passes through the coil, it becomes a magnet, turns and winds up a spiral hair spring until the force caused by the current balances the resisting force of

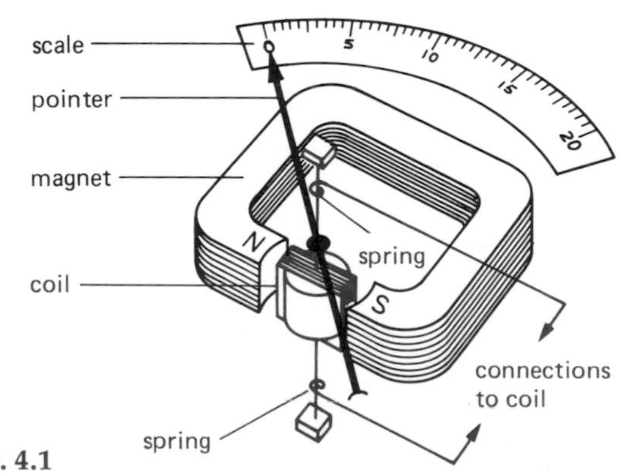

Fig. 4.1

the wound-up spring. The greater the current, the more does the coil turn and the larger the pointer deflection. The scale is linear, i.e. the divisions are evenly spaced.

Two important properties are the *resistance* of the coil and the *current to give a full scale deflection* (f.s.d.). If, for example, the coil resistance is 1000 Ω and the f.s.d. current is 100 μA (0.1 mA), a p.d. = 1000 × 0.1 = 100 mV will cause a f.s.d. The meter is both a microammeter reading up to 100 μA and a millivoltmeter reading to 100 mV.

Connecting a meter into a circuit to make a measurement should cause minimum disturbance to the conditions which existed before.

Ammeters and shunts

An ammeter is inserted in *series* in a circuit. It must have a *low* resistance compared with the rest of the circuit, otherwise it changes the current to be measured.

To convert the previous microammeter to an ammeter reading to 1 A, a low value resistor called a *shunt* is connected in *parallel* with the meter. It must have a value S which, when the current is 1 A, allows only 100 μA (0.0001 A) to go through the meter and provides a bypass itself for the remaining 0.9999 A.

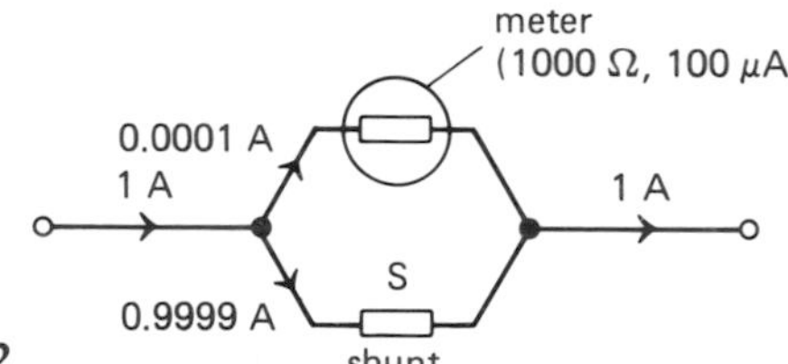

Fig. 4.2

From Fig. 4.2 we have

$$\text{p.d. across meter} = \text{p.d. across shunt}$$
$$\therefore 0.0001 \times 1000 = 0.9999 \times \text{S} \qquad (\text{from } V = IR)$$
$$\therefore \text{S} = \frac{0.0001 \times 1000}{0.9999} \approx 0.1\,\Omega$$

The resistance of the ammeter (meter + shunt) is less than 0.1 Ω.

Voltmeters and multipliers

A voltmeter is connected in *parallel* with the component, e.g. a resistor, across which the p.d. is to be measured. It should have a *high* resistance compared with the resistor (at least 10 times greater), otherwise the total resistance of the whole circuit is reduced by the 'loading' effect of the voltmeter and the required p.d. changes.

To convert the same microammeter to a voltmeter reading 0 – 1 V, a high value resistor, called a *multiplier*, is connected in *series* with the meter. It must have a value M which, when a p.d. of 1 V is applied across multiplier and meter together, restricts the current to 100 μA (0.0001 A) and a f.s.d. is recorded.

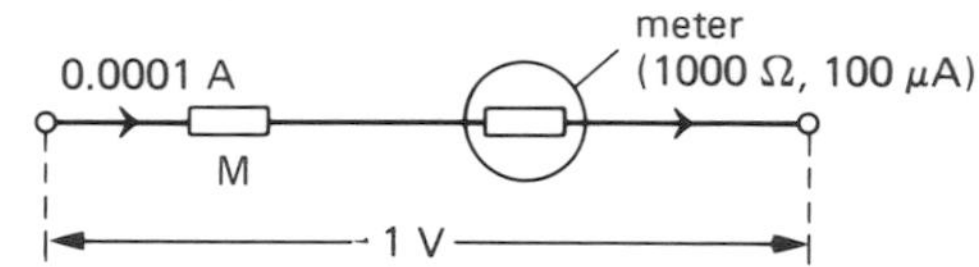

Fig. 4.3

From Fig. 4.3 we have

p.d. across multiplier and meter at f.s.d.

$$= 0.0001\,(\text{M} + 1000) = 1 \quad (\text{from } V = IR)$$
$$\therefore \text{M} + 1000 = 1/0.0001 = 10\,000$$
$$\therefore \text{M} = 9000\,\Omega$$

The resistance of the voltmeter (meter + multiplier) is 10 000 Ω.

Sensitivity of a voltmeter

(a) Definition. The *sensitivity* (or quality) of a voltmeter is stated in ohms per volt (Ω/V). The larger it is, the smaller is the current taken and the less is the disturbance to the circuit. It is calculated from

$$\text{sensitivity} = \frac{\text{resistance of meter + multiplier}}{\text{p.d required to give f.s.d.}}$$

For example, for the voltmeter above

sensitivity = 10 000 Ω/1 V = 10 000 Ω/V.

To extend its range to 10 V, a multiplier of 99 000 Ω would be required to make the resistance of the voltmeter (meter + multiplier) 100 000 Ω and so keep the full scale current at 100 μA. The sensitivity is still 10 000 Ω/V. To read up to 100 V this voltmeter would need to have a resistance of 1 MΩ.

(b) Example. In Fig. 4.4, V_1 and V_2 are two voltmeters of sensitivity 10 kΩ/V used on their 10 V range. The resistance of each is therefore 100 kΩ.

V_1 gives a fairly accurate reading of the p.d. (4.5 V) which existed across R_2 (10 kΩ) before V_1 was connected; its resistance is 10 times greater than R_2.

V_2 gives a misleading reading of the p.d. across R_4 (100 kΩ) because its resistance is the same as R_4. The

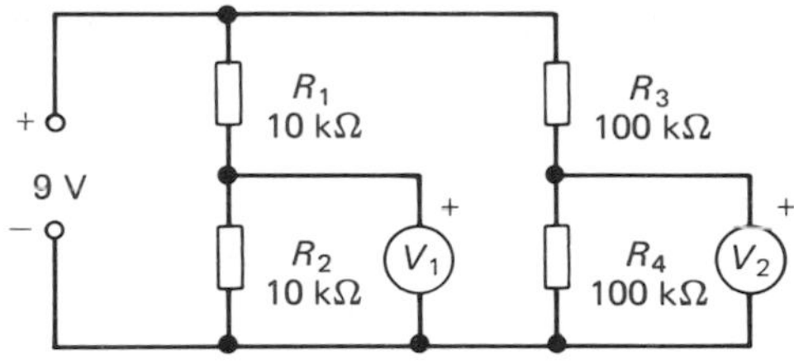

Fig. 4.4

combined resistance of R_4 and V_2 in parallel is 50 kΩ. The loading effect of V_2 is large and it records the p.d. across R_4 as 3 V instead of 4.5 V.

Electronic voltmeters (p. 97) have much higher resistances, e.g. 10 MΩ, than moving coil types.

Measurement of resistance

(a) Ammeter-voltmeter method. The resistance R of a conductor can be found from $R = V/I$ using the circuit of Fig. 4.5. A gives the sum of the currents in

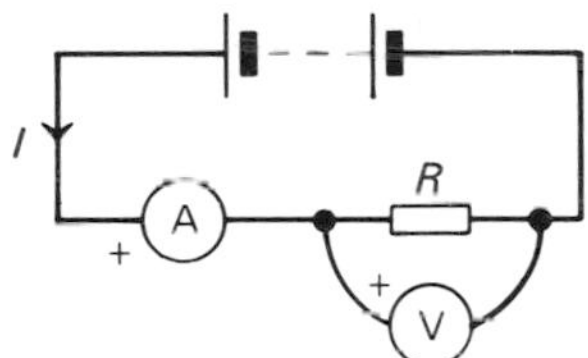

Fig. 4.5

R and V, which can be taken as the current I through R if the resistance of V is much greater than that of R. If not, V is connected across A and R in series so that A records the exact current I in R and the reading on V nearly equals the p.d. V across R if the resistance of A is much lower than R.

(b) Wheatstone bridge method. This is a better method because it is a 'null' method which does not depend on the accuracy of a meter.

Four resistors P, Q, R, S are joined as in Fig. 4.6. If R is the unknown resistor, S must be known as must P and Q or the ratio P/Q. One or more of P, Q and S are adjusted until there is no deflection on the galvanometer G, i.e. the bridge is balanced. There is

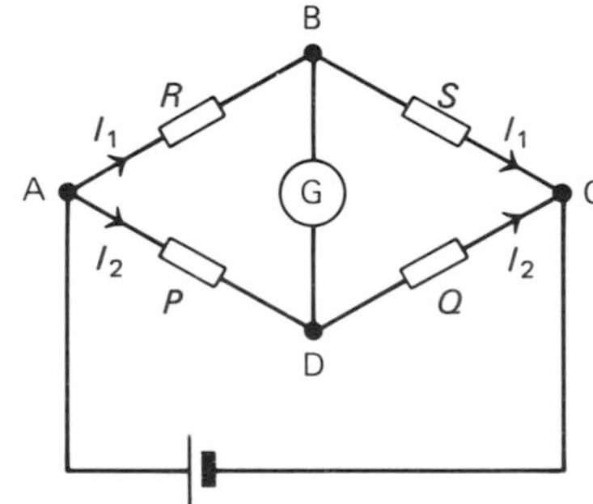

Fig. 4.6

then no current through G and the p.d. across BD must be zero.

$$\therefore \text{p.d. across AB} = \text{p.d. across AD}$$

$$\therefore I_1 \times R = I_2 \times P \qquad (1)$$

Also, p.d. across BC = p.d. across DC

But, current through R = current through S = I_1

and, current through P = current through Q = I_2

$$\therefore I_1 \times S = I_2 \times Q \qquad (2)$$

Dividing (1) by (2)

$$\frac{R}{S} = \frac{P}{Q} \text{ or } R = S \times \frac{P}{Q}$$

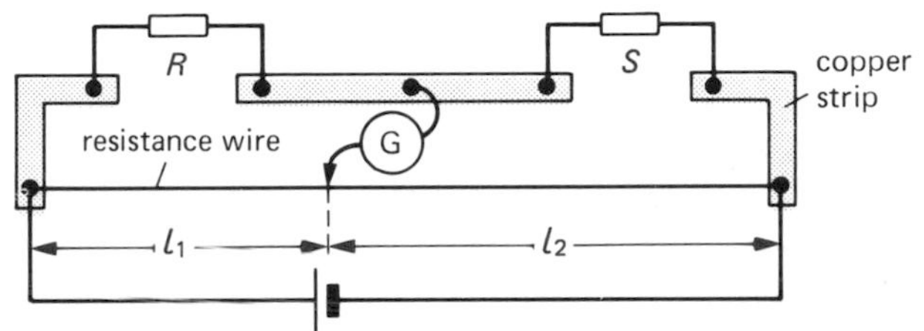

Fig. 4.7

The *metre bridge* is a practical form of Wheatstone bridge in which P and Q are replaced by a resistance wire of uniform cross-section area, 1 m long, Fig. 4.7. The ratio P/Q is altered by changing the position on the wire of a movable contact. The resistance of the wire is directly proportional to its length and so at balance $R = S \times l_1/l_2$.

Questions

1. A meter has a resistance of 50 Ω and a f.s.d. current of 1 mA. How can it be adapted to measure (i) currents up to 1 A, (ii) p.ds up to 10 V?

2. If a voltmeter has a sensitivity of 1000 Ω/V, what is its resistance on (i) the 1 V range, (ii) the 5 V range? On which range will it give a more accurate measurement of the p.d. across a certain resistor? Why?

3. What is the f.s.d. current of a voltmeter with a sensitivity of (i) 2000 Ω/V, (ii) 20 000 Ω/V?

4. In Fig. 4.8 what is the reading on V if it has a resistance on the scale used of (i) 20 kΩ, (ii) 2 kΩ?

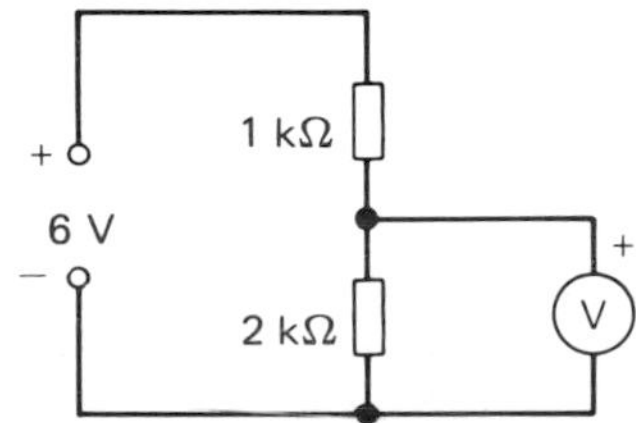

Fig. 4.8

5 Potential divider

A potential divider divides a voltage into a number of equal parts so that its output voltage is some fraction of the input voltage.

Using two fixed resistors

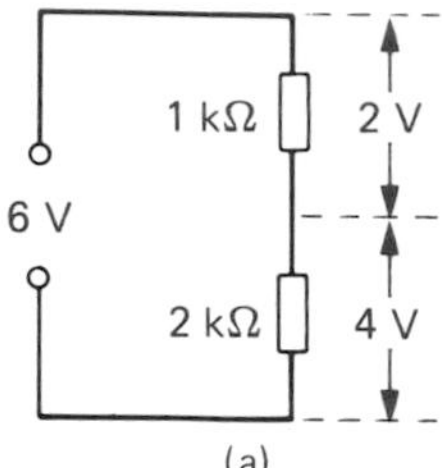

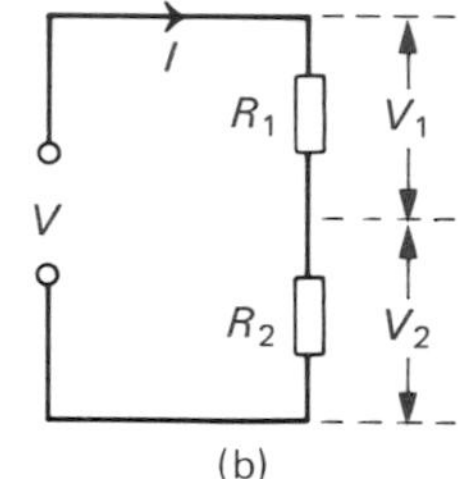

Fig. 5.1

In the circuit of Fig. 5.1a the 1 kΩ and 2 kΩ resistors divide the 6 V input voltage into three (1 + 2) equal parts. One part (2 V) appears across the 1 kΩ and two parts (4 V) across the 2 kΩ resistor. The division is in the ratio of the two resistances, i.e. 1 kΩ/2 kΩ = 1/2 since the same current passes through each.

For the more general circuit of Fig. 5.1b we can say

$$V_1 = IR_1 \quad (1)$$
$$V_2 = IR_2 \quad (2)$$
$$V = V_1 + V_2 = I(R_1 + R_2) \quad (3)$$

Dividing (1) by (3)

$$\frac{V_1}{V} = \frac{IR_1}{I(R_1 + R_2)} = \frac{R_1}{R_1 + R_2}$$

$$\therefore \quad V_1 = \frac{R_1}{R_1 + R_2} \cdot V$$

Similarly from (2) and (3) we get

$$V_2 = \frac{R_2}{R_1 + R_2} \cdot V$$

Note also from (1) and (2) that

$$\frac{V_1}{V_2} = \frac{R_1}{R_2}$$

Using a variable resistor

If all three connections on a variable resistor are used, it acts as a potential divider or potentiometer ('pot') in which the ratio R_1/R_2 is readily changed.

In Fig. 5.2 the resistance between A and B represents R_1 and that between B and C represents R_2. A continuously variable output voltage, from 0 to V,

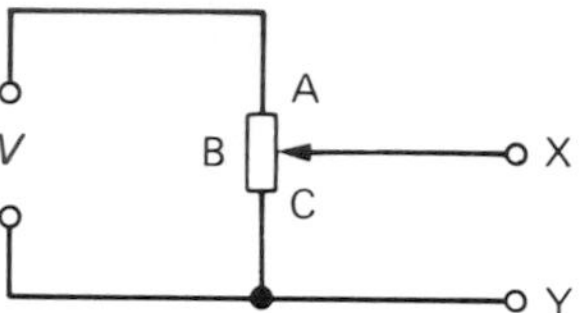

Fig. 5.2

is available between X and Y, depending on the position of the sliding contact. It will be $V/2$ when $R_1 = R_2$, which, in a linear pot (p. 25), occurs when AB = BC.

Effect of load on output voltage

When a potential divider drives current through another circuit, the latter 'loads' the potential divider and the output voltage is less than the calculated value. The difference is small if the resistance R_L of the load is at least *ten times greater* than that of the part of the potential divider across which it is connected. In other words the current drawn by the load should not exceed 1/10 of the current through the potential divider. (As we saw when considering voltmeters (p. 14) the same problem arises.)

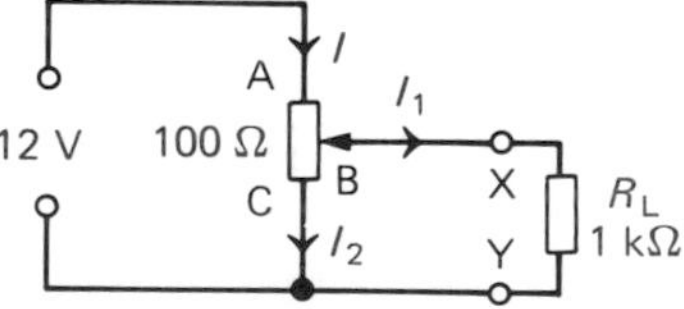

Fig. 5.3

For example, if the potential divider in Fig. 5.3 has a resistance of 100 Ω between A and C then I_1 through the 1 kΩ load will always be at least ten times smaller than I_2 through the potential divider. Therefore when B is midway between A and C, the output voltage between X and Y will be 6 V. If however R_L is 50 Ω, the output voltage is only 4 V (see Question 3).

Questions

1. For the potential divider in Fig. 5.1b calculate V_2 when V, R_1 and R_2 have the values in the table below.

	(a)	(b)	(c)	(d)
V	3 V	5 V	6 V	9 V
R_1	2 kΩ	30 kΩ	100 kΩ	100 Ω
R_2	1 kΩ	20 kΩ	50 kΩ	200 Ω

2. In Fig. 5.2 if BC = 2/3 AC and V = 9 V, what is the output voltage across XY?

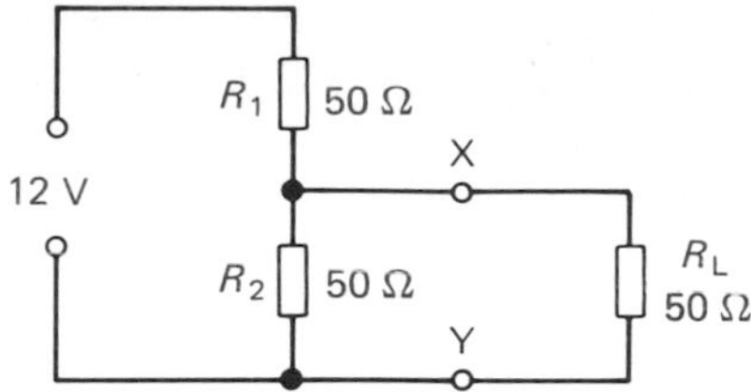

Fig. 5.4

3. What is the p.d. between X and Y in Fig. 5.4?

6 Electric power

Power

(a) Expression for power. The power of a device is 1 watt (1 W) if it changes energy at the rate of 1 joule per second (1 J/s) from one form to another. Every electronic component has a maximum power rating which should not be exceeded. For example a $\frac{1}{2}$ W resistor can safely change $\frac{1}{2}$ J of electrical energy per second into heat without damage.

The power P of a component in *watts* is calculated from

$$P = V \times I$$

where V is the p.d. across it in *volts* and I is the current through it in *amperes*. If it has resistance R in *ohms*, then since $V = IR$ we also have

$$P = (IR) \times I = I^2R$$

If $R = 10\,\Omega$ and $I = 1$ A, $P = 1^2 \times 10 = 10$ W but if $I = 2$ A then $P = 2^2 \times 10 = 40$ W. Doubling I quadruples P.

(b) Example. To find the maximum safe current which can be passed through, for instance, a 10 kΩ $\frac{1}{4}$ W resistor, we have

$$R = 10\,\text{k}\Omega = 10^4\,\Omega \text{ and } P = 0.25\,\text{W}$$

But $P = I^2R \quad \therefore I^2 = P/R$

$$\therefore \quad I^2 = 0.25/10^4 = 0.25 \times 10^2/(10^4 \times 10^2)$$

$$= 25/10^6$$

$$\therefore \quad I = \sqrt{25/10^6} = 5/10^3\,\text{A} = 5\,\text{mA}$$

Power transfer

(a) Internal resistance. Power supplies have some resistance themselves, called *internal* or *source resistance*. It causes the energy loss which occurs inside a battery (i.e. the 'lost' volts, p. 9) when a current is driven through a load in a circuit. The greater the current the greater is the loss and the smaller is the terminal p.d. of the battery. Before we can calculate the maximum power a supply can deliver to a load of resistance R we need to know the e.m.f. E of the supply and its internal resistance r.

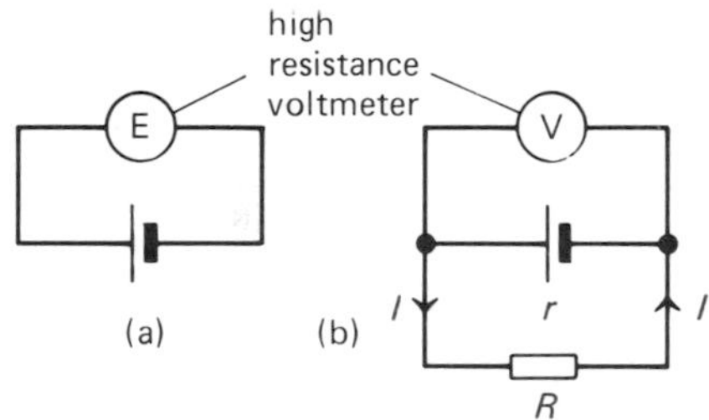

Fig. 6.1

In Fig. 6.1a the high resistance voltmeter measures E (since the cell is on open circuit) and in Fig. 6.1b it measures the terminal p.d. V which drives I through R, i.e. $V = IR$. The 'lost' volts $v = E - V = Ir$ since the current all round the circuit is I. But $E = V + v$ (p. 9).

$$\therefore \quad E = IR + Ir = I(R + r)$$

$$\therefore \quad I = E/(R + r)$$

$$\therefore \quad P = I^2R = E^2R/(R + r)^2$$

For example if $E = 1.5\,V$, $V = 1.2\,V$ and $R = 4\,\Omega$, then from $V = IR$ we get $I = V/R = 1.2/4 = 0.3\,A$. Also $v = E - V = Ir$, therefore $r = v/I = 0.3/0.3 = 1\,\Omega$ and $P = I^2R = 0.3^2 \times 4 = 0.36\,W$.

(b) Maximum power theorem. This states that the maximum power is delivered to a load *when its resistance equals the internal resistance of the power supply*, i.e. when $R = r$. Putting $r = R$ in the equation for P we get

$$\text{maximum power in } R = E^2R/(R + R)^2$$
$$= E^2R/4R^2$$
$$= E^2/4R$$

In the above example maximum power transfer would occur when $R = r = 1\,\Omega$ and would be $(1.5)^2/4 = 0.56\,W$. (The power wasted in the cell is $E^2/4r$, i.e. the same as that in R since $R = r$. The *efficiency* of the power transfer process is therefore 50%, only half of the total power available being supplied to R.)

Questions

1. Calculate the maximum safe current through a resistor of

(a) $100\,\Omega$ 1 W (b) $100\,\Omega$ 4 W (c) $1.8\,k\Omega$ $\frac{1}{4}$ W.

2. If the p.d. across a resistor is 9 V and it carries a current of 3 mA, what is (a) its resistance, (b) the power delivered to it?

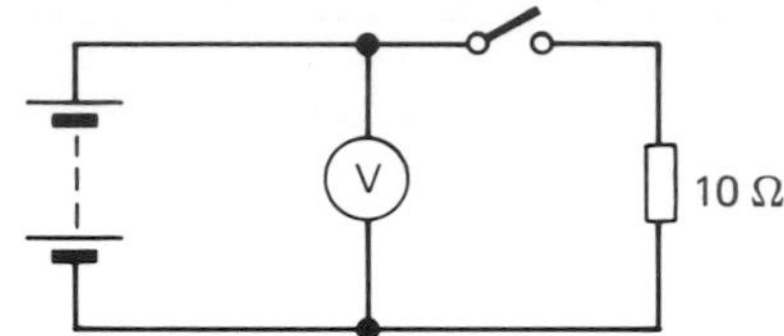

Fig. 6.2

3. (a) In Fig. 6.2, V is a high resistance voltmeter. With the switch open it reads 4.0 V and when the switch is closed it falls to 3.2 V. Calculate the e.m.f. and internal resistance of the battery.
(b) By what resistor must the $10\,\Omega$ resistor be replaced to give maximum power output? What will be the value of that power output? (*L.*)

7 Alternating current

Direct and alternating currents

A *direct current* (d.c.) flows in one direction only; batteries produce d.c. The *waveform* of a current is a graph whose shape shows how the current varies with time. Those in Fig. 7.1 are for steady and varying d.c.

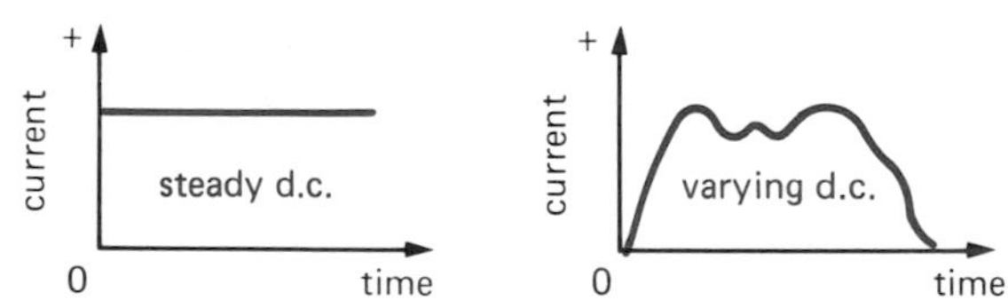

Fig. 7.1

An *alternating current* (a.c.) is one that continually changes direction; car and power station alternators (p. 83) produce a.c. Fig. 7.2 shows the simplest a.c. waveform, it has a sine wave or sinusoidal shape. The current rises from zero to a maximum in one direction (+), falls to zero again before becoming a maximum in the opposite direction (−) and then rises to zero once more and so on. The circuit symbol for an a.c. power supply is ~.

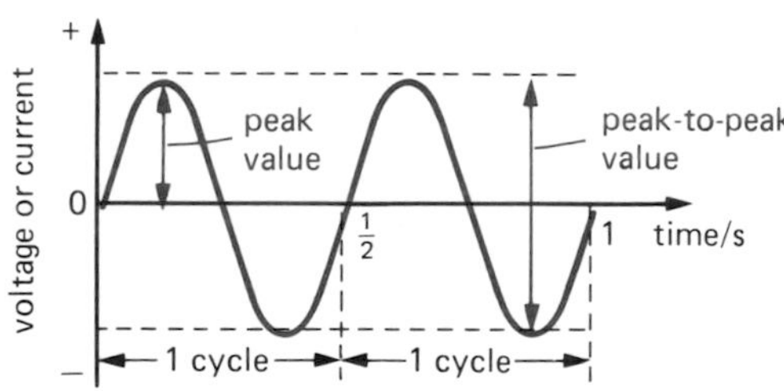

Fig. 7.2

Electric heaters and lamps work off either a.c. or d.c. but most electronic systems require d.c. The pointer of a moving coil meter is deflected one way by d.c.; a.c. makes it move to and fro about the scale zero if the direction changes are slow enough, otherwise no deflection occurs.

Frequency of a.c.

The *frequency* f of a.c. is the number of complete alternations or cycles made in 1 second and the unit is the *hertz* (Hz). The frequency of the a.c. in Fig. 7.2 is 2 Hz; that of the electricity mains supply in many countries (including the U.K.) is 50 Hz. Larger units are the kilohertz (1 kHz = 1000 Hz = 10^3 Hz) and the megahertz (1 MHz = 1000 kHz = 10^6 Hz).

The *period* T is the time for which one cycle of the a.c. lasts. If f = 2 Hz, there are 2 cycles per second, so T = 1/2 s. Smaller units are the millisecond (1 ms = 1/1000 s = 10^{-3} s and the microsecond (1 μs = 1/1000 ms = 10^{-6} s). In general $T = 1/f$. A useful conversion table is given below

f	1 Hz	1 kHz	1 MHz
T	1 s	1 ms	1 μs

The *peak value* or *amplitude* of an a.c. is its maximum positive or negative value. For a sine wave it is half the peak-to-peak value.

Audio frequency (a.f.) currents have frequencies from 20 Hz or so to about 20 kHz. They produce an audible sound in a loudspeaker.

Radio frequency (r.f.) currents have frequencies above 20 kHz. They produce radio waves from an aerial.

Root-mean-square values

Since the value of an alternating quantity changes, the problem arises of what value to take to measure it. The average value of a sine wave over a complete cycle is zero; the peak value might be used. However, the *root-mean-square* (r.m.s.) value is chosen because many calculations can then be done as they would for d.c.

The r.m.s. (or effective) value of an alternating current (or voltage) is the steady direct current (or voltage) which would give the same heating effect.

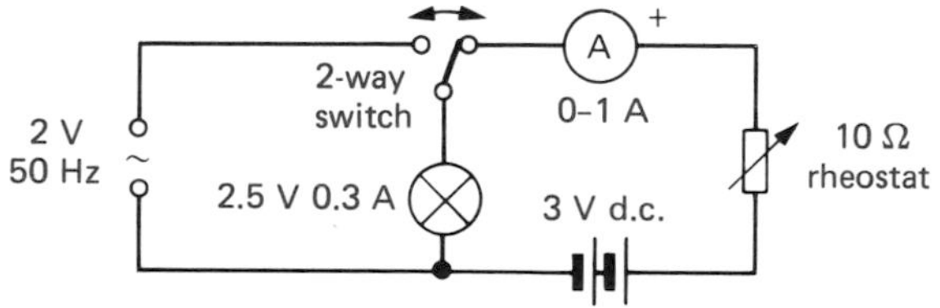

Fig. 7.3

For example, if the lamp in the circuit of Fig. 7.3 is lit first by a.c. (by moving the 2-way switch to the left) and its brightness noted, then if 0.3 A d.c. produces the same brightness (when the switch is moved to the right and the rheostat adjusted), the r.m.s. value of the a.c. is 0.3 A. A lamp designed to be fully lit by a current of 0.3 A d.c. will also be fully lit by a.c. of r.m.s. value 0.3 A.

It can be shown that for a sine wave a.c.

$$\text{r.m.s. value} = \frac{\text{peak value}}{\sqrt{2}} \approx 0.7 \times \text{peak value}$$

The r.m.s. voltage of the U.K. mains supply is 240 V: the peak value is much higher and is given by

$$\text{peak value} = \frac{\text{r.m.s. value}}{0.7} = \frac{240}{0.7} = 340\text{ V}$$

The value given for an alternating current or voltage is always assumed to be the r.m.s. one, unless stated otherwise. The power P of a device on an a.c. supply is given by the same expression as for d.c., i.e. $P = V \times I$ where V and I are r.m.s. values.

If the average value of a sine wave over *half* a cycle is required, it can be found from

$$\text{r.m.s. value} = 1.1. \times \text{average value}$$

Waveforms

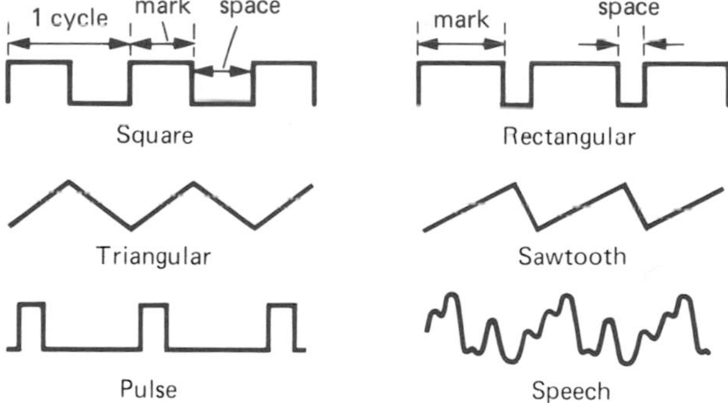

Fig. 7.4

(a) Types. Some of the many types of repetitive waveform occurring in electronics (apart from the sine wave) are shown in Fig. 7.4. The term *mark-to-space ratio* is used in connection with rectangular and square waves. It is given by

$$\text{mark-to-space ratio} = \frac{\text{mark time}}{\text{space time}}$$

For a square wave the ratio is 1 since the mark and space times are equal.

(b) Harmonics. All waveforms, except sine waves, can be shown to consist of

(i) a *fundamental*, which is a sine wave having the same frequency f as the original waveform, and
(ii) *harmonics*, which are sine waves with frequencies that are multiples of the fundamental and with amplitudes that are usually smaller. The second harmonic has frequency $2f$, the third $3f$ and so on.

A sine wave is the only waveform that consists of one frequency (it produces a pure tone in the a.f. range). All others can be analysed into a number of sine waves of different frequencies and amplitudes.

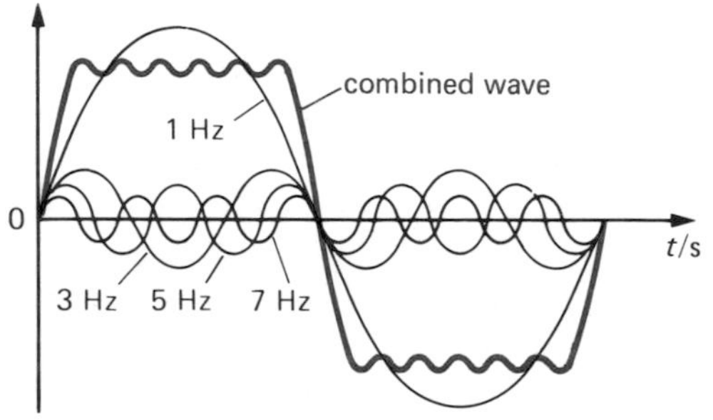

Fig. 7.5

Conversely, they can be built up by adding a number of sine waves together. For example, a triangular wave consists of the fundamental with an infinite number of *even* harmonics. A square wave consists of the fundamental and an infinite number of *odd* harmonics. Fig. 7.5 shows that a roughly square wave is obtained with frequencies of 1, 3, 5 and 7 Hz.

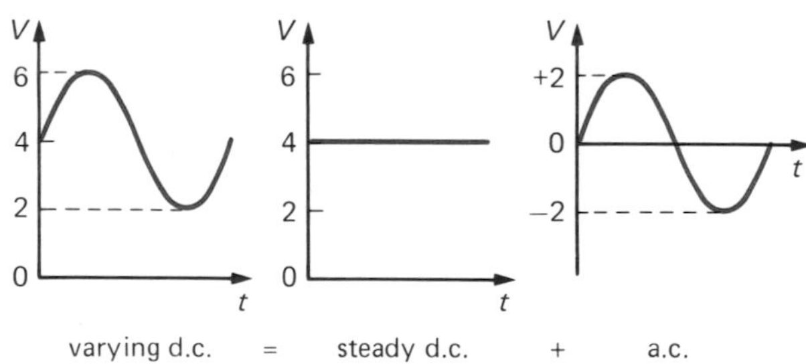

Fig. 7.6

(c) Varying d.c. waveform. Sometimes it is useful to consider that a varying d.c. consists of a steady d.c. plus an a.c., Fig. 7.6.

Questions

1. The waveform of an alternating voltage is shown in Fig. 7.7. What is (a) the period, (b) the frequency, (c) the peak voltage, (d) the r.m.s. voltage?

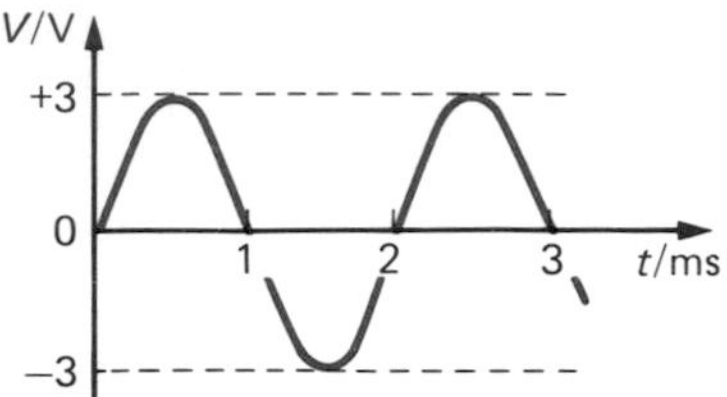

Fig. 7.7

2. An a.c. supply lights a lamp with the same brightness as does a 12 V battery. What is (a) the r.m.s. voltage, (b) the peak voltage, (c) the power of the lamp if it takes a current of 2 A on the a.c. supply?

3. What are the mark-to-space ratios of the waves in Fig. 7.8a,b?

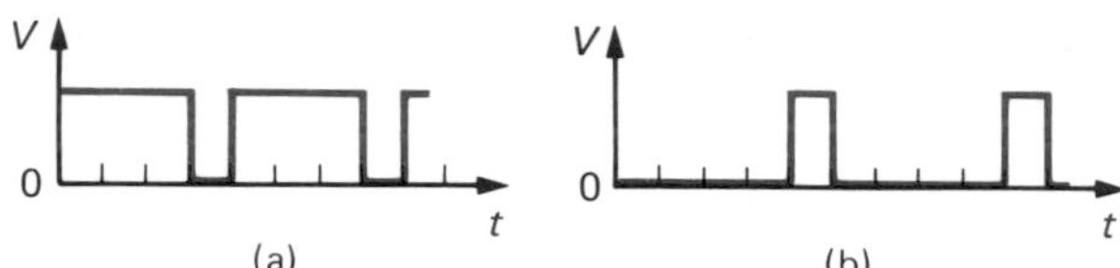

Fig. 7.8

4. What values of *steady* direct voltage and *peak* alternating voltage when added give the varying direct voltage of Fig. 7.9?

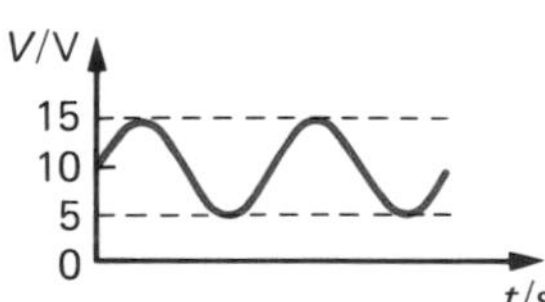

Fig. 7.9

8 Progress questions

1. (a) An electric current in a metal wire can be considered as a drift of free electrons in one direction through the wire.

(i) Explain why the 'free electrons' are so called.
(ii) What causes the drift of free electrons in the wire?
(iii) Describe the difference in the movement of the free electrons when there is no current flowing.

(b) Explain why current cannot be made to flow in an insulator. *(O.L.E. part qn.)*

2. In the circuit of Fig. 8.1 all lamps are rated 6 V 0.2 A. The battery and ammeters have negligible resistance. State the reading, in amperes, which would be indicated on A_1, A_2, A_3, A_4. *(A.E.B. 1982 Control Tech.)*

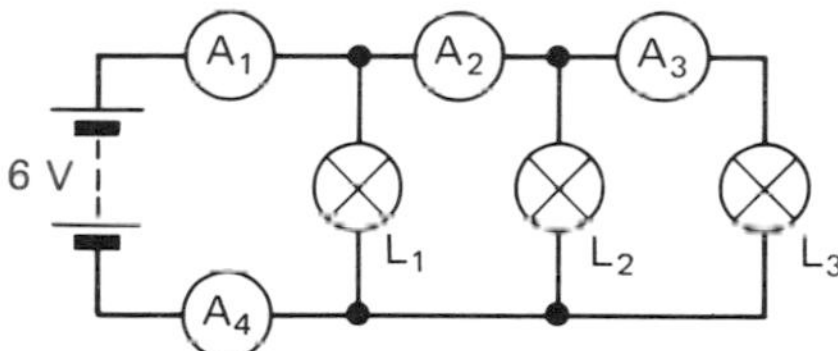

Fig. 8.1

3. An ammeter and a voltmeter are connected to a circuit comprising a 12 V battery and one or more 2 kΩ resistors. The meter readings are shown alongside the first circuit in Fig. 8.2a. What are the readings of the meters in each of the other circuits (b) to (e)? *(A.E.B. 1982 Electronics)*

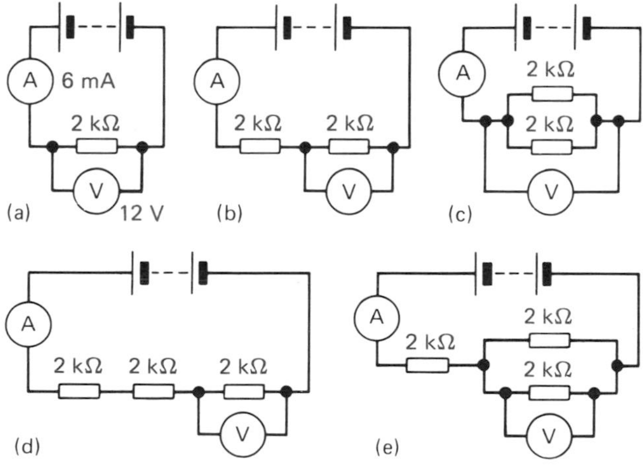

Fig. 8.2

4. How does the resistance of (a) a metal wire, and (b) a slice of pure (intrinsic) semiconductor, change as the temperature rises? Fig. 8.3 shows graphs of current against time for (i) a metal wire, and (ii) a slice of semiconductor across which a constant p.d. is applied. Explain the shape of each graph. *(L.)*

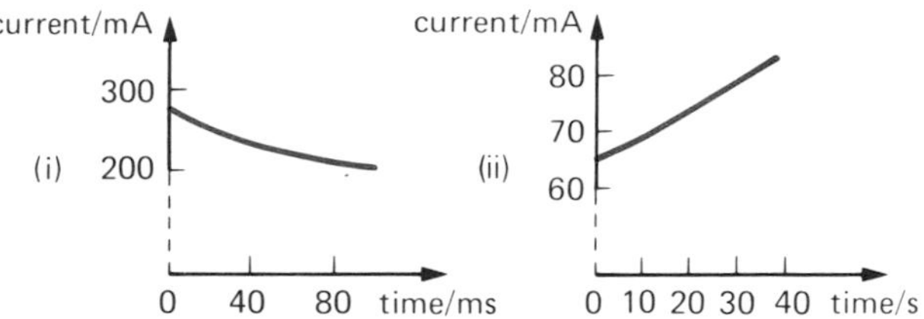

Fig. 8.3

5. A conductor has a length of 5 m and a cross-sectional area of 0.1 mm². If the resistivity of the material is 45×10^{-8} Ωm, calculate the resistance of the conductor.

What length of this conductor is needed to form a 1 Ω resistor? *(C.)*

6. A meter has a resistance of 5 Ω and gives a full scale deflection for a current of 2 mA. How can it be converted to a voltmeter reading up to 10 V? *(C.)*

7. The diagrams in Fig. 8.4a,b show two ways of measuring the p.d. across and the current through a resistor. State, giving your reasons, which arrangement is to be preferred if (i) $R = 2\,\Omega$ and $S = 4\,\Omega$, (ii) $R = 500\,\text{k}\Omega$ and $S = 1\,\text{M}\Omega$. *(L.)*

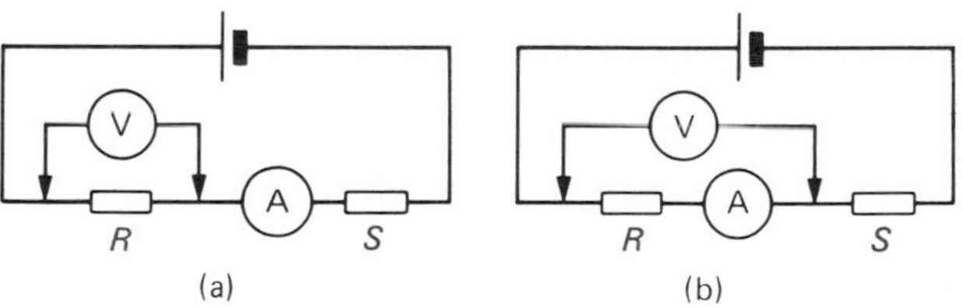

Fig. 8.4

8. Give a description of the Wheatstone bridge and an explanation of the theory when the bridge is balanced.

How would you investigate the variation of the resistance of a wire resistor with temperature? Draw a sketch-graph to illustrate this variation between 0 and 100°C. How would you find the *temperature coefficient of resistance* of the material from the graph?

A tungsten filament has a resistance of 320 Ω at 0°C. An experiment is performed in which this filament is heated and its resistance measured at various temperatures. The results are given in the table:

R (in Ω)	1780	2230	2705	3145
T (in °C)	880	1150	1455	1700

Plot a graph of these results and use it to find the temperature coefficient of resistance of tungsten over the temperature range 880–1700°C. *(C.)*

9. The circuit in Fig. 8.5 is designed to allow a variable voltage to be applied across the 500 Ω resistor by using a 1 kΩ linear potentiometer. What is the value of the p.d. across the 500 Ω resistor when the potentiometer slider Y is (a) at the end X, (b) at the mid-point of the potentiometer, (c) at the end Z?
Comment on these values. (*L.*)

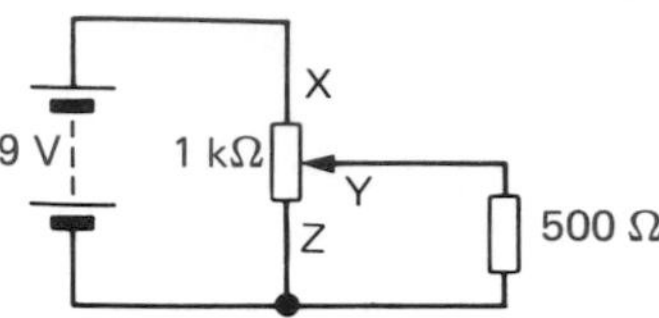

Fig. 8.5

10. In Fig. 8.6, V is a high resistance voltmeter and A is an ammeter of negligible resistance. When the switch S is open the voltmeter reads 6 V and when switch S is closed the voltmeter reads 5 V.

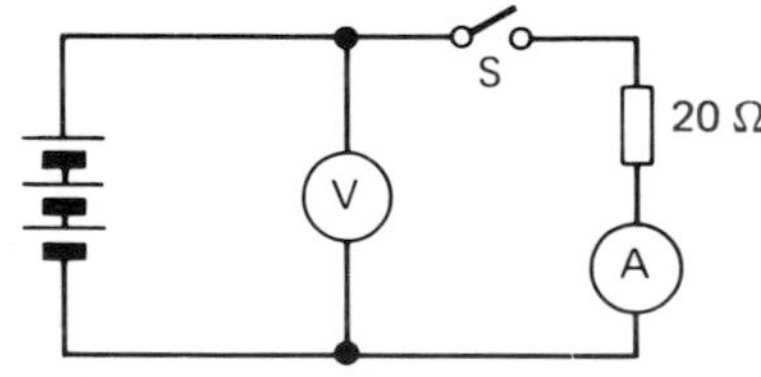

Fig. 8.6

(a) Calculate
(i) the e.m.f. and the internal resistance of the battery,
(ii) the ammeter reading when switch S is closed,
(iii) the power consumption in the 20 Ω resistor with switch S closed.
(b) The 20 Ω resistor is then replaced by another resistor *R*.
(i) State the value of *R* required to obtain maximum power from the battery.
(ii) Calculate the maximum power output. (*C.*)

11. For the circuit in Fig. 8.7, calculate
(a) the resistance of the circuit,
(b) the electric current through the 6 Ω resistor,
(c) the p.d. between the cell terminals. (*C.*)

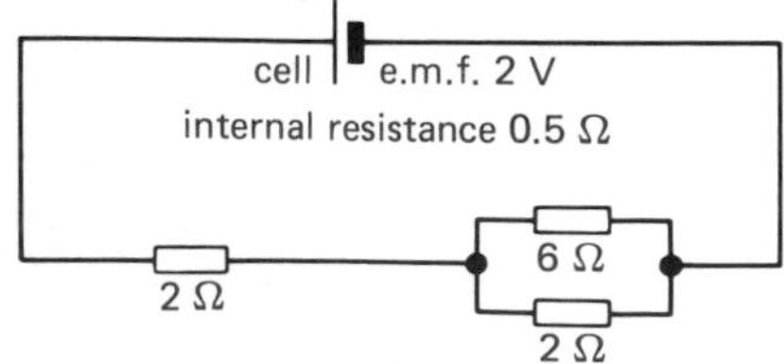

Fig. 8.7

12. If in the circuit of Fig. 8.8 the voltmeter has a resistance of 8 kΩ calculate the p.d. across the 2 kΩ resistor with and without the voltmeter, V, across it, Comment upon the results. (*L.*)

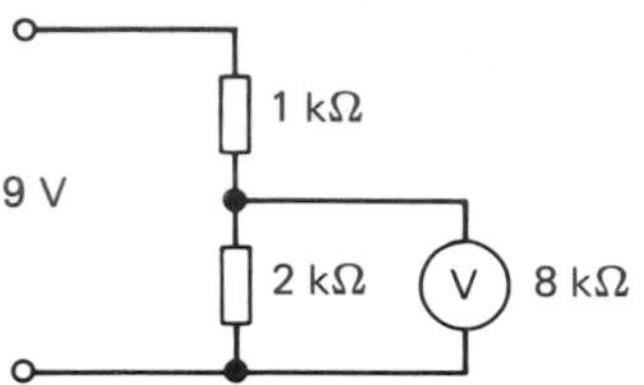

Fig. 8.8

13. (a) Define the volt. Distinguish between the e.m.f. of a cell and the p.d. across its terminals.
(b) State Ohm's law and describe how it may be verified for a metallic conductor.
(c) Two identical cells each of e.m.f. *E* are connected as shown in Fig. 8.9. What is the e.m.f. of the combination in each case?

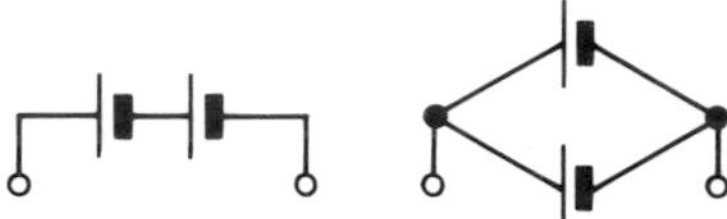

Fig. 8.9

(d) Two cells each of e.m.f. 1.8 V and internal resistance 0.5 Ω pass a current through a 2 Ω resistor. Calculate the p.d. across the resistor when the cells are (i) in series, (ii) in parallel. (*C.*)

14. Calculate the r.m.s. voltage and frequency of the signal shown in Fig. 8.10. (*L.*)

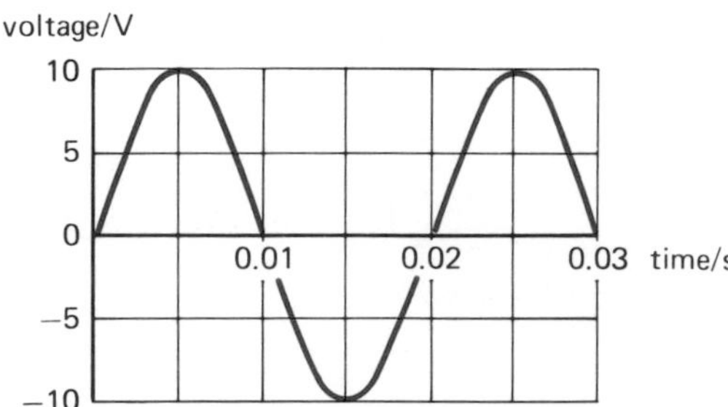

Fig. 8.10

Components and circuits

9 Resistors

About resistors

Basically resistors are used to limit the current in circuits. When choosing one, three factors to be considered, apart from the value, are:

(i) the *tolerance*: exact values cannot be guaranteed by mass-production methods but this is not a disadvantage because in most electronic circuits the values of resistors are not critical. A resistor with a stated (called nominal) value of 100 Ω and a tolerance of ± 10% could have any value between 90 and 110 Ω;

(ii) the *power rating*: this is the maximum power which can be developed in a resistor without damage occurring by overheating. For most electronic circuits 0.25 or 0.5 W ratings are adequate. The greater the physical size of a resistor the greater is its rating; and

(iii) the *stability*: this is the ability to keep the same value with changes of temperature and with age.

Fixed resistors

Four types are shown in Fig. 9.1 with their methods of construction and properties.

Since exact values of fixed resistors are unnecessary in most electronic circuits, only certain 'preferred' values are made. The number of values depends on the tolerance. The table overleaf shows the values required to give maximum coverage with minimum overlap for three tolerances. For example, a 22 Ω ± 10% resistor may have any value between (22 + 2.2) ≈ 24 Ω and (22 − 2.2) ≈ 20 Ω. The next higher value of 27 Ω (for ± 10% tolerance) more or less covers the range 24 Ω to 30 Ω.

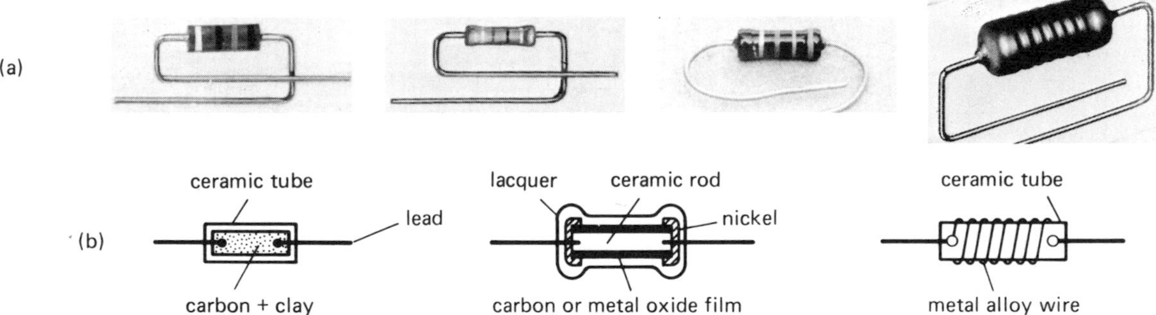

Fig. 9.1

Property \ Type	Carbon composition	Carbon film	Metal oxide	Wirewound
Maximum value	20 MΩ	10 MΩ	100 MΩ	270 Ω
Tolerance	± 10%	± 5%	± 2%	± 5%
Power rating	0.125–1 W	0.25–2 W	0.5 W	2.5 W
Stability	poor	good	very good	very good
Use	general	general	accurate work	low values

Tolerance	Preferred values (multiples × 10)							
± 5%	1.0	1.1	1.2	1.3	1.5	1.6	1.8	2.0
± 10%	1.0		1.2		1.5		1.8	
± 20%	1.0				1.5			
± 5%	2.2	2.4	2.7	3.0	3.3	3.6	3.9	4.3
± 10%	2.2		2.7		3.3		3.9	
± 20%	2.2				3.3			
± 5%	4.7	5.1	5.6	6.2	6.8	7.5	8.2	9.1
± 10%	4.7		5.6		6.8		8.2	
± 20%	4.7				6.8			

Resistors with ± 10% tolerance belong to the E12 series (it has 12 basic values). Those with ± 5% tolerance form the E24 series.

Resistance codes

(a) Colour code. The examples in Fig. 9.2 show how the first three coloured bands on a resistor give its resistance in ohms and the fourth band the tolerance. To decide which is the first band, remember that the fourth band, if it is present, will be either gold or silver, these being colours not used for the first band.

(b) Printed code. The code is printed on the resistor and consists of letters and numbers. It is also used on circuit diagrams and on variable resistors. The examples below show how it works. R means × 1, K means × 10^3 and M means × 10^6 and the position of the letter gives the decimal point.

Tolerances are indicated by adding a letter at the end: J = ±5%, K = ±10%, M = ±20%. For example 5K6K = 5.6 kΩ ±10%.

Number	Colour
0	black
1	brown
2	red
3	orange
4	yellow
5	green
6	blue
7	violet
8	grey
9	white

r a i n b o w

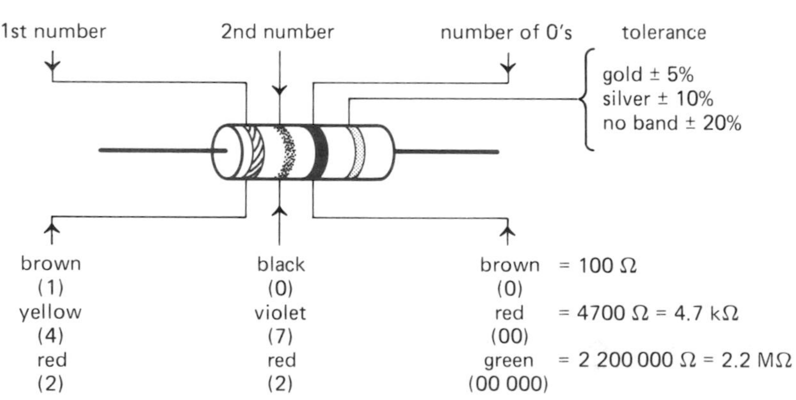

Note. The E96 series of resistors has six bands—the first four for the value, the fifth for tolerance and the sixth for temperature stability.

Fig. 9.2

Printed code examples

Value	0.27 Ω	3.3 Ω	10 Ω	220 Ω	1 kΩ	68 kΩ	100 kΩ	4.7 MΩ
Mark	R27	3R3	10R	220R	1K0	68K	100K	4M7

Variable resistors

(a) Rotary type, Fig. 9.3a. The construction is shown in Fig. 9.3b. Connections are made to each end of the track by terminal tags A and B. Tag C is connected to the sliding contact or wiper which varies the resistance between C and the end tags when it is rotated by the spindle. For power ratings up to 2 W, the track is made either of carbon or of cermet (ceramic and metal oxide); above 2 W it is wirewound.

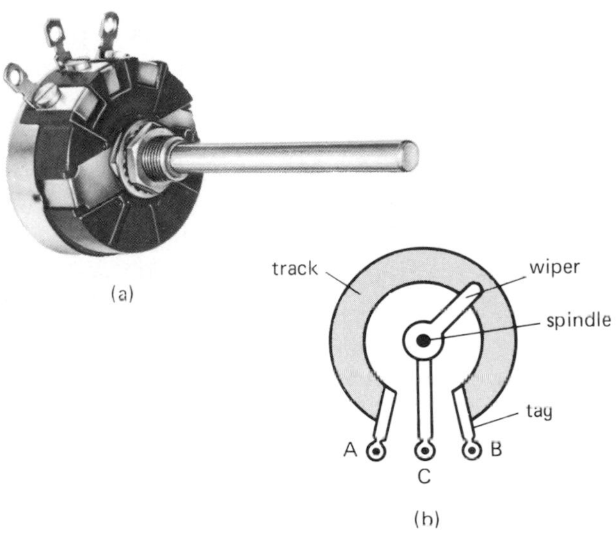

Fig. 9.3

If the track is 'linear' equal changes of resistance occur when the wiper is rotated through equal angles. In a 'log' track the resistance change for equal angular rotations is greater at the end of the track than at the start. Common values are 10 kΩ, 50 kΩ, 100 kΩ, 500 kΩ and 1 MΩ.

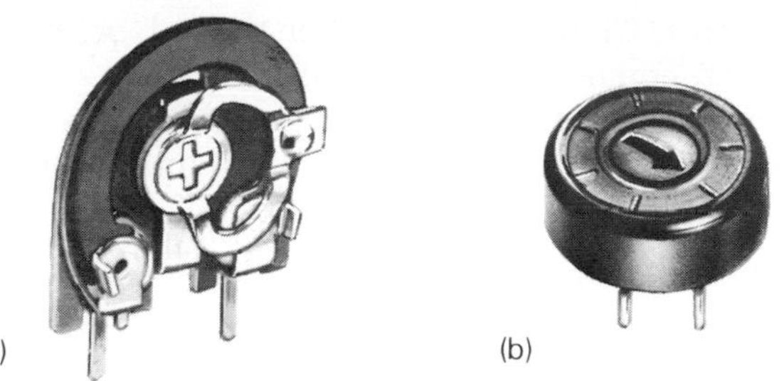

Fig. 9.4

(b) Preset type, Fig. 9.4a,b. These have carbon or cermet tracks and when adjustment is necessary, a screwdriver is used. Ratings vary from 0.25 to 1 W.

Questions

1. What is the value and tolerance of R_1, R_2 and R_3 colour-coded in the table below.

Band	1	2	3	4
R_1	brown	black	red	silver
R_2	yellow	violet	orange	gold
R_3	green	blue	yellow	none

2. What is the colour code for the following:
(a) 150 Ω ± 10% (b) 10 Ω ± 5% (c) 3.9 kΩ ± 10% (d) 10 kΩ ± 10% (e) 330 kΩ ± 20% (f) 1 MΩ ± 10%?

3. What are the values and tolerances of resistors marked:
(a) 2K2M (b) 270KJ (c) 1M0K (d) 15RK?

4. What are the printed codes for:
(a) 100 Ω ± 5% (b) 4.7 kΩ ± 20% (c) 100 kΩ ± 10% (d) 56 kΩ ± 20%?

5. What E12 preferred values would you use if you calculated that a circuit needed resistors having values of:
(a) 1.3 kΩ (b) 5.0 kΩ (c) 72 kΩ (d) 350 kΩ?

10 Capacitors

About capacitors

A capacitor stores electric charge. Basically it consists of two metal plates separated by an insulator called the *dielectric*.

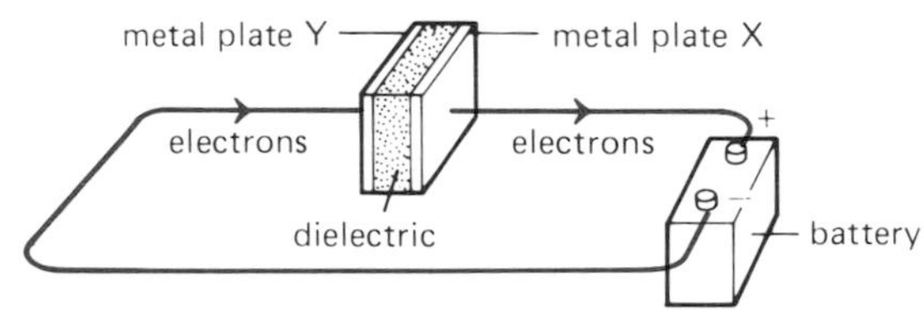

Fig. 10.1

(a) Charging. When connected to a battery, Fig. 10.1, the positive of the battery attracts electrons from plate X and the negative repels electrons to plate Y. Positive charge (deficit of electrons) builds up on X and an equal negative charge (excess of electrons) builds up on Y. During the charging, there is a brief flow of electrons round the circuit from X to Y. Charging stops when the p.d. between X and Y equals (and opposes) the e.m.f. of the battery. The process takes time, i.e. the response of a capacitor to a change of p.d. is not immediate.

If the battery is removed, the charge may take a long time to leak away unless a conductor is connected across it.

(b) Capacitance. The capacitance C of a capacitor measures its charge-storing ability. It is 1 *farad* (F) if it stores a charge of 1 coulomb when the p.d. across it is 1 volt. If the charge is 6 C when the p.d. is 2 V, then $C = 6\ \mathrm{C}/2\ \mathrm{V} = 3\ \mathrm{F}$. In general for charge Q and p.d. V

$$C = \frac{Q}{V} \quad \text{or} \quad Q = VC$$

Smaller more convenient units are the microfarad ($1\,\mu\mathrm{F} = 10^{-6}\,\mathrm{F}$), the nanofarad ($1\,\mathrm{nF} = 10^{-9}\,\mathrm{F} : 1\,\mu\mathrm{F} = 1000\,\mathrm{nF}$) and the picofarad ($1\,\mathrm{pF} = 10^{-12}\,\mathrm{F} : 1\,\mathrm{nF} = 1000\,\mathrm{pF}$).

C is large when the area of the plates is large, the plate separation is small and certain dielectrics are used.

(c) Energy stored. A charged capacitor stores electrical energy. For charge Q and p.d. V, the energy W stored is

$$W = \tfrac{1}{2}QV = \tfrac{1}{2}CV^2 \quad (\text{since } Q = VC)$$

where W is in joules if Q is in coulombs and V in volts.

In a photographic flash unit a capacitor discharges through a lamp and its energy is changed to heat and light.

Practical capacitors

When choosing a capacitor two factors to be considered, apart from its value and tolerance, are

(i) the *working voltage*: this is the maximum voltage (d.c. or peak a.c.) it can withstand before the dielectric breaks down (it is often marked on it), and

(ii) the *leakage current*: no dielectric is a perfect insulator but the loss of charge by leakage through it should be small.

(a) Fixed capacitors. Non-polarized types (symbol –▮▮–) can be connected either way round. Polarized types (symbol +–▯▮–) have a positive and a negative terminal and must be connected so that there is d.c. through them in the correct direction (to maintain the dielectric by electrolytic action).

Five fixed capacitors which use different dielectrics are shown in Fig. 10.2. Values may be marked on them as for resistors, e.g. 4μ7 for 4.7 μF or 2n2 for 2.2 nF. Some polyesters use the resistor colour code, see Fig. 10.3.

The 'Swiss roll' method of construction used for polarized capacitors is shown in Fig. 10.4. Some other types are also made in this way but a very thin metal film is deposited on each side of a flexible strip of the dielectric to act as the plates. In mica and

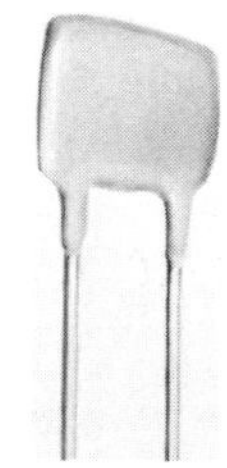

Fig. 10.2

Type / Property	Non-polarized			Polarized (electrolytic)	
Property	Polyester	Mica	Ceramic	Aluminium	Tantalum
Values	0.01–10 μF	1 pF–0.01 μF	10 pF–1 μF	1–100 000 μF	0.1–100 μF
Tolerance	± 20%	± 1%	− 25 to + 50%	− 10 to + 50%	± 20%
Leakage	small	small	small	large	small
Use	general	high frequency	decoupling	low frequency	low voltage

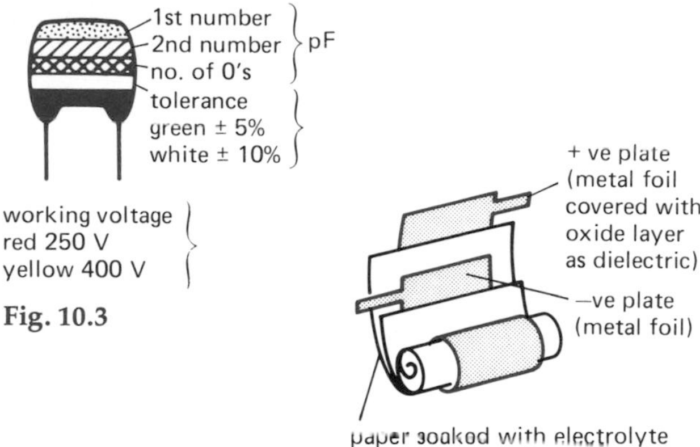

Fig. 10.3

Fig. 10.4

ceramic capacitors the plates consist of a deposit of silver on a thin sheet of mica or ceramic.

(b) Variable capacitors. They are used to tune radio receivers. Their value is varied (e.g. up to 500 pF) by altering the overlap between a fixed set of metal plates and a moving set, separated by a dielectric of air. Often two or more are 'ganged', Fig. 10.5a.

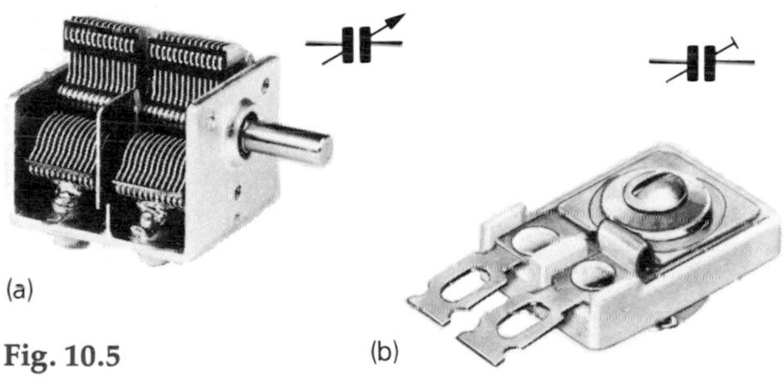

Fig. 10.5

Small variable capacitors called *trimmers* or *presets* are used to make fine, infrequent changes to the capacitance of a circuit. Fig. 10.5b shows a type in which metal foils and mica sheets are compressed more or less by a screw to change the value.

Capacitor networks

(a) Parallel. In Fig. 10.6a the p.d. across each capacitor is the same (but the charges are different

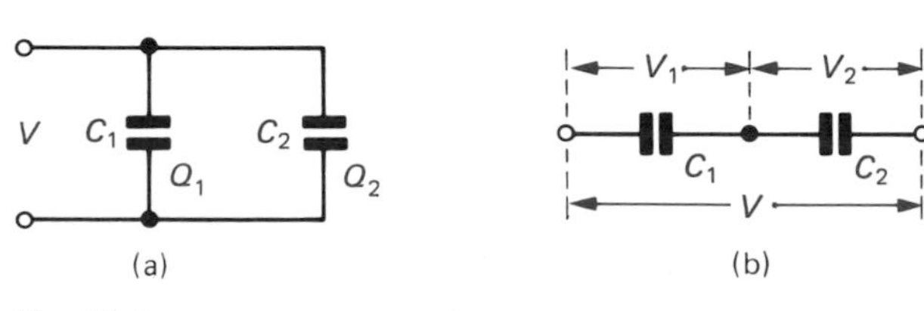

Fig. 10.6

unless $C_1 = C_2$). The total charge $Q = Q_1 + Q_2$; it therefore follows that the combined capacitance C is

$$C = C_1 + C_2$$

(b) Series. In Fig. 10.6b each capacitor has the same charge (but the p.ds are different unless $C_1 = C_2$). Since the total p.d. $V = V_1 + V_2$, it follows that the combined capacitance C is given by

$$\frac{1}{C} = \frac{1}{C_1} + \frac{1}{C_2} \quad \text{or} \quad C = \frac{C_1 \times C_2}{C_1 + C_2}$$

(c) Example. What is the combined capacitance of the network in Fig. 10.7?

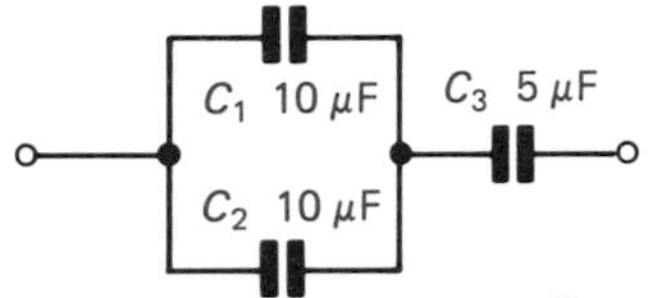

Fig. 10.7

The capacitance of C_1 and C_2 in parallel is $C_1 + C_2 = 10 + 10 = 20\,\mu\text{F}$.

This is in series with $C_3 = 5\,\mu\text{F}$ and their combined capacitance C is

$$C = \frac{5 \times 20}{5 + 20} = \frac{100}{25} = 4\,\mu\text{F}$$

Capacitors in a.c. circuits

(a) Action. In Fig. 10.8 the lamp lights only when S is in the a.c. position, i.e. C blocks d.c. but allows a.c. to pass. With d.c. there is a brief current which charges C. With a.c., as its direction changes each half cycle, C is charged, discharged, charged in the opposite direction and discharged again 50 times a second. No current actually passes through C but it *seems* to because electrons flow to and fro in the wires joining the plates to the a.c. supply.

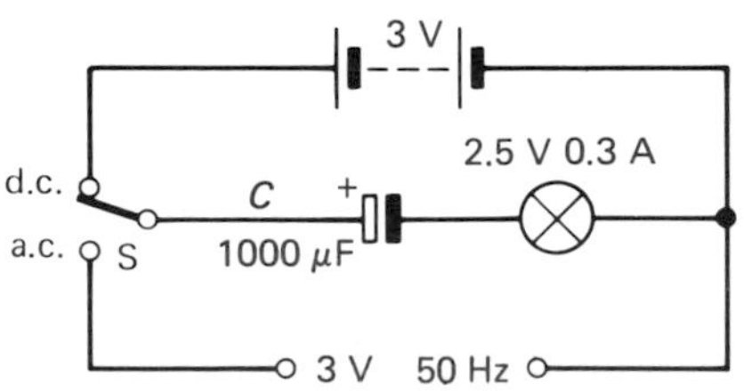

Fig. 10.8

(b) Capacitive reactance. The opposition of a capacitor to a.c. is measured by its capacitive reactance X_C, given by

$$X_C = \frac{1}{2\pi fC}$$

where X_C is in ohms if f is in hertz and C in farads. This expression agrees with the facts that if C is large, electron flow is large and if f is high, electrons flow rapidly on and off the plates. Large C and f thus lead to a large current, i.e. small opposition to the a.c. For example, if $C = 1000\,\mu\text{F} = 10^{-3}$ F and $f =$ 50 Hz, then $X_C = 1/(2\pi \times 50 \times 10^{-3}) = 3.2\,\Omega$. X_C would decrease if f increased (as Fig. 10.9 shows) or C increased.

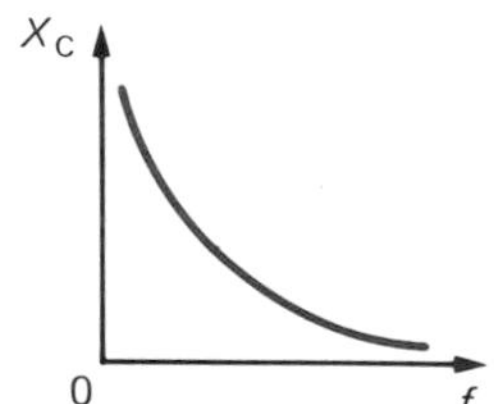

Fig. 10.9

X_C is to a capacitor in an a.c. circuit as R is to a resistor in a d.c. circuit and if an r.m.s. voltage V is applied to a capacitor of reactance X_C, the r.m.s. current I is given by

$$I = \frac{V}{X_C} \qquad \left(\text{compare } I = \frac{V}{R} \text{ for d.c.}\right)$$

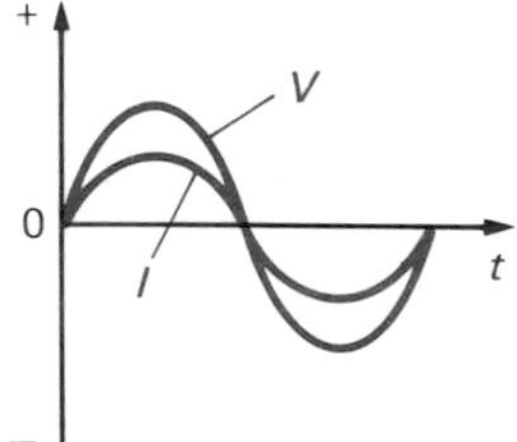

Fig. 10.10

(c) Phase shift. For a resistor in an a.c. circuit the current and p.d. reach their peak values at the same time, i.e. they are in *phase*, Fig. 10.10. For a capacitor, at the start of the cycle of applied p.d. there is a 'rush' of current I which falls to zero when the p.d. V across the capacitor equals the applied p.d., i.e. I is a maximum when $V = 0$ and vice versa, Fig. 10.11a. There is a phase shift of $\frac{1}{4}$ cycle between I and V with I ahead, i.e. it reaches its peak value first.

(d) Power. The power P taken by a capacitor from an a.c. supply can be found at any instant by multiplying (taking + and − signs into account) the values of V and I at that instant since $P = I \times V$.

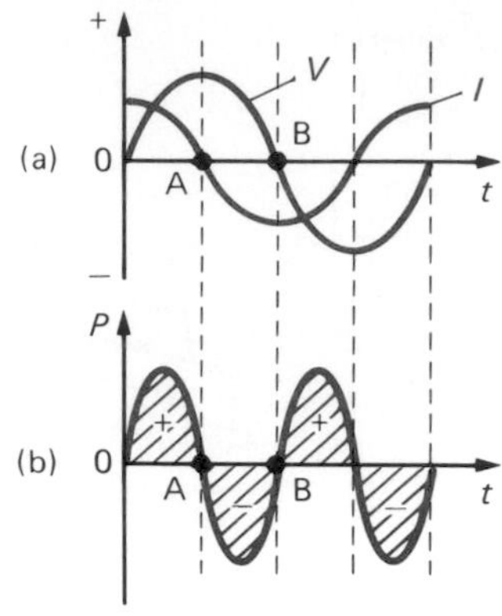

Fig. 10.11

This can be done using the V–t and I–t graphs of Fig. 10.11a (remembering that $+ \times + = +$, $+ \times - = -$ and $- \times - = +$). The P–t graph so obtained is shown in Fig. 10.11b. It is a sine wave like the V–t and I–t graphs but with twice the frequency. The average power taken by the capacitor over a cycle is zero since the graph is symmetrical about the t-axis. A capacitor is said to be a 'wattless' component.

To explain this behaviour we consider that during the first quarter-cycle 0A (V and I are both positive, making P positive), power is drawn from the supply and energy is stored in the charged capacitor. In the second quarter-cycle AB (V is positive, I is negative so P is negative), the capacitor discharges and returns its stored energy to the supply.

Coupling and decoupling

Capacitors are used to separate a.c. from d.c. For example, if the output from one circuit, X, contains d.c. and a.c. (e.g. is varying d.c.) but only the a.c. is wanted as the input to another circuit Y, the circuits can be *coupled* by a suitable capacitor offering low reactance at the frequencies involved, Fig. 10.12a.

Capacitors are also used to prevent a.c. passing through a circuit or component, i.e. it *decouples* them. In Fig. 10.12b C has a low reactance at the frequencies involved and decouples R by acting as a bypass for a.c. while d.c. goes through R.

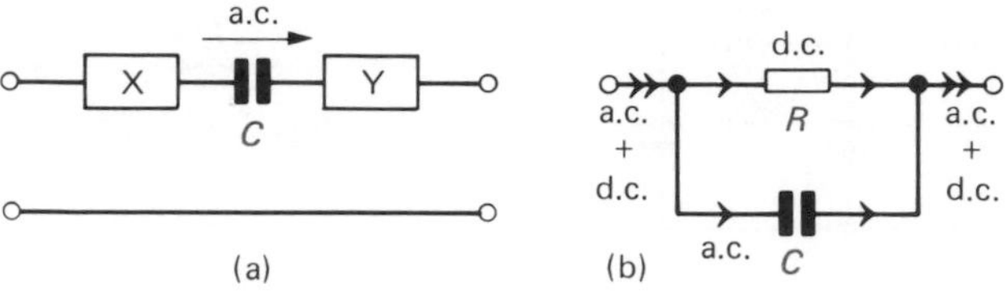

Fig. 10.12

Uses of capacitors

1. To separate a.c. from d.c. by coupling or de-coupling.
2. To control current in an a.c. circuit (p. 28).
3. To smooth the output of a power supply by storing charge (p. 89).
4. To tune a radio receiver (p. 181).
5. To control the frequency of an oscillator (p. 37).
6. In time delay circuits (p. 68).

Questions

1. (a) Calculate Q if $V = 10\,\text{V}$ and $C = 100\,000\,\mu\text{F}$.
(b) What is C if $Q = 12\,\mu\text{C}$ and $V = 6\,\text{V}$?
(c) Find V if $C = 10\,\mu\text{F}$ and $Q = 50\,\mu\text{C}$.

2. If 500 V is applied across a $2\,\mu\text{F}$ capacitor, calculate (a) the charge and (b) the energy stored.

3. What is the combined capacitance of (a) $2.2\,\mu\text{F}$ and $4.7\,\mu\text{F}$ in parallel, (b) two $100\,\mu\text{F}$ capacitors in series?

4. Calculate the reactance of a $1\,\mu\text{F}$ capacitor at frequencies of (a) 1 kHz, (b) 1 MHz.

5. A constant voltage variable frequency a.c. supply is connected to a $2\,\mu\text{F}$ capacitor.
(a) Draw a graph to show how the current 'through' the capacitor varies with the frequency.
(b) If the supply is 6 V r.m.s. and the frequency is 100 Hz what is the current?

6. Why should a capacitor with a working voltage of 250 V not be used on a 240 V a.c. supply?

11 Inductors

About inductors

An inductor is a coil of wire with a core of air or a magnetic material. Four types with their symbols are shown in Fig. 11.1. Inductors have *inductance* (symbol L); they oppose changing currents as we will now see.

In the circuit of Fig. 11.2, if the rheostat is first adjusted so that the lamps are equally bright when S is closed, the resistance of the rheostat then equals the resistance (due to its coil) of the iron-cored inductor. When S is opened and closed again, the lamp in series with the inductor lights up a second or two *after* that in series with the rheostat. The inductor opposes the *rise* of the d.c. from zero to its steady value.

If the 3 V battery is replaced by a 3 V a.c. supply, the lamp in series with the inductor never lights (the current is changing all the time), unless the inductance is reduced by removing the iron core.

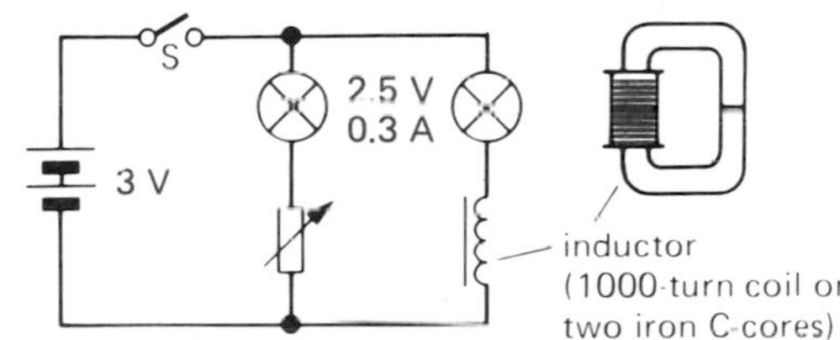

Fig. 11.2

To explain the behaviour of an inductor we need to know about electromagnetic induction.

Electromagnetic induction

When a conductor is in a *changing* magnetic field, an e.m.f. is produced in it. This can be shown by pushing a magnet into a coil, one pole first, Fig. 11.3a,

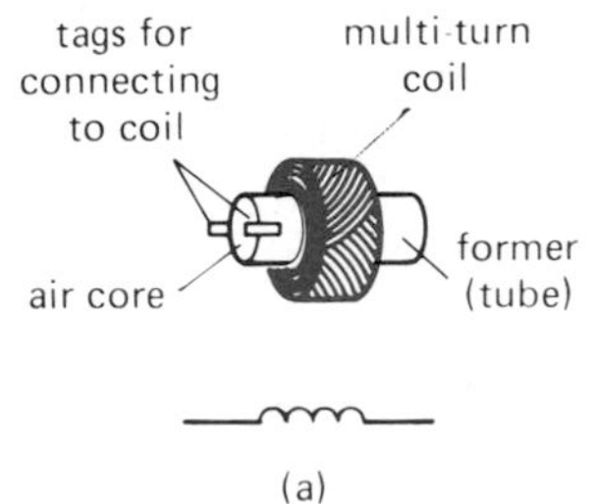

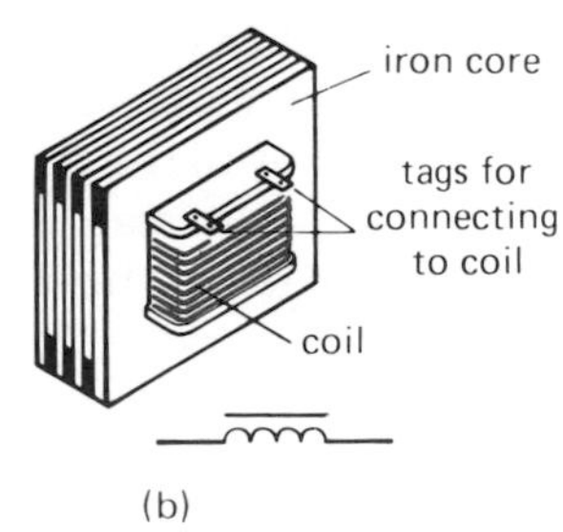

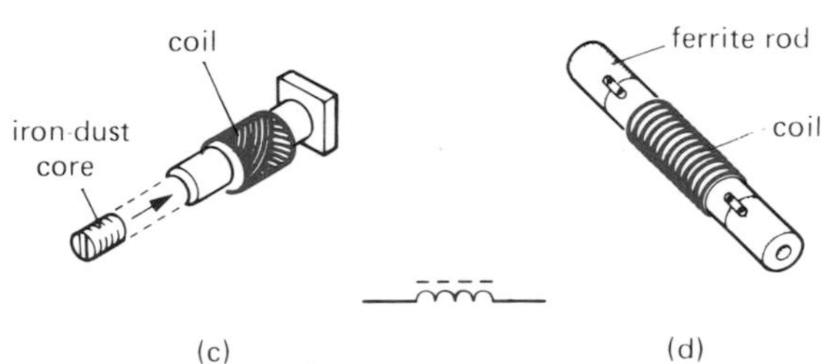

Fig. 11.1

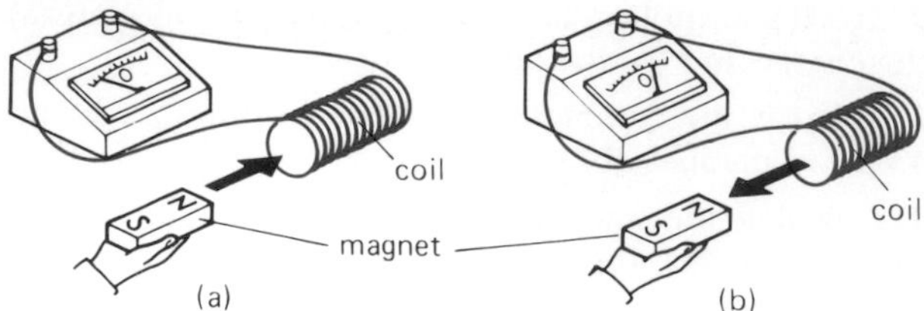

Fig. 11.3

holding it still inside the coil and then withdrawing it, Fig. 11.3b. A centre-zero galvanometer in series with the coil shows that there is a current when the magnet is *moving* but not when it is at rest. The current is in opposite directions when the magnet enters and leaves the coil. The result is the same if the coil is moved towards and away from the stationary magnet.

Experiments show that the induced e.m.f. (which causes the current in the coil)

(i) increases when the rate at which the magnetic field changes also increases, e.g. it increases if the magnet is moved faster. This is *Faraday's law*, and

(ii) always opposes the change causing it, e.g. in Fig. 11.3a the end of the coil nearest the magnet becomes a N pole (due to the induced current), in Fig. 11.3b it becomes a S pole. In both cases it tries to oppose the motion of the magnet. This is *Lenz's law*.

How inductors work

A changing magnetic field can be produced by an electric current if its value alters.

(a) In d.c. circuits. When a direct current increases in a coil from zero to its steady value, the accompanying magnetic field builds up to its final shape. During the process the field is changing and induces an e.m.f. in the coil *itself* which opposes the change causing it, i.e. the rising current that is trying to establish the field. The opposition delays the rise of current, Fig. 11.4a.

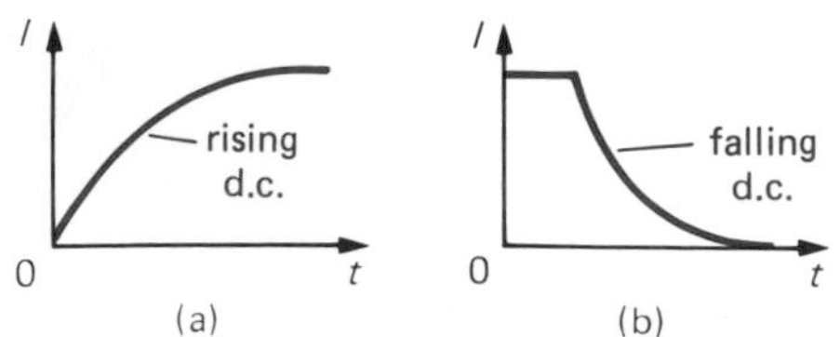

Fig. 11.4

When the current is switched off, the field collapses rapidly and induces a large e.m.f. in the coil which opposes the collapsing field caused by the falling current. It tries to keep the current flowing longer, so delaying its fall to zero, Fig. 11.4b. This effect causes sparking at switch contacts, the energy for which comes from that stored in the magnetic field round the coil.

(b) In a.c. circuits. Since an alternating current is changing all the time, the magnetic field it produces is also changing continuously. There is therefore always an e.m.f. in the coil and permanent opposition to the a.c. on this account.

Inductance

An inductor has an inductance of 1 *henry* (H) if a current changing in it at the rate of 1 ampere per second induces an e.m.f. of 1 volt. If the induced e.m.f. is 2 V, the inductance is 2 H. The millihenry (1 mH = 10^{-3} H) and the microhenry (1 μH = 10^{-6} H) are more convenient sub-units.

In general the inductance (also called the self-inductance) of a coil increases if

(i) its cross-sectional area is large and its length small,

(ii) it has a large number of turns, and

(iii) it has a core of magnetic material.

It can be shown that the energy W stored in the magnetic field of an inductor of inductance L carrying a current I is

$W = \frac{1}{2}LI^2$ (for a capacitor $W = \frac{1}{2}CV^2$)

where W is in joules if L is in henrys and I in amperes.

Inductors in a.c. circuits

(a) Inductive reactance. In a d.c. circuit when the current is constant, the opposition of an inductor is due entirely to the resistance of the copper wire used to wind it. In an a.c. circuit the current is changing all the time and opposition arises not only from the resistance of the coil but also because of its inductance. The opposition due to the latter is called the *inductive reactance* X_L of the inductor and its value is calculated from

$X_L = 2\pi fL$ (compare $X_C = 1/(2\pi fC)$)

where X_L is in ohms if f is in hertz and L in henrys.

X_L increases if either f or L increase since in both cases the e.m.f. induced by the changing current (and magnetic field) would be greater. Fig. 11.5 shows the linear relationship between X_L and f for a given L; compare it with Fig. 10.9 for X_C and f. As an example, if $f = 50\,\text{Hz}$ and $L = 10\,\text{H}$, then $X_L = 2\pi \times 50 \times 10 = 3.1\,\text{k}\Omega$: if $f = 500\,\text{Hz}$, $X_L = 31\,\text{k}\Omega$.

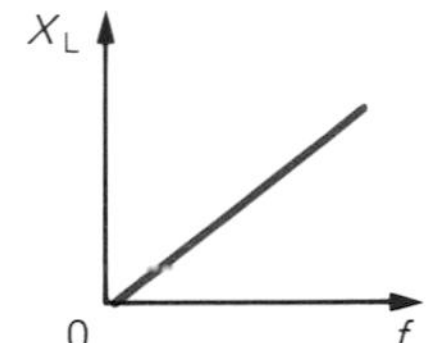

Fig. 11.5

If an r.m.s. voltage V is applied to an inductor of reactance X_L, the r.m.s. current I is given by

$$I = \frac{V}{X_L} \quad \left(\text{compare } I = \frac{V}{X_C} \text{ for a capacitor}\right)$$

(b) Phase shift. The current 'through' a capacitor in an a.c. circuit *leads* the p.d. across it (see Fig. 10.11a). In an inductor the current I lags behind the p.d. V, Fig. 11.6. The phase shift arises from the fact that when the current starts to flow, although it is small at first, it is increasing at its fastest rate, therefore so too is the build-up of the magnetic field. As a result the e.m.f. induced in the inductor has its maximum value and opposes the applied p.d.

(c) Power. Like a perfect capacitor, a perfect inductor, i.e. one with zero resistance, is a 'wattless' component. Its power against time (P–t) graph is obtained from Fig. 11.6 (as for a capacitor) and is the same as Fig. 10.11b.

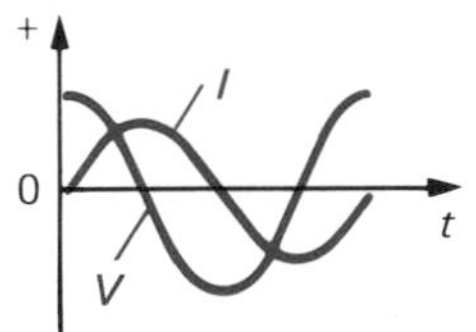

Fig. 11.6

Again the average power taken from the a.c. supply over a cycle is zero. In this case the energy drawn from the supply on the first quarter-cycle is stored in the magnetic field of the inductor. During the second quarter-cycle, the current and magnetic field decrease and the e.m.f. induced in the inductor makes it act as a generator returning the energy stored to the supply. In practice an inductor has resistance and some energy is taken from the supply on this account and not replaced.

Current control by an inductor in a.c. circuits

If the current in an a.c. circuit is to have a certain value, the necessary 'opposition' can be provided by a resistor but it wastes electrical energy as heat. The opposition of an inductor to a.c. is due to its reactance X_L and its resistance R. Although both are measured in ohms they cannot be added directly (because of phase shifts) to give the combined opposition. This is called the *impedance* Z of the inductor and is given by

$$Z - \sqrt{R^2 + X_L^2}$$

For example, if the resistance R of an inductor coil $= 50\,\Omega$ and X_L at a certain frequency $= 120\,\Omega$, then $Z = \sqrt{50^2 + 120^2} = \sqrt{2500 + 14\,400} = \sqrt{16\,900} = 130\,\Omega$ at that frequency. The r.m.s. current I is found from

$$I = \frac{V}{Z}$$

where V is the r.m.s. voltage.

If an inductor is used to control current in an a.c. circuit instead of a resistor only part of the required opposition is resistive and so less heat is produced since the reactance is 'wattless'.

Current control in an a.c. circuit is also possible using a capacitor, the opposition being its reactance X_C. Usually, however, it is only used for high fre quency currents. At low frequencies C would be inconveniently large to give low values of X_C.

Uses of inductors

(a) Air-cored types. These have small inductances (e.g. 1 mH) and are used at high frequencies, either in radio tuning circuits above 2 MHz (p. 181) or as r.f. 'chokes' to stop radio frequency currents taking certain paths in a circuit. Their reactance is large to radio frequencies but small to low frequencies, enabling them to separate r.f. from a.f. signals.

(b) Iron-cored types. In a current-carrying coil with an iron core, the core becomes magnetized and the strength of this magnetic field is several hundred times greater than that due to the coil alone. For a typical iron-cored inductor $L = 10\,\text{H}$.

Iron cores are laminated, that is, they consist of flat sheets which are coated thinly with an insulator. The laminations thus offer a high resistance to currents, called *eddy currents*, that would otherwise be produced in the core and cause energy losses as

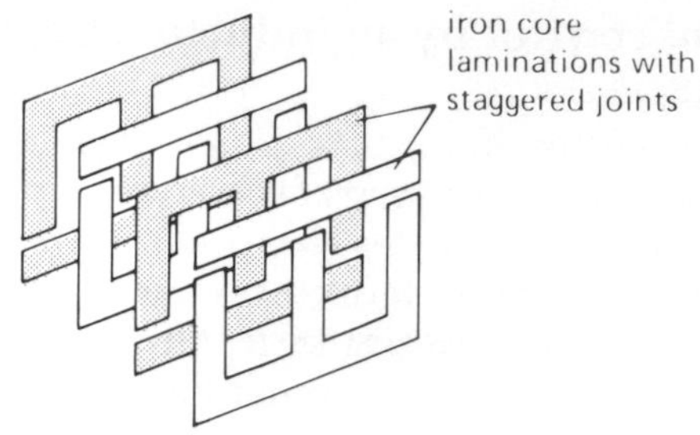

Fig. 11.7

heat. E- and I-shaped laminations are shown in Fig. 11.7.

Iron-cored inductors are used as low frequency smoothing chokes in power supplies (p. 90). They are also used in fluorescent lamps (p. 77) to control a.c.

(c) Iron-dust and ferrite types. They are used at high frequencies. The iron-dust type is in the form of a powder, coated with an insulator and pressed to give a magnetic core of high resistance which reduces eddy current losses. Ferrite cores are made from non-conducting, magnetic materials.

Both types are used in radio tuning circuits up to about 2 MHz. They enable a small coil to give the required inductance, which may be varied by screwing the core in or out of the coil, Fig. 11.1c. The aerial of a radio receiver is often a coil on a ferrite rod, Fig. 11.1d.

Questions

1. What property do all inductors have as well as inductance?

2. At a certain frequency of a.c. a resistor, a capacitor and an inductor each offer the same 'opposition' to the a.c. How does the opposition of each change (if it does) when the frequency of the a.c. increases?

3. What is the inductance of an inductor in which a current changing at the rate of $1\,\mathrm{A\,s^{-1}}$ induces an e.m.f. of 100 mV?

4. Calculate the inductive reactance of (a) a 15 H smoothing choke at 100 Hz, (b) a 1 mH r.f. choke at 1 MHz.

5. When 9 V is applied to an inductor the current through it is 3 A if the supply is d.c. and 0.1 A if it is a.c. What is (a) the resistance R, (b) the impedance Z, (c) the reactance X_L (ignore R compared with Z) and (d) the inductance L of the inductor if the a.c. has frequency 50 Hz?

12 CR and LR circuits

Capacitor charging in a CR circuit

In the circuit of Fig. 12.1, when S is in position 1, C charges through R from the supply. The microammeter measures the charging current I and the voltmeters record the p.ds V_C and V_R across C and R respectively at different times t.

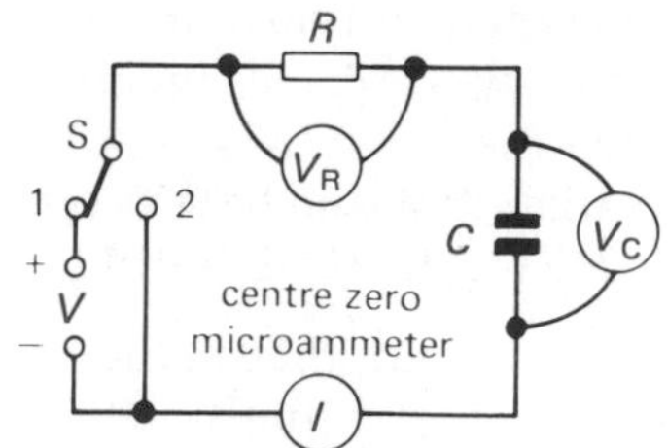

Fig. 12.1

Graphs like those in Fig. 12.2a and b can be plotted from the results and show that

(i) I has its maximum value at the start and decreases more and more slowly to zero as C charges up,

(ii) V_C rises rapidly from zero and slowly approaches the supply voltage V which it equals when C is fully charged, and

(iii) V_R behaves like I.

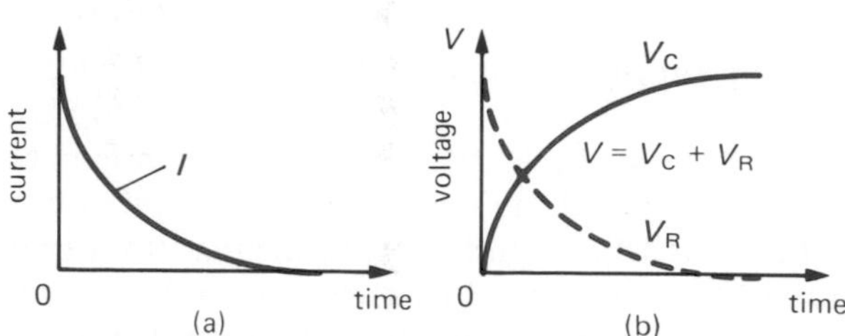

Fig. 12.2

All three graphs are *exponential* curves, I and V_R being 'decaying' ones and V_C a 'growing' one. In general we can say that at any time,

$$I = \frac{V - V_C}{R}, \; V_C = V - V_R \text{ and } V_R = IR$$

When charging starts, $V_C = 0$ therefore $I = V/R$ and $V_R = V$.
When charging stops, $V_C = V$ therefore $I = 0$ and $V_R = 0$.

Capacitor discharging in a CR circuit

In Fig. 12.1 when S is moved from position 1 to position 2, C discharges through R. If graphs of I, V_C, and V_R are plotted as before, they are again exponential curves like those in Figs. 12.3a and b.

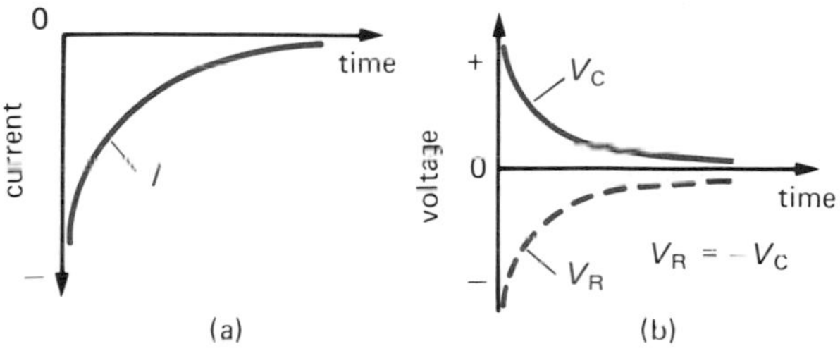

Fig. 12.3

They show that

(i) I, the discharge current, has its maximum value at the start but is in the opposite direction to the charging current (as is V_R), and
(ii) V_C and V_R fall as C discharges and are equal and opposite at all times.

In general we can say that at any time

$$I = \frac{V_C}{R} \text{ and } V_R = IR$$

Time constant of CR circuit

The charging and discharging of a capacitor through a resistor do not occur instantaneously. The *time constant* is a useful measure of how long these processes take in a particular CR circuit.

(a) Charging. If a capacitor of capacitance C is charged at a constant rate through a resistor of resistance R by a *steady current* I, it would be fully charged with charge Q and p.d. V after a time t where $t = CR$ seconds if C is in farads and R in ohms. To prove this we use

$$Q = It = \frac{V}{R} \times t \qquad \text{(from } V = IR\text{)}$$

also $$Q = VC$$

therefore $$\frac{V}{R} \times t = VC \qquad \therefore t = CR$$

In fact, as we have seen, the charging current decreases exponentially with time, Fig. 12.2a, and it can be shown that

(i) in CR seconds, called the *time constant*, the p.d. V_C across the capacitor rises to only 0.63 of its final value, and

(ii) in $5CR$ seconds the capacitor is, in effect, fully charged and V_C has the final value.

For example, if $C = 500\,\mu\text{F} = 500 \times 10^{-6}\,\text{F}$ and $R = 100\,\text{k}\Omega = 10^5\,\Omega$, then $CR = 500 \times 10^{-6} \times 10^5 = 50\,\text{s}$. Hence, if the charging p.d. $V = 9\,\text{V}$ then after 50 s, $V_C = 0.63 \times 9 \approx 2/3 \times 9 = 6\,\text{V}$. After 100 s, V_C rises by 2/3 of the p.d. remaining across C after 50 s, i.e. by $2/3$ of $(9-6) = 2/3 \times 3 = 2\,\text{V}$, so making $V_C = 6 + 2 = 8\,\text{V}$. After 250 s, $V_C \approx 9\,\text{V}$. See Fig. 12.4.

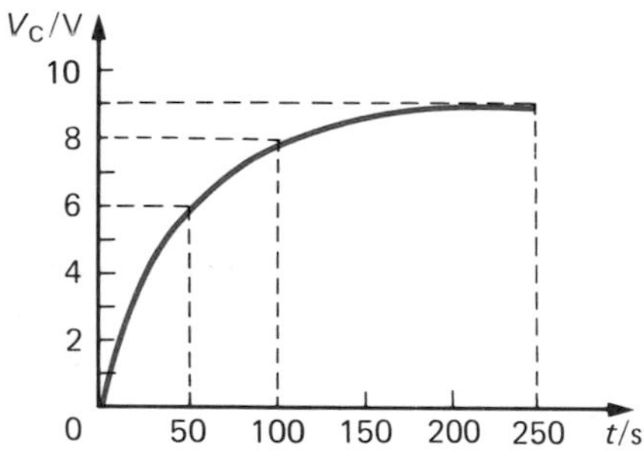

Fig. 12.4

(b) Discharging. The time constant CR is useful here too and is the time for V_C to fall by 0.63 ($\approx 2/3$) of its value at the start of discharging. For example, considering the discharge of C in **(a)**, we can say that after 50 s, V_C will fall by 2/3 of 9 V, i.e. by 6 V, from 9 V to 3 V. In the next 50 s it falls by 2/3 of 3 V, i.e. by 2 V, making $V_C = 1\,\text{V}$. After about 250 s ($5CR$), $V_C = 0$.

Worked example

In the circuit of Fig. 12.1, $C = 10\,\mu F$, $R = 10\,k\Omega$ and $V = 10\,V$. If S is switched to position 1, what will be (a) *the maximum current through R,* (b) *the maximum charge on C?*

(a) Maximum current I occurs at the instant S is closed, i.e. before the p.d. builds up across C and opposes the voltage V of the supply. It is given by

$$I = V/R = 10\,V/10\,k\Omega = 1\,mA$$

(b) Maximum charge Q occurs when the charging current has fallen to zero and the p.d. across C equals V, i.e.

$$Q = VC = 10\,V \times 10\,\mu F = 100\,\mu C$$

Capacitors and p.d. changes

A capacitor tends to hold constant the charge and so also the p.d. between its plates. A rise or fall of potential on one plate creates a rise or fall of potential on the other. Hence an a.c. voltage which changes potential rapidly is passed on from one plate to the other. A d.c. voltage on the other hand gives the capacitor time to alter its charge to suit the new p.d. and is blocked. The time needed to adjust to different p.ds depends on the time constant of the circuit.

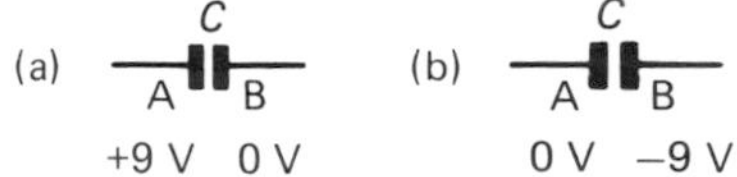

Fig. 12.5

In Fig. 12.5a suppose C is charged to a p.d. of 9 V with plate A having a potential of +9 V and plate B being at 0 V. If the potential of plate A suddenly falls to 0 V, the p.d. cannot change instantaneously since charging and discharging take time. Therefore the potential of plate B must also drop by 9 V, to −9 V, to maintain the p.d. of 9 V across it, Fig. 12.5b.

CR coupled circuits

Parts of a circuit are often coupled (joined) by a capacitor and resistor. The output of the circuit may then not have the same waveform as the input, depending on the time constant of the coupling. The effect can be studied with an input of square waves since it contains many frequencies (p. 20).

(a) Capacitor coupling. The coupling circuit is given in Fig. 12.6a and the square wave input of period T in Fig. 12.6b. When the input rises from 0 to V (i.e. AB in Fig. 12.6b), the output p.d. V_R, which is taken from across R, rises immediately to V because the p.d. across C cannot change suddenly, i.e. $V_C = 0$. C starts to charge through R, V_C rises and V_R falls and the rate at which they do so depends on the time constant CR.

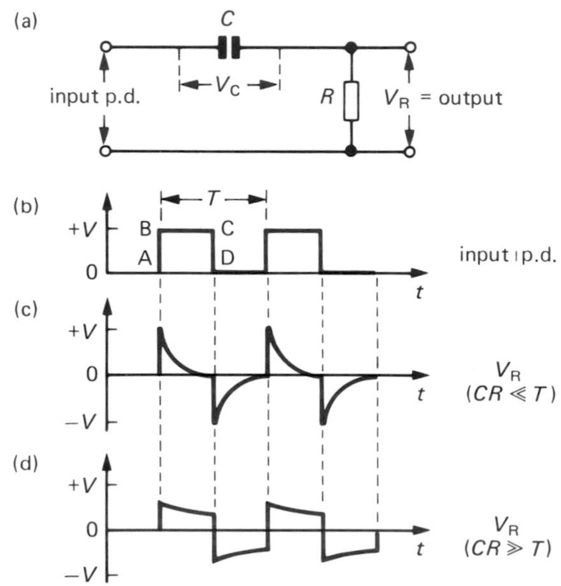

Fig. 12.6

If CR is much *smaller* than T, C has time to charge and discharge before the input reverses again. During charge, V_R varies as in Fig. 12.2b and during discharge (CD in Fig. 12.6b), it varies as in Fig. 12.3b. The complete V_R waveform is shown in Fig. 12.6c and is very different from the input, i.e. distortion has occurred.

If CR is much *greater* than T, C charges and discharges slowly and V_R more or less follows the input. Its waveform settles down to that in Fig. 12.6d, i.e. very little distortion occurs. This coupling circuit is used to connect the stages of an audio amplifier, the time constant being about 10 times greater than the period of the input waveform (e.g. 1/100 s for a 1 kHz input).

Also note that V_R is an alternating p.d. whereas the input p.d. is direct. If we think of the input as being steady d.c. plus a.c., then only the a.c. passes through C as the output.

(b) Resistor coupling. In this case, Fig. 12.7a, the output is the p.d. across C, i.e. V_C, and the waveforms are as in Fig. 12.7b, c and d. When CR is

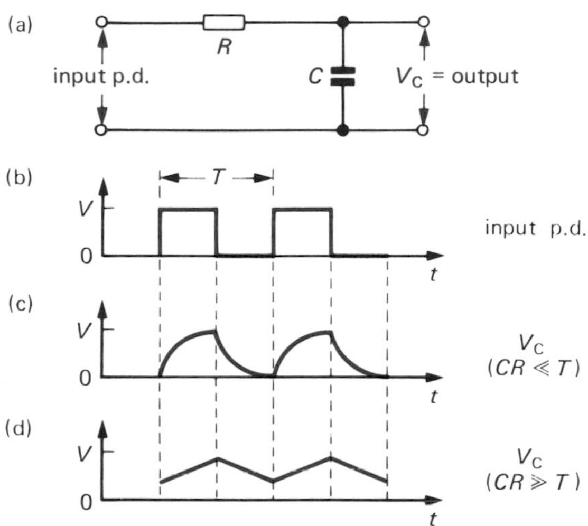

Fig. 12.7

much greater than T, V_C does not change much and the input p.d. changes are smoothed in the output. This kind of coupling is used for smoothing power supplies (p. 89) and for detection in radio receivers (p. 181).

LR circuit

The time constant can also be used as a measure of the time taken by the current to rise or fall in a circuit containing inductance L and resistance R, Fig. 12.8. It is given in seconds by L/R if L is in henrys and R in ohms.

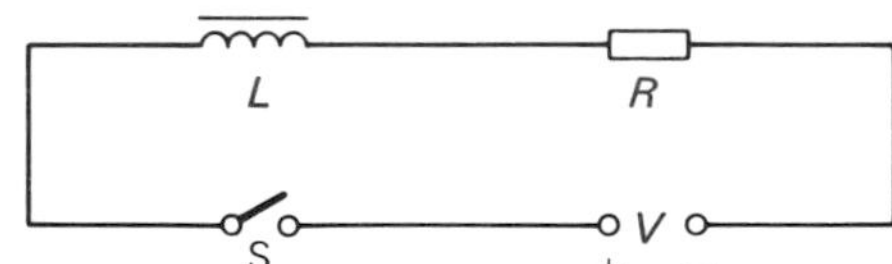

Fig. 12.8

When S is closed the current I rises to 0.63 (about 2/3) of its final steady value in L/R seconds. In $5L/R$ seconds I will have its final value of V/R, which does not depend on L, Fig. 12.9.

When S is opened, R becomes very large, making L/R very small and causing the rapidly collapsing magnetic field to induce a large 'back' e.m.f. in L.

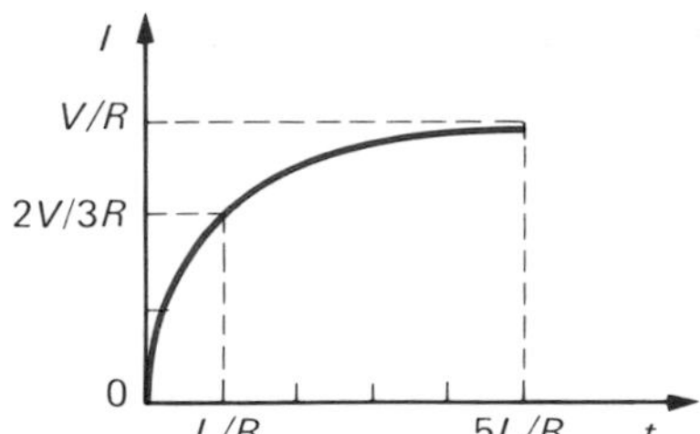

Fig. 12.9

Questions

1. What p.d. is reached across a 100 μF capacitor when it is charged by a constant current of 20 μA for 1 minute?

2. What is the time constant of a circuit in which (a) C = 1 μF and R = 1 MΩ, (b) C = 100 μF and R = 50 kΩ?

3. In the circuit of Fig. 12.10, C starts to charge up through R when S is closed. What is the p.d. (approximately) across

(a) the capacitor after (i) 1 s, (ii) 2 s, (iii) 5 s, and
(b) the resistor after (i) 1 s, (ii) 2 s, (iii) 5 s?

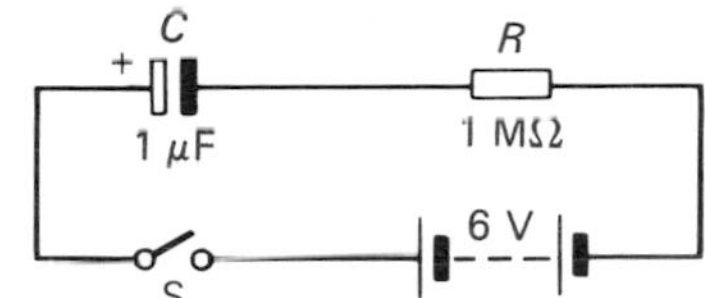

Fig. 12.10

4. In the circuit of Fig. 12.11, C starts to discharge through R when S is closed. What is the p.d. (approximately) across

(a) the capacitor after (i) 5 s, (ii) 10 s, (iii) 25 s and
(b) the resistor after (i) 5 s, (ii) 10 s, (iii) 25 s?

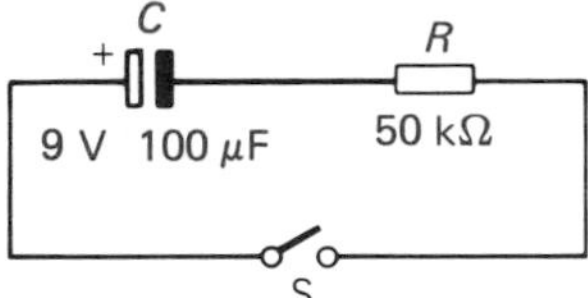

Fig. 12.11

5. A 1.5 V cell is connected to a 1000 μF capacitor in series with a 150 Ω resistor.

(a) What is the maximum current which flows through the resistor during charging?
(b) What is the maximum charge on the capacitor?
(c) How long does the capacitor take to charge to 1.0 V?

6. A capacitor is charged to 6 V so that one plate, A, is at +6 V and the other plate, B, at 0 V. What is the potential of plate B if plate A is suddenly connected to (a) 0 V, (b) +12 V, (c) −6 V? (Assume in each case that initially A is at +6 V and B at 0 V.)

13 LCR circuits

LCR series circuit

(a) Impedance. If an a.c. supply of r.m.s. voltage V and frequency f is applied to a series circuit having components of inductance L, capacitance C and resistance R, Fig. 13.1, each offers some opposition

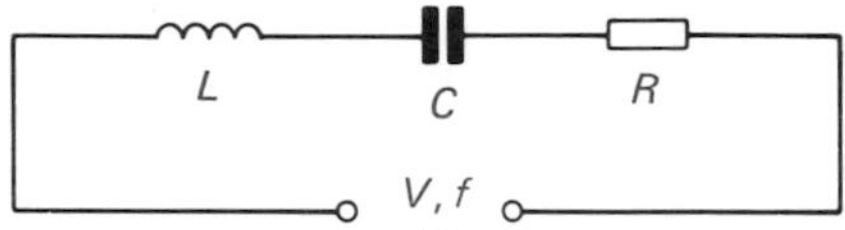

Fig. 13.1

to the current. The total opposition is called the *impedance* (symbol Z) and is measured (like resistance and reactance) in ohms. It can be shown that

$$Z = \sqrt{R^2 + (X_L - X_C)^2} \qquad (1)$$

where $X_L = 2\pi fL$ and $X_C = 1/(2\pi fC)$

The r.m.s. current I in the circuit is given by

$$I = \frac{V}{Z} \qquad \text{(compare } I = \frac{V}{R} \text{ for d.c.)}$$

(b) Example. If $L = 2.0\,\text{H}, C = 10\,\mu\text{F}, R = 100\,\Omega, V = 24\,\text{V}$ and $f = 50\,\text{Hz}$, calculate X_L, X_C, Z and I.

We have

$$X_L = 2\pi \times 50 \times 2.0 = 200\pi = 630\,\Omega$$

$$X_C = 1/(2\pi \times 50 \times 10 \times 10^{-6}) = 320\,\Omega$$

$$\therefore \quad Z = \sqrt{100^2 + (630 - 320)^2} = 330\,\Omega$$

$$\text{Hence} \quad I = \frac{V}{Z} = \frac{24}{330} = 0.073\,\text{A} = 73\,\text{mA}$$

Resonant circuits

A resonant circuit consists of a capacitor and inductor in series or in parallel and is used for frequency selection.

(a) Series. In the circuit of Fig. 13.2a, when the frequency of the a.c. has a certain value f_o, called the *resonant frequency*, $X_L = X_C$. From equation (1) it follows that Z then has a minimum value, equal to R which in this case is the small resistance of the inductor coil. The *current* I is a maximum at resonance, given by V/R.

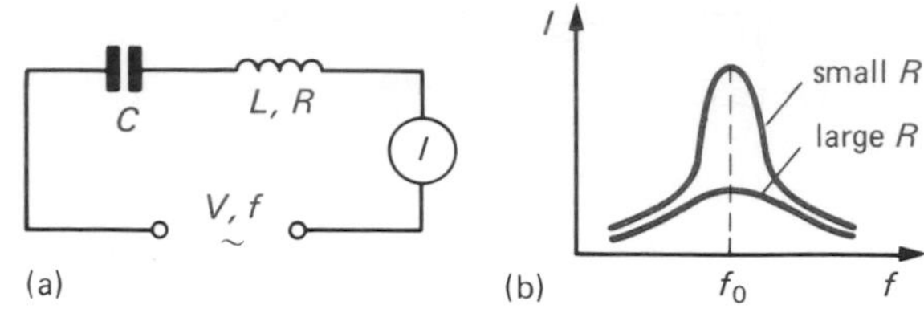

Fig. 13.2

The expression for f_o is obtained from $X_L = X_C$, that is

$$2\pi f_o L = \frac{1}{2\pi f_o C}$$

$$\text{or} \quad 4\pi^2 f_o^2 LC = 1$$

$$\therefore \quad f_o = \frac{1}{2\pi\sqrt{LC}}\ \text{Hz}$$

where L is in henrys and C in farads.

The graph of I against f for the circuit is shown in Fig. 13.2b. The greatest response occurs at f_o but the selectivity, i.e. the sharpness of the peak, falls off as R increases.

(b) Parallel. For this circuit, Fig. 13.3a, f_o is given by the same expression but Z is a maximum and I a minimum at resonance. The *voltage* developed across the circuit at resonance is thus a maximum (for the same I at other frequencies it would be smaller due to Z being smaller). Fig. 13.3b shows the response curve, this time of Z against f.

(c) Uses. Tuning in a radio receiver (p. 181) is done using resonant circuits in which C (or L) is varied until f_o equals the frequency of the wanted signal.

Note. As in all cases of resonance, a 'driving force' (here the applied p.d. V) is coupled to a 'driven system' (the *LCR* circuit) which responds to the correct driving frequency.

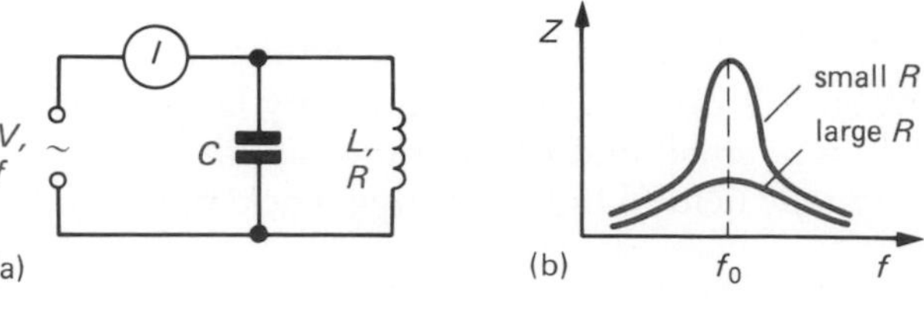

Fig. 13.3

Oscillatory circuit

If a charged capacitor C discharges through an inductor L in a circuit of low resistance, Fig. 13.4a, an a.c. of constant frequency and decreasing amplitude, called a damped oscillation, is produced which eventually dies off, Fig. 13.4b. Its frequency f, known as the *natural frequency* of the circuit, has the same value as the resonant frequency, that is, $f = 1/(2\pi\sqrt{LC})$.

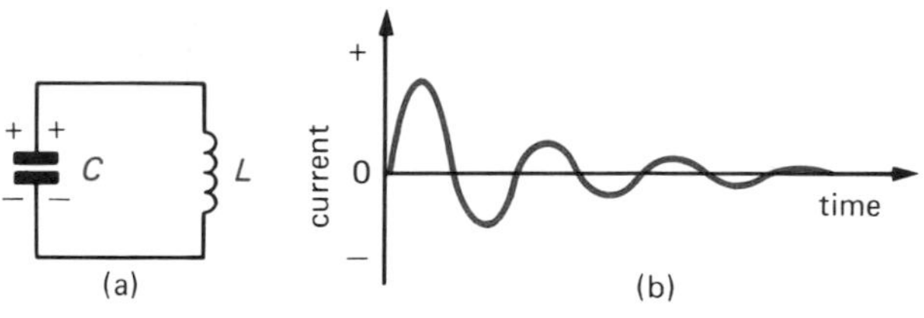

Fig. 13.4

The oscillations occur because the energy stored initially in C is transferred to the magnetic field of L which, when C has discharged, collapses and tries to keep the current going. C charges in the opposite direction. This is repeated but energy is gradually lost due to the resistance of L.

The LC circuit is the basis of oscillators (p. 116) for producing a.c. of constant amplitude, i.e. undamped oscillations.

Questions

1. Calculate Z if $R = 6\Omega$, $X_L = 58\Omega$ and $X_C = 50\Omega$.

2. What is the resonant frequency if $L = 1\,\text{mH}$ and $C = 1\,\text{nF}$?

14 Transformers

About transformers

A transformer changes (transforms) an alternating p.d. from one value to another. It consists of two coils, called the *primary* and the *secondary* which are not connected electrically. The windings are either one on top of the other or are side-by-side on an iron, iron-dust or air core. Transformer symbols are given in Fig. 14.1.

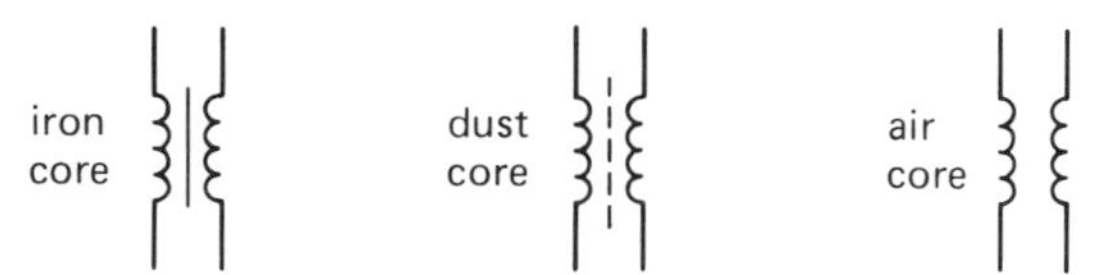

Fig. 14.1

A transformer works by electromagnetic induction. An alternating p.d. is applied to the primary and produces a changing magnetic field which passes through the secondary, thereby inducing an alternating p.d. in it. A practical arrangement for ensuring as much as possible of the magnetic field links the secondary is shown in the iron-cored transformer of Fig. 14.2; the secondary is wound on top of the primary.

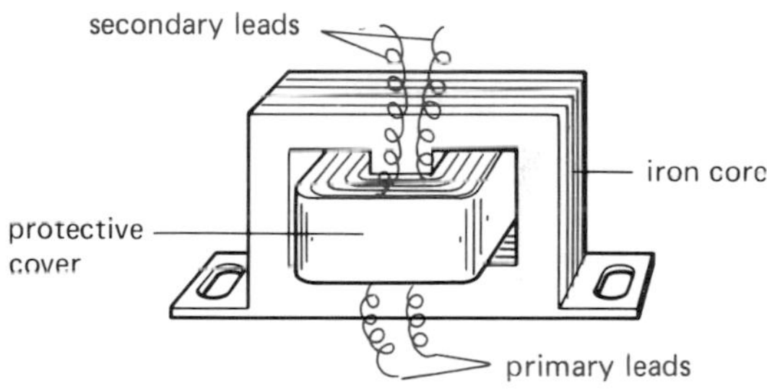

Fig. 14.2

Transformer equations

(a) Turns ratio. It can be shown that if a transformer is 100% efficient at transferring electrical energy from primary to secondary (many are nearly so), then:

$$\frac{\text{secondary p.d.}}{\text{primary p.d.}} = \frac{\text{turns on secondary}}{\text{turns on primary}}$$

In symbols,

$$\frac{V_s}{V_p} = \frac{n_s}{n_p}$$

If n_s is twice n_p, the transformer is a *step-up* one and V_s will be twice V_p. In a *step-down* transformer there are fewer turns on the secondary than on the primary and V_s is less than V_p. The ratio n_s/n_p is the *turns ratio.*

(b) Power. If the p.d. is stepped-up by a transformer the current is stepped-down in proportion and vice versa. This must be so if we assume that all the electrical energy given to the primary appears in the secondary. Hence

$$V_p \times I_p = V_s \times I_s$$

where I_p and I_s are the primary and secondary currents. A transformer (unlike an amplifier, p. 104) gives no power gain.

Worked example

If the transformer in Fig. 14.3 is 100% efficient, calculate (a) *the p.d. V_s across the 4 Ω resistor and* (b) *the primary current I_p.*

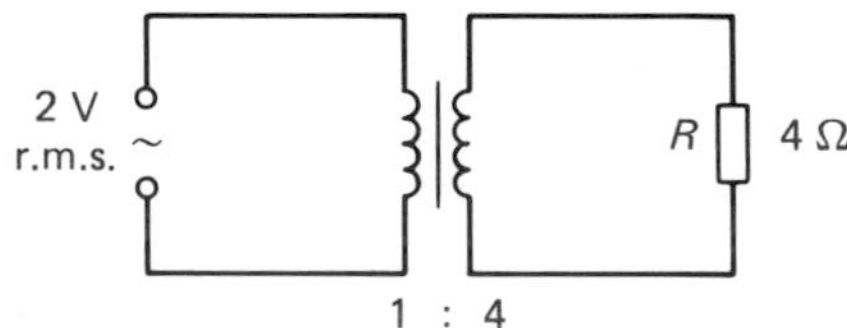

Fig. 14.3

(a) $V_s/V_p = n_s/n_p = 4/1$

$\therefore \quad V_s = 4 \times V_p = 4 \times 2 = 8\,\text{V}$

(b) $I_s = V_s/4 = 8/4 = 2\,\text{A}$

But $I_p \times V_p = I_s \times V_s$

$\therefore \quad I_p \times 2 = 2 \times 8$

$\therefore \quad I_p = 8\,\text{A}$

Types of transformer

(a) Mains, Fig. 14.4a. The primary is connected to the a.c. mains supply (240 V 50 Hz in U.K.) and the secondary may be step-up or step-down or there may be one or more of each. They have laminated iron cores and are used in power supplies (p. 87).

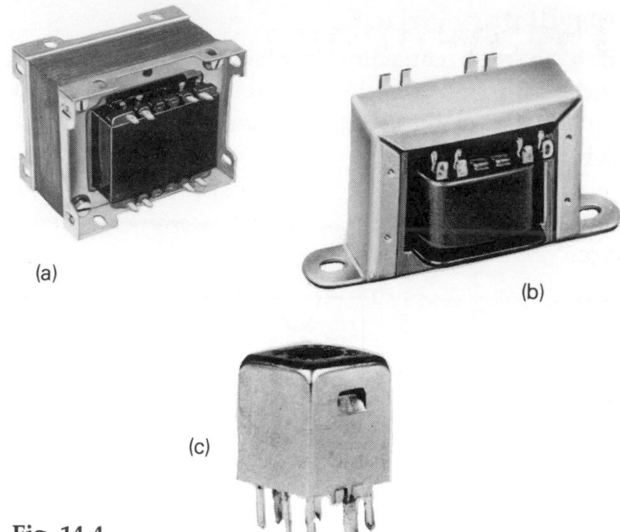

Fig. 14.4

(b) Audio frequency, Fig. 14.4b. It also has a laminated iron core and acts as a matching transformer to ensure maximum power transfer from, for example, an amplifier to a loudspeaker (p. 173).

(c) Radio frequency, Fig. 14.4c. It has an iron-dust core and forms part of the tuning circuit in a radio (p. 181), being surrounded by a small metal 'screening' can to stop radiation from it getting to other parts of the circuit.

Question

1. A 12 V lamp is operated from a 240 V a.c. mains step-down transformer.

(a) What is the turns ratio?

(b) How many turns are on the primary if the secondary has 80 turns?

(c) What is the primary current if the current in the lamp is 2 A?

15 Switches

About switches

In a mechanical switch, metal contacts have to be brought together or separated to make or break a circuit.

A switch is rated according to (i) the *maximum current* it can carry and (ii) its *working voltage.* These both depend on whether it is to be used in a.c. or d.c. circuits. For example, one with an a.c. rating of 250 V 1.5 A, has a d.c. rating of 20 V 3 A. If these values are exceeded the life of the switch is shortened. This is due to overheating while it carries current or to vaporization of the contacts due to sparking as a result of the current trying to keep flowing in the air gap when it is switched off. In general sparking lasts longer with d.c. than with a.c. because a.c. falls to zero twice per cycle.

Switches have different numbers of *poles* and *throws.* The poles (P) are the number of separate circuits the switch makes or breaks at the same time. The throws (T) are the number of positions to which each pole can be switched. The symbols for various types are given in Fig. 15.1. In a SPDT (single pole double throw) switch there are two positions for the switch (B or C) and only one circuit (that joined to A) is switched.

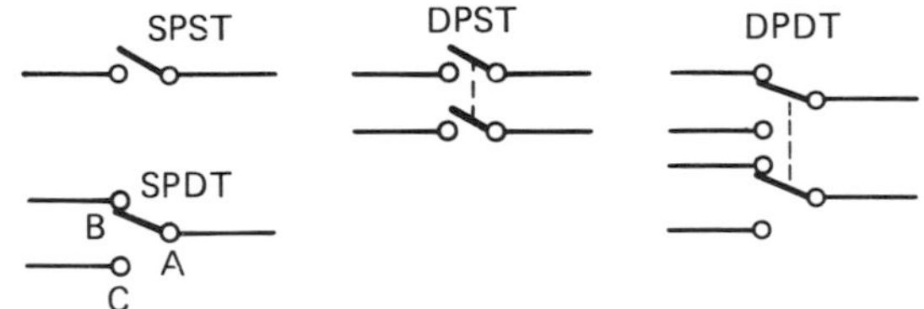

Fig. 15.1

Types of switch

(a) Push-button. The one in Fig. 15.2a is a 'push-on, release-off' type; its symbol is given in Fig. 15.2b and that for the 'push-off, release-on' variety in Fig. 15.2c. 'Push-to-change-over' switches are also made; their symbol is shown in Fig. 15.2d.

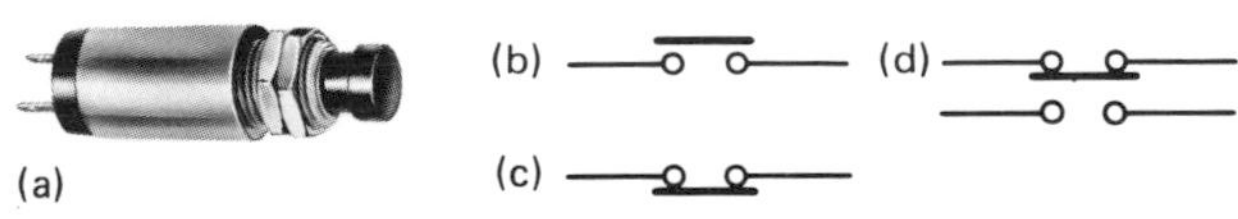

Fig. 15.2

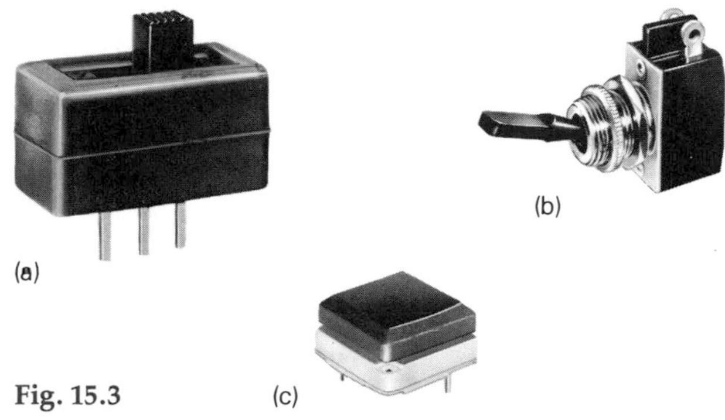

Fig. 15.3

(b) Slide. The slide switch in Fig. 15.3a is a 'change-over' SPDT type.

(c) Toggle. This type is often used on equipment as a power supply 'on-off' switch, either in the SPST form shown in Fig. 15.3b or as a SPDT, DPST or DPDT type.

(d) Keyboard. The one shown in Fig. 15.3c is a SPST push-to-make momentary type which can be mounted on a printed circuit board (p.c.b.).

(e) Rotary wafer. One or more insulating plastic discs or wafers are mounted on a twelve-position spindle as shown in Fig. 15.4a. The wafers have metal contact strips on one or both sides and rotate between a similar number of fixed wafers with springy contact strips. The contacts on the wafers can be arranged to give switching that is 1 pole 12 throw, 2 pole 6 throw, 3 pole 4 throw, 4 pole 3 throw (as in Fig. 15.4b) or 6 pole 2 throw.

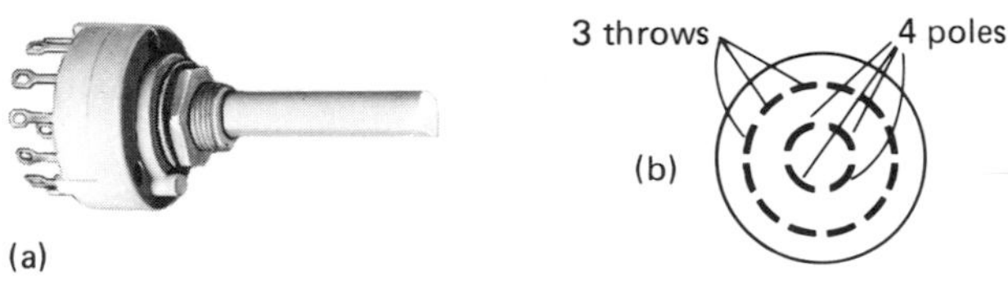

Fig. 15.4

Two switch circuits

Two circuits in which switches are used to perform logical tasks are given overleaf.

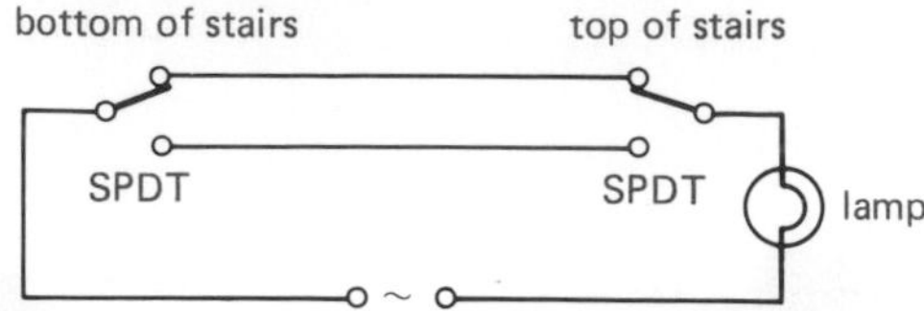

Fig. 15.5

(a) Two-way control of staircase lighting. The light can be switched on or off either by the SPDT switch at the top or the bottom of the stairs.

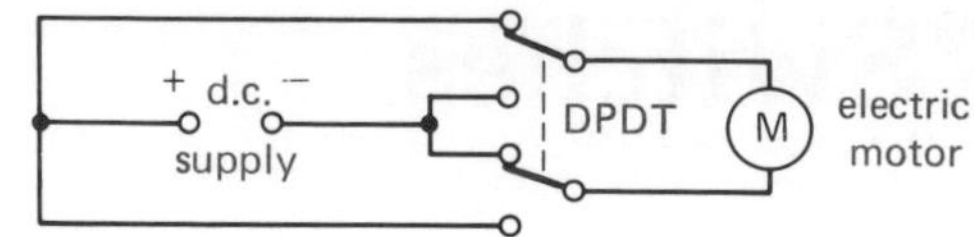

Fig. 15.6

(b) Reversing polarity of supply to electric motor. By changing the polarity of the supply to the motor (a permanent magnet type, p. 52) its direction of rotation is reversed by the DPDT switch.

16 Progress questions

1. A resistor may have four coloured bands on it (a, b, c and d in Fig. 16.1). What do each of these bands indicate? What is the maximum current (approximately) that can be conducted safely by a 1 kΩ $\frac{1}{2}$W resistor? *(O. and C.)*

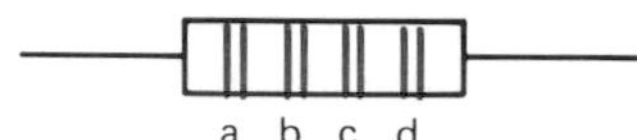

Fig. 16.1

2. A number of 1000 Ω resistors are available.
(a) State the colour coding for a 1000 Ω resistor.
(b) By means of diagrams, show how the resistors could be connected to make: (i) a combination of resistance 3000 Ω; (ii) a combination of resistance 500 Ω; (iii) a combination of resistance 1250 Ω.
(c) Each of the 1000 Ω resistors will safely dissipate a maximum power of 0.5 W. Show how they could be connected to make a combination of resistance 1000 Ω and power rating 2 W. Explain. *(O.L.E.)*

3. A manufacturer's catalogue lists capacitors as 0.1 μF, 40 V d.c. What does this mean? What charge is stored by one such capacitor when it has a p.d. of 20 V across it? *(O. and C.)*

4. A 2 μF capacitor is charged to a p.d. of 100 V and immediately discharged through a 5 Ω resistor. Calculate
(a) the charge taken by the capacitor,
(b) the heat generated in the resistor. *(C.)*

5. For the circuit in Fig. 16.2 calculate
(a) the capacitance between X and Z,
(b) the charge on the 5 μF capacitor,
(c) the p.d. between X and Y. *(C.)*

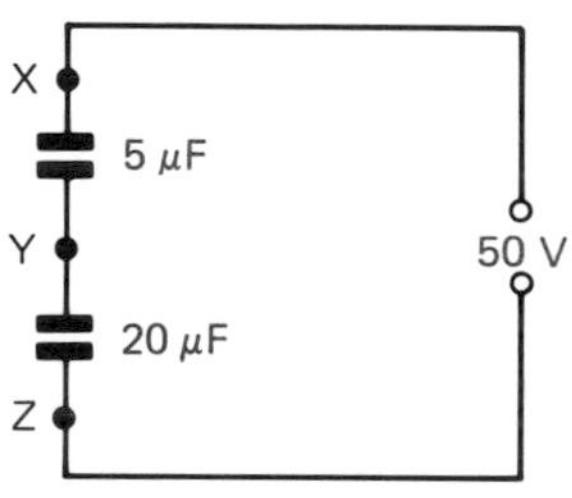

Fig. 16.2

6. If an a.c. signal of peak p.d. 5 V and frequency 1000 Hz is applied to the circuit in Fig. 16.3, calculate
(a) the r.m.s. current through the 200 Ω resistor,
(b) the reactance of the 8 μF capacitor,
(c) the peak current through the 8 μF capacitor. *(C.)*

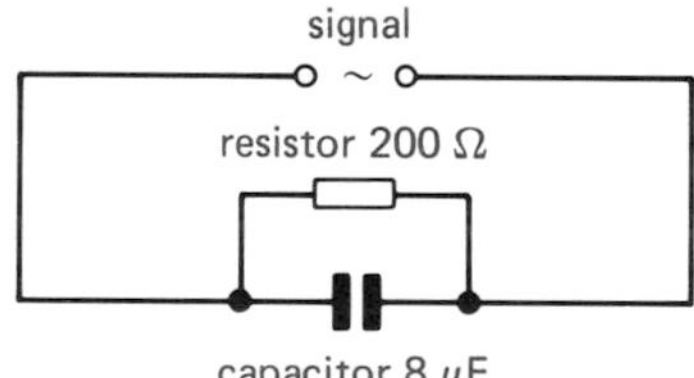

Fig. 16.3

7. Draw a labelled diagram to show the structure of a low-frequency choke. What is meant by the *self-inductance* of a choke? On what factors does its value depend?
A choke (of negligible resistance) has a self-inductance of 0.5 H. Draw a graph to show how its reactance varies with frequency between 0 and 1000 Hz. This choke is connected across a 2 V peak-to-peak generator of negligible internal resistance. The generator frequency is 50 Hz. Draw a graph to show how the current and voltage in the circuit vary in one cycle, and explain their phase relationship to one another.
What is the r.m.s. current I in the circuit at 50 Hz? At what frequency will the r.m.s. current be $\frac{1}{2}I$? *(C.)*

8. Calculate, for the circuit of Fig. 16.4,
(a) the maximum current,
(b) the maximum charge which can be stored in the capacitor,
(c) the time constant of the circuit.
Hence draw sketch maps of the current against time and

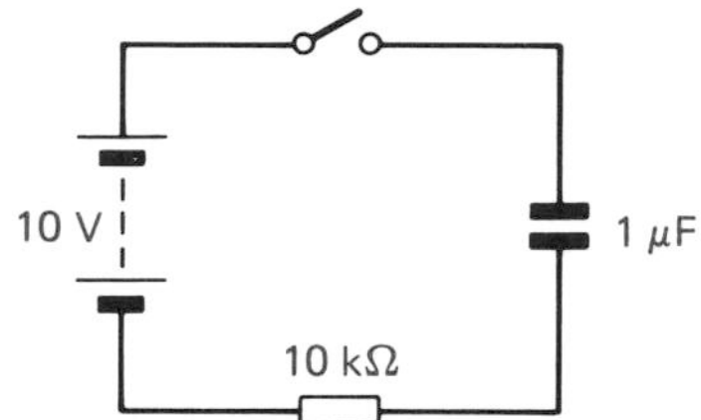

Fig. 16.4

the charge stored against time. Take $t = 0$ as the moment when the switch is closed. Label all axes. (*L.*)

9. For the circuit in Fig. 16.5 sketch a graph showing
(a) how the voltage V across the capacitor varies with time when S is switched from position 1 to position 2,
(b) the maximum value of V, and
(c) the approximate value of V after a time equal to the time constant.

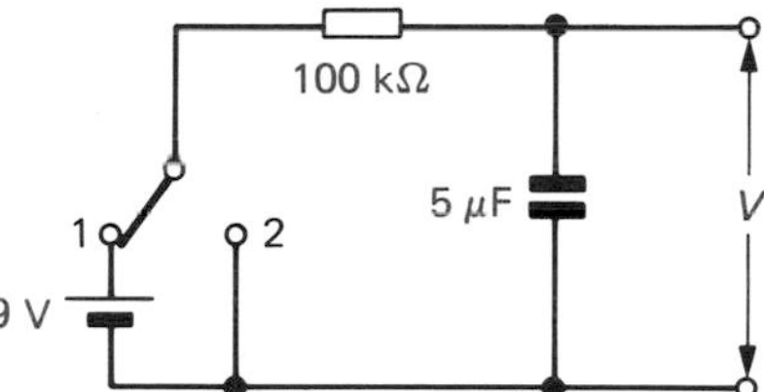

Fig. 16.5

10. In Fig. 16.6 three circuits are shown each with an input waveform. Copy the input waveforms and below each draw the output waveform V_o. (*L. part qn.*)

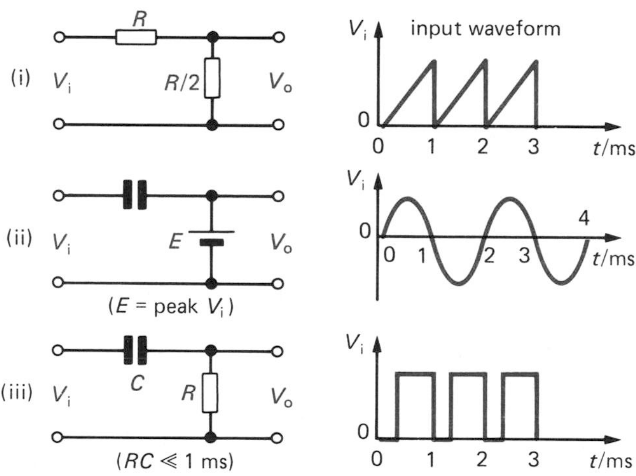

Fig. 16.6

11. (a) What is meant by the self-inductance of a coil? Define the henry and explain what one means by a self-inductance of 0.1 H.

Fig. 16.7

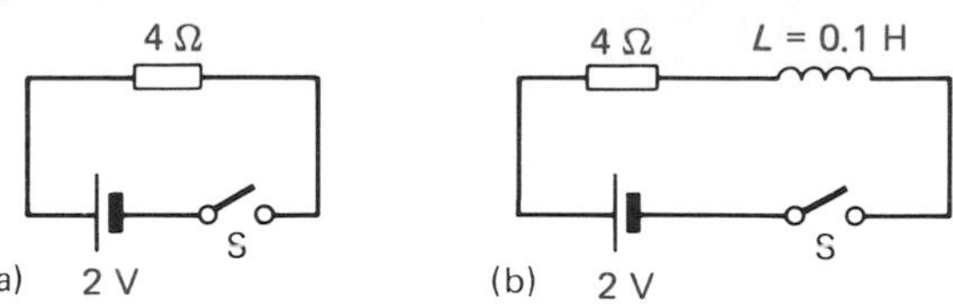

(b) Draw a graph over a time scale of 1 s of the current in the circuit shown in (i) Fig. 16.7a, (ii) Fig. 16.7b, when the position of switch S is altered every 0.25 s (that is S is on for 0.25 s then off for 0.25 s, etc.).
How would the current variation alter in the circuit in Fig. 16.7b if the value of the inductance were made larger? (*C.*)

12. Calculate the value of the capacitor C to make the circuit in Fig. 16.8 have a resonant frequency of 1 kHz. Explain briefly what is meant by the *impedance* of a circuit. (*C.*)

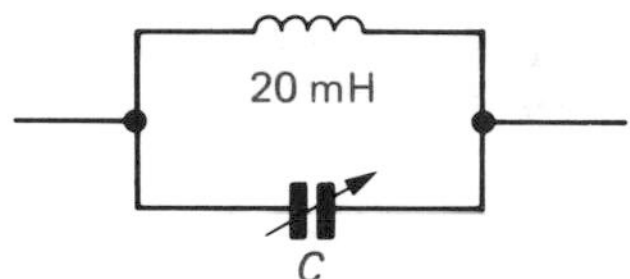

Fig. 16.8

13. What is the tuning range in MHz of a circuit containing a 1 μH inductor in parallel with a 100–500 pF variable capacitor?

14. The output of an oscillator is a sinusoidal voltage whose frequency can be varied but whose amplitude is constant and equal to 1 V r.m.s. The output is connected to a 1 μF capacitor in series with a 10 kΩ resistor and the frequency is varied from 1 Hz to 1 MHz.
What will be the *approximate* value of (i) the current flowing from the oscillator, and (ii) its phase relative to the output voltage for a frequency of (a) 1 Hz, (b) 1 MHz? (*O. and C.*)

15. A supplier's stocklist has the following entry:
Transformer: primary 240 V 50 Hz
secondaries 20 V, 0.5 A; 20 V, 0.5 A
Explain carefully what the entry means. If the transformer is 100% efficient, what is the step-down ratio for one of the secondary windings? (*O. and C.*)

16. A transformer has a primary of 100 turns and a secondary of 200 turns. An alternating potential difference of 40 V r.m.s. is applied across its primary and a load of 160 Ω is placed across its secondary, Fig. 16.9. Calculate the p.d. across the load and the current in the primary. State any assumptions you have made. (*L.*)

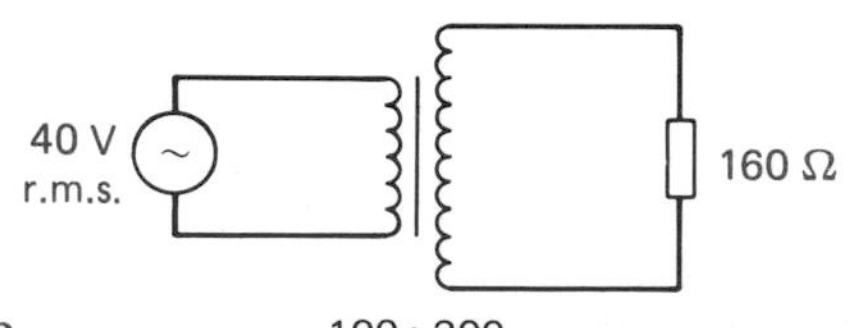

Fig. 16.9

Transducers

17 Microphones

Transducers

Transducers change energy from one form to another. Those in which electrical energy is the input or output will be considered; they enable electronic systems to communicate with the outside world, Fig. 17.1.

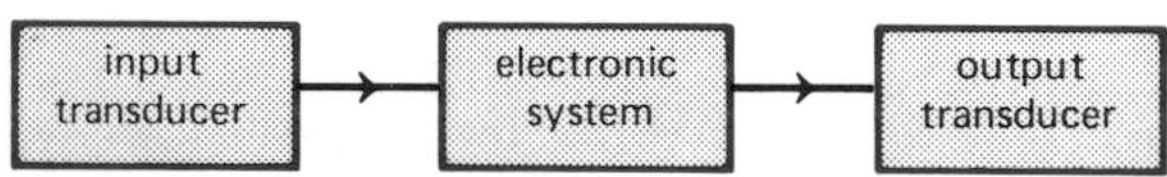

Fig. 17.1

A transducer must be matched to the system with which it is designed to work. Sometimes this means ensuring there is maximum voltage transfer between transducer and system, at other times we have to transfer maximum power. In all cases, as we will see later (p. 113), the impedance of the transducer is a critical factor (rather than its resistance since many transducers have inductance or capacitance and we are generally dealing with a.c.).

About microphones

A microphone changes sound into electrical energy: its symbol is given in Fig. 17.2.

Fig. 17.2

Apart from its impedance, two other important properties are:

(i) the *frequency response*: ideally this should cover the audio frequency range from 20 Hz or so to about 20 kHz and the sensitivity should be the same for the full range, i.e. sounds of the same intensity but different frequency should produce the same output, and

(ii) the *directional sensitivity*: a microphone may be more sensitive, i.e. give a bigger output, for sound coming from one or more directions than from others. This is shown by polar diagrams like those in Fig. 17.3a,b,c for uni-, bi- and omni-directional microphones respectively. In these, the length of the line from the origin 0 to any point P on the diagram is a measure of the sensitivity in the direction 0P.

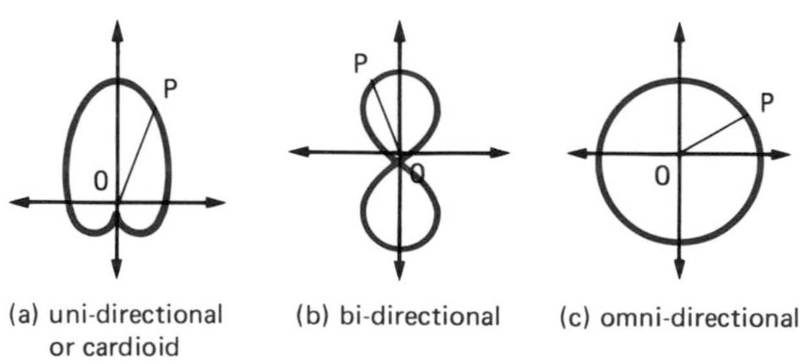

Fig. 17.3

Microphones fall into three groups.

Velocity-dependent microphones

The output voltage depends on the *velocity* of vibration of the moving parts which in turn depends on the sound received.

(a) Moving-coil or dynamic type. This is a popular type because of its good quality reproduction, robustness and reasonable cost. One is shown in Fig. 17.4a. It consists of a small coil of many turns of fine wire wound on a tube (the former) which is attached to a light disc (the diaphragm) as in Fig. 17.4b. When sound strikes the diaphragm, the coil moves in and out of the circular gap between the

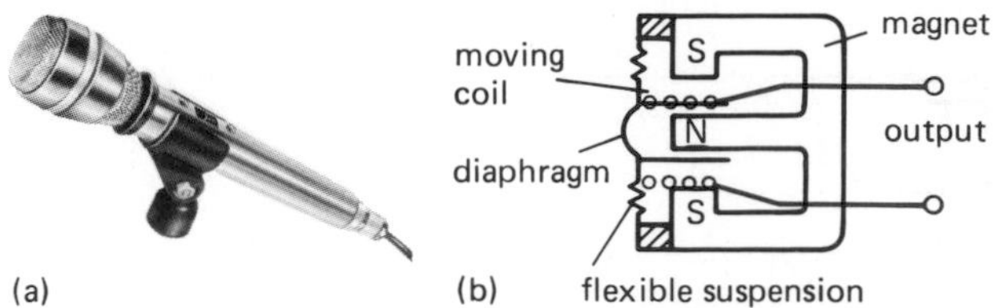

Fig. 17.4

poles of a strong permanent magnet. Electromagnetic induction occurs and the alternating e.m.f. induced in the coil (typically 1 to 10 mV) has the same frequency as the sound.

A moving-coil microphone has an impedance of two or three hundred ohms. Uni- and near-omni-directional models are available.

(b) Ribbon type. The moving part is a thin corrugated ribbon of aluminium foil suspended between the poles of a powerful magnet, Fig. 17.5. Sound falling on either side of the ribbon makes it vibrate at the same frequency and a small e.m.f. is induced in it.

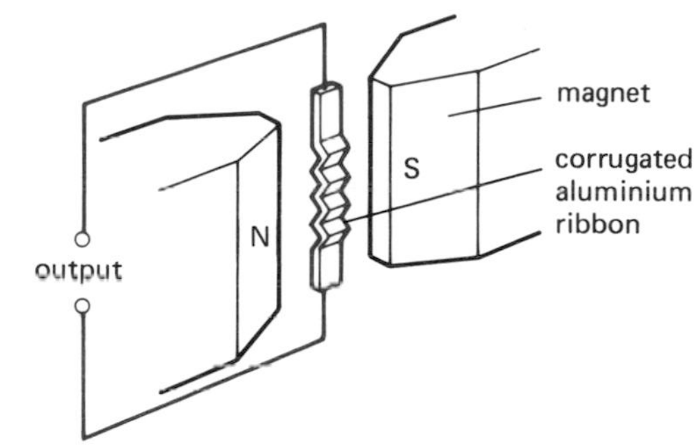

Fig. 17.5

Ribbon microphones are bi-directional and have a very low impedance (e.g. 0.2Ω). Their performance is similar to that of the moving-coil type but they can be damaged by blasts of air.

Amplitude-dependent microphones

In these the output depends on the amount by which the sound waves displace the diaphragm from its rest position, i.e. on the *amplitude* of the motion.

(a) Capacitor type. It is used in broadcasting studios, for public address systems and for concerts where the highest quality is necessary. It consists of two capacitor plates, A and B in Fig. 17.6a. B is fixed while the metal foil disc A acts as the movable diaphragm. Sound waves make the diaphragm vibrate and as its distance from B varies, the capacitance changes. A small battery in the microphone causes the resulting charging and discharging current to produce varying p.d. across a resistor. This p.d. provides the input to a small amplifier in the stem of the microphone.

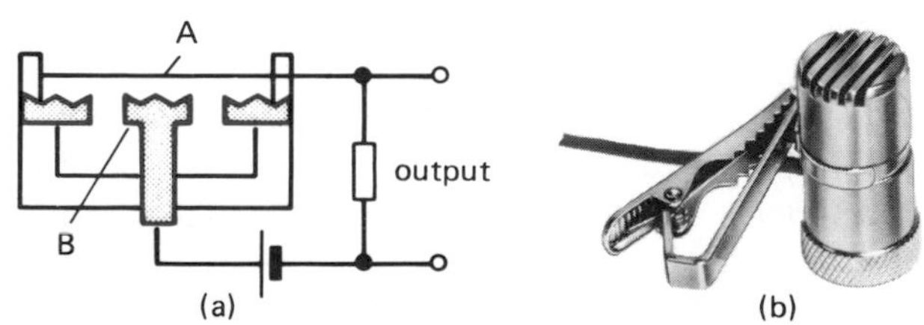

Fig. 17.6

Capacitor microphones, like the tie model in Fig. 17.6b, may be uni- or omni-directional.

(b) Crystal type. Its action depends on the *piezoelectric effect.* In this, an electric charge and so also a p.d. is developed across opposite faces of a slice of a crystal when it is bent, because of the displacement of ions. The effect is given by natural crystals such as Rochelle salt (sodium potassium tartrate) and quartz and by synthetic ceramic crystals of lead zirconium titanate. When sound waves set the diaphragm vibrating, the crystal bends to and fro, Fig. 17.7, and an alternating p.d. is produced between metal electrodes deposited on two of its faces.

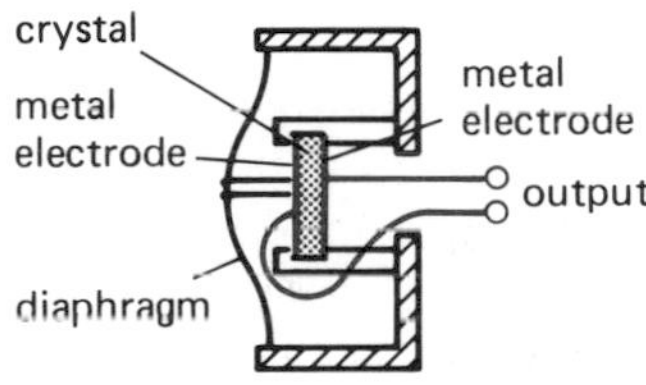

Fig. 17.7

The crystal microphone has a **high output** (e.g. 100 mV), is usually omni-directional and has a very high impedance (several megohms). Its performance is not as good as the previous types but it is inexpensive and much used in cassette recorders.

Carbon microphone

It is used in telephone handsets but is not a high quality device. The principle of the latest type is shown in Fig. 17.8.

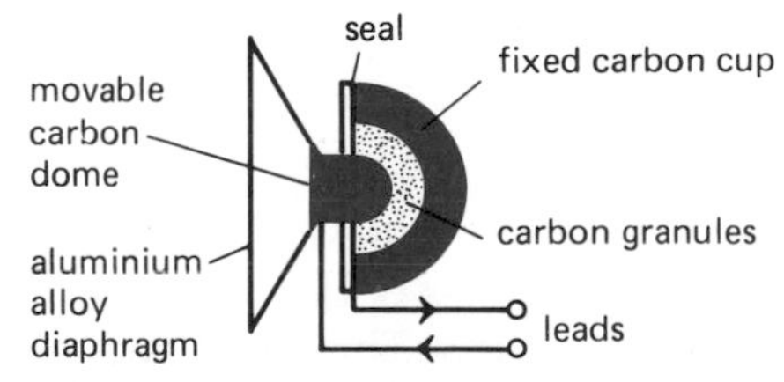

Fig. 17.8

Sound makes the diaphragm vibrate and vary the pressure on the carbon granules between the movable carbon dome, which is attached to the diaphragm, and the fixed carbon cup at the back. An increase of pressure squeezes the granules together and reduces their electrical resistance. Reduced pressure increases the resistance. *Current* through the microphone from a supply (of 50 V at the telephone exchange) varies accordingly to give a varying d.c. which can be regarded as a steady d.c. plus a.c. having the frequency of the sound.

Current variations represent information in a carbon microphone, in other types, e.g. crystal, the representation is in the form of a varying voltage. In the first case a *current analogue* is produced, in the second it is a *voltage analogue*.

Questions

1. What does a transducer do?

2. Name *four* types of microphone. Give *one* use for each.

3. What is the piezoelectric effect?

18 Loudspeakers, headphones and earpieces

Introduction

Loudspeakers, headphones and earpieces change electrical energy into sound. Three of their important properties are:

(i) the *frequency response* which, as with microphones, should be uniform over the audio frequency range,
(ii) the *impedance*, which must be known for matching purposes, and
(iii) the *power rating*, which should not be exceeded or damage may occur.

Loudspeakers

(a) Moving-coil type. Most loudspeakers in use today are of this type, shown in Fig. 18.1. Their construction, Fig. 18.2a, is basically the same as that of the moving-coil microphone, in fact a loudspeaker can be used as a low impedance microphone. The loudspeaker symbol is given in Fig. 18.2b.

Fig. 18.1

When audio frequency a.c. (e.g. from an amplifier) passes through the coil of the speaker, it vibrates in or out between the poles of the magnet depending on the current direction. The paper cone, which is fixed to the coil, also vibrates and passes the motion

flexible rim
frame
metal case
coil on former
End-on-view of magnet
paper cone
coil leads
magnet
(a)
(b)

Fig. 18.2

on to the large area of air touching it. A sound wave of the same frequency as the a.c. in the coil is thus produced. (The motion of the coil is due to the magnetic field caused by the current in it, interacting with the field of the magnet.)

The loudspeaker coil has inductive reactance X_L and resistance R and its impedance $Z = \sqrt{R^2 + X_L^2}$. X_L, and therefore Z, varies with the frequency of the a.c. but is usually given for 1 kHz. Common values are 4 Ω, 8 Ω and 15 Ω, the maximum being about 80 Ω.

(b) Crystal type. This type depends on the reverse piezoelectric effect (p. 43), i.e. an alternating p.d. applied across opposite faces of the crystal makes it vibrate and produce sound of the same frequency as the a.c. In speaker systems with two or more speakers it is used as a *tweeter* to handle high (treble) frequencies since it rejects low (bass) frequencies and has a typical frequency response of 3.5 kHz to 35 kHz. It has a high impedance, of the order of 1 kΩ.

(c) Efficiency. The small loudspeakers used for radio and TV are only 5 to 10% efficient at changing electrical energy into sound. For large speakers at low power levels the value is near 50%.

Headphones

(a) Moving-coil type. High quality headphones as used for listening to stereo are moving-coil (dynamic) types with, in a typical case, a frequency response of 20 Hz to 20 kHz and an impedance of 8 Ω per earphone, Fig. 18.3a. Often each earphone is fitted with its own volume control. The headphone symbol is given in Fig. 18.3b.

Fig. 18.3

(b) Magnetic type. Headphones for general use, e.g. mono listening, work on a different principle; they have a higher impedance, e.g. 1 kΩ per earphone, and a smaller frequency response, e.g. 30 Hz to 15 kHz. Their action can be explained from Fig. 18.4. Current passes through the coils of an

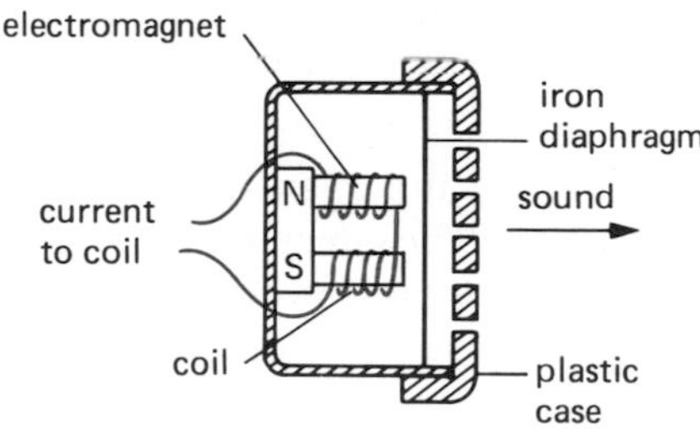

Fig. 18.4

electromagnet and attracts the iron diaphragm more, or less, depending on the value of the current. As a result the diaphragm vibrates and produces sound.

Telephone handset receiver

The construction of the latest telephone handset receiver, called the 'rocking armature' type, is shown in Fig. 18.5. (An armature is a piece of iron which is made to move by an electromagnet.)

The coils are wound in opposite directions on the two S poles of the magnet so that if the current goes

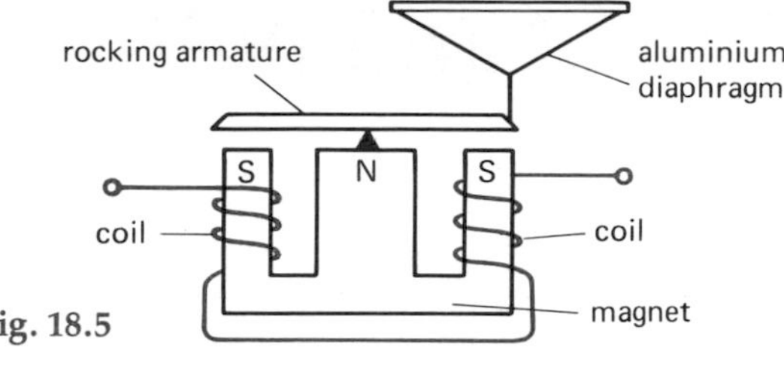

Fig. 18.5

round one in a clockwise direction, it goes round the other anticlockwise. Current in one direction therefore makes one S pole stronger and the other weaker so causing the iron armature to rock on its pivot towards the stronger S pole. When the current reverses, the S pole which was the stronger before will now be the weaker and the armature rocks the other way. These movements of the armature are passed on to the aluminium diaphragm, making it vibrate and produce sound of the same frequency as the a.c. speech current in the coil.

Earpieces

Earpieces like that in Fig. 18.6a are used in deaf aids and transistor radios. The *magnetic* type works on the moving-coil principle and has an impedance of

Fig. 18.6

8 Ω. The *crystal* type depends on the reverse piezoelectric effect and has an impedance of several megohms. The symbol for an earpiece is shown in Fig. 18.6b.

Questions

1. A manufacturer's catalogue gives the following information about a certain loudspeaker.

Frequency response : 25 Hz to 19 kHz
Impedance : 8 Ω
Power (r.m.s.) : 5 W

Explain each statement.

2. (a) The lowest frequency to which a certain loudspeaker, with a paper cone of diameter 5 inches, responds is 25 Hz. For a similar 8 inch diameter speaker, the lowest frequency is 18 Hz. What conclusion can you draw from these figures?
(b) Why are loudspeakers with an elliptical-shaped cone often used in television sets, record players and transistor radios?

19 Heat and light sensors

Thermistors

Thermistors or *therm*al res*istors* are semiconductor devices whose use as transducers is due to the fact that their resistance changes markedly when their temperature changes. They are used for the measurement and control of temperature, being heated either externally or internally by the current they carry. Rod-, disc- and bead-shaped varieties are shown in Fig. 19.1 along with the thermistor symbol. There are two types.

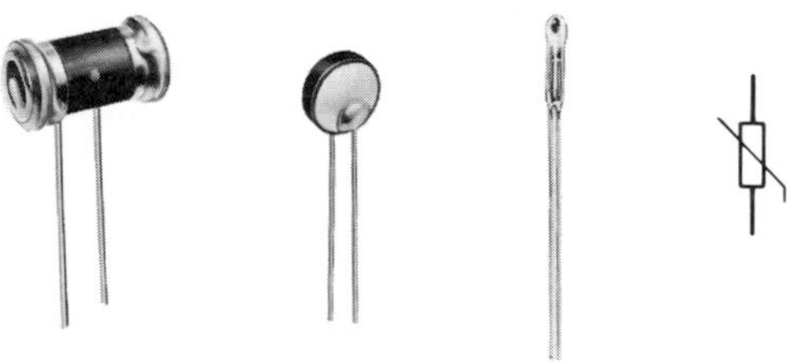

Fig. 19.1

(a) n.t.c. types (standing for *n*egative *t*emperature *c*oefficient, p. 12) are commonest. Their resistance *decreases* as the temperature increases as shown by the typical resistance-temperature graph of Fig. 19.2. They are made from oxides of nickel, manganese and other elements which, after heat treatment, have a ceramic-like appearance.

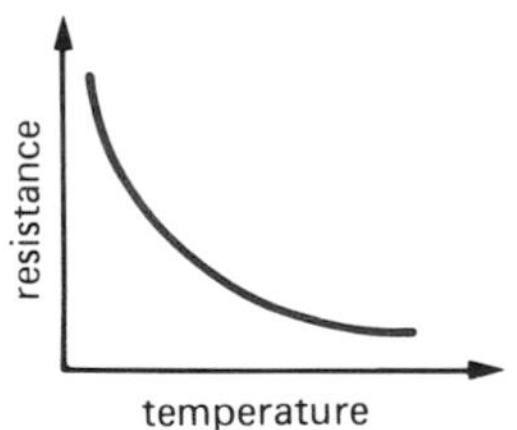

Fig. 19.2

The circuit for a simple electronic thermometer using a small bead thermistor is shown in Fig. 19.3a.

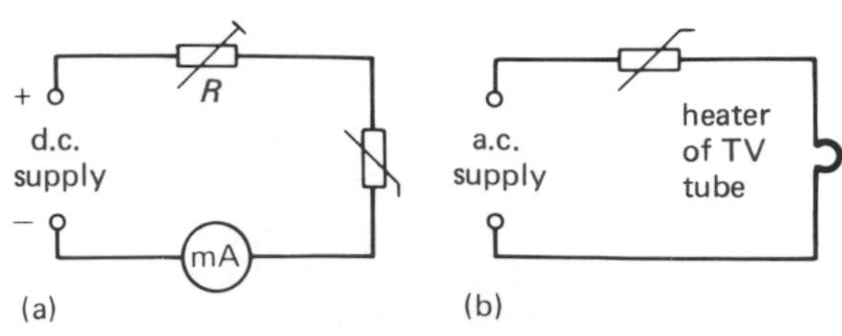

Fig. 19.3

Temperature changes cause the resistance of the thermistor to change and the current through the milliammeter mA changes. By adjusting the preset resistor R to give a suitable deflection at room temperature and then using several known temperatures, the meter can be calibrated to read °C directly.

In Fig. 19.3b the heater of the TV tube (p. 48) has a low resistance when cold and would, but for the disc thermistor, allow a dangerous current surge when the tube is switched on. As the current warms up the thermistor and the heater, the rise in heater resistance (a metal) is compensated by the fall in thermistor resistance (a semiconductor).

(b) p.t.c. types (standing for *p*ositive *t*emperature *c*oefficient). Their resistance *increases* sharply above a certain temperature. They are used mainly to prevent damage in circuits which might experience a large temperature rise, e.g. one in which an electric motor becomes overloaded.

Light dependent resistor (LDR)

This type of transducer changes light into electrical energy and is also called a photoconductive cell. Its action depends on the fact that the resistance of certain semiconductors, such as cadmium sulphide, decreases as the intensity of the light falling on them increases. The effect (also given by infrared and ultraviolet radiation) is due to light supplying the energy to set free electrons from atoms of the semiconductor, so increasing its conductivity, i.e. reducing its resistance.

A popular LDR (the ORP12) is shown in Fig. 19.4 with the LDR symbol. There is a 'window' over the grid-like metal electrodes to allow light to fall on a thin layer of cadmium sulphide. Its resistance varies from about 10 MΩ in the dark to 1 kΩ or so in daylight.

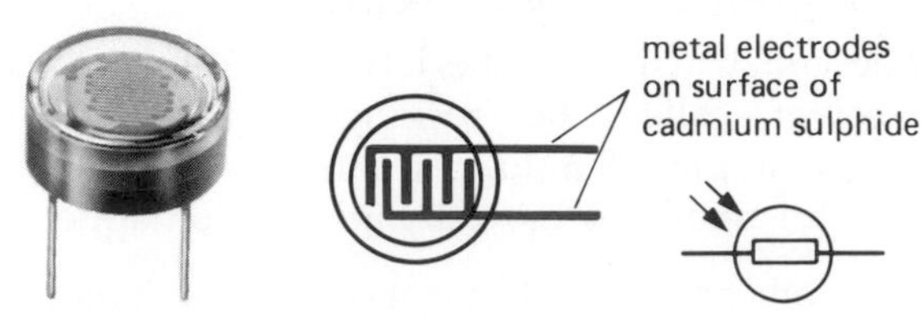

Fig. 19.4

They have many uses, e.g. in photographic exposure meters, but their response to light changes is slow compared with a photodiode (p. 61), typically 100 milliseconds compared with 1 microsecond. This makes them unsuitable for fast counting operations.

Questions

1. A lamp in series with a thermistor is just alight. Does it get brighter if the thermistor is cooled and is (a) an n.t.c., (b) a p.t.c., type?

2. A lamp in series with an LDR is just alight. What happens if the intensity of the light on the LDR (a) increases, (b) decreases?

20 Digital displays

These are transducers for changing electrical energy into light and are used as numerical indicators in calculators, digital watches, cash registers, weighing machines and other measuring instruments. They consist of seven segments arranged as a figure eight so that the digits 0 to 9 (and sometimes the letters A to F) can be displayed when different segments are energized, Fig. 20.1.

0 1 2 3 4 5 6 7 8 9

Fig. 20.1

Five types are in common use.

Light-emitting diode (LED) display

The segments are light-emitting diodes or LEDs (p. 60) made from the semiconductor gallium arsenide phosphide. When a typical current of 10 mA passes through the diode it emits red or green light, depending on its composition.

Each segment has an anode lead at which the current enters and a cathode lead at which it leaves. All seven anodes (or cathodes) are joined together to form a common anode (or cathode), Fig. 20.2. Such displays are often designed to work on a 5 V d.c. supply with a current limiting resistor (e.g. 330 Ω) in series with each segment.

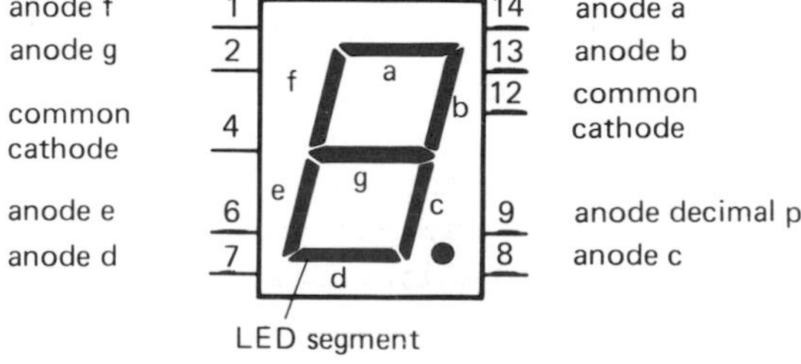

Fig. 20.2

Filament display

This was one of the first seven-segment displays and is commonly used in petrol filling station pumps. Like an ordinary electric filament lamp (p. 77), the segments consist of short coiled lengths of tungsten wire (in small glass tubes) which get white hot and emit light when current passes through them. By operating at lower temperatures, e.g. 1500 °C, their working life exceeds 100 000 hours. Each segment needs about 10 mA at 5 V.

Gas discharge display (GDD)

Each segment of a GDD consists of a small glass tube containing mainly neon gas at low pressure which produces a bright orange glow when a current passes through it. Although about 170 V is required to start the display, the current taken by a segment may only be 0.2 mA or so.

Fluorescent vacuum display

The display is a blue-green colour, popular for calculators. The light is produced when electrons, emitted from an electrically heated filament (p. 48) are accelerated in an evacuated glass tube and strike a fluorescent screen (p. 48). The filament needs, in a typical case, a current of 50 mA at 2 V a.c. or d.c. and an accelerating p.d. of 15 V is also required.

Liquid crystal display (LCD)

These are much used in digital watches where their very small current needs (typically 5 μA for all segments on) prolongs battery life. Liquid crystals are organic (carbon) compounds which exhibit both solid and liquid properties. A 'cell' with transparent electrodes on opposite faces, containing a thin layer of liquid crystal, Fig. 20.3a, and on which light falls, goes 'dark' when a p.d. is applied across the electrodes. The effect is due to molecular rearrangement within the liquid crystal.

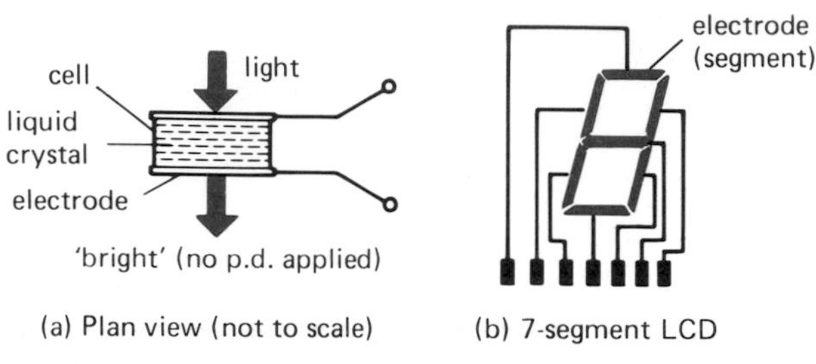

Fig. 20.3

Only the liquid crystal under those electrodes to which the p.d. is applied goes 'dark'. The display has a silvered background which reflects back incident light and it is against this continuously visible background (except in darkness when it has to be illuminated) that the numbers show up as dark segments, Fig. 20.3b.

Thousands of tiny LCDs are used to form the picture elements (pixels) of the screen in one type of black-and-white pocket television receiver.

Questions

1. Explain the abbreviations (a) LED, (b) GDD, (c) LCD.
2. Give one advantage and one disadvantage of (a) LED, (b) liquid crystal, displays.

21 Cathode ray tube (CRT)

Introduction

A cathode ray tube changes electrical energy into light and is used in oscilloscopes (p. 99), television receivers (p. 184), computer visual display units (VDUs), radar and other electronic systems. It has three parts:

(i) an *electron gun* for producing a narrow beam of high-speed electrons (a cathode ray) in an evacuated glass tube,
(ii) a *beam deflection system* (electrostatic or magnetic), and
(iii) a *fluorescent screen* at the end of the tube which emits light where the beam falls on it.

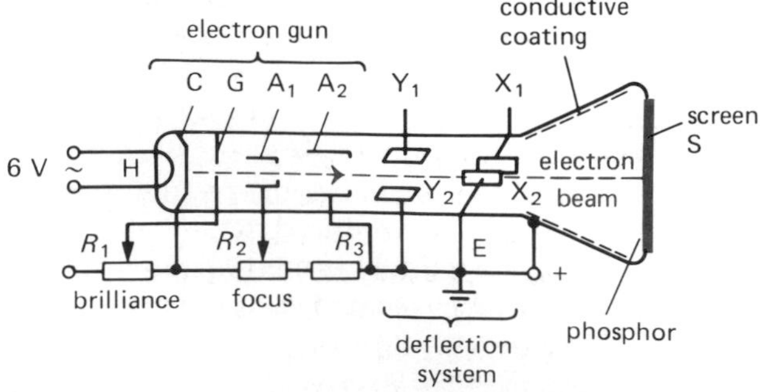

Fig. 21.1

A CRT with electrostatic deflection is shown in Fig. 21.1; a voltage divider provides appropriate voltages from a high voltage power supply.

Electron gun

The *heater*, H, is a tungsten wire inside, but electrically insulated from the hollow cylindrical nickel *cathode* C. When current passes through H (usually from a 6 V a.c. supply) it raises the temperature of C to dull red heat, causing a special mixture on the tip of C to emit electrons copiously. The process is called *thermionic emission*.

The negatively charged electrons are accelerated towards the *anodes*, A_1 and A_2 (metal discs or cylinders), which have positive potentials with respect to C, A_2 more so than A_1. The *grid* G, another hollow metal cylinder, has, relative to C, a negative potential which can be varied by R_1. The more negative it is, the fewer is the number of electrons emerging from the hole in the centre of G. The spot produced on the *screen* S by the beam, after it has shot through the central holes in A_1 and A_2, is then less bright. R_1 is the *brilliance* control.

Focusing of the beam to give a small luminous spot on S is achieved by changing the p.d. between A_1 and A_2, i.e. by altering the *focus* control R_2. Typical voltages for a small CRT are 1 kV for A_2, 200 to 300 V for A_1 and −50 V to 0 V for G.

There must be a return path for the electrons, from S to C, otherwise unwanted negative charge would build up on S. This does not happen because when struck by electrons, S emits a more or less equal number of *secondary* electrons. These are attracted to and collected by a conducting coating (e.g. of graphite) on the inside of the tube near S and which is connected to A_2. From there the circuit is completed through the power supply to C.

It is common practice to earth A_2 (and the conducting coating), thus preventing earthed objects (e.g. people) near S upsetting the electron beam. The other electrodes in the gun are then negative with respect to A_2 but the electrons are still accelerated through the same p.d. between C and A_2.

Beam deflection system

In electrostatic deflection (used in oscilloscopes) the beam from A_2 passes between two pairs of metal plates, Y_1Y_2 and X_1X_2, at right angles to each other. If one plate of each pair is at a positive potential with respect to the other, the beam moves towards it. The Y-plates, which are farther from the screen, are horizontal and deflect the beam vertically, i.e. along the y-axis. The X-plates are vertical and produce horizontal deflections, i.e. along the x-axis.

To minimize the defocusing effect which the voltages on the deflecting plates might have on the beam, one plate of each pair is at the same potential as A_2, i.e. earth. In practice X_2, Y_2 and A_2 are connected internally.

In magnetic deflection (used in TV receiver tubes whose need for a very large deflecting angle of 110° would require extremely high electrostatic deflecting voltages) two pairs of current-carrying coils are placed outside the tube at right angles to each other, Fig. 21.2. The magnetic field produced by one pair causes deflections in the y-direction and by the other pair in the x-direction.

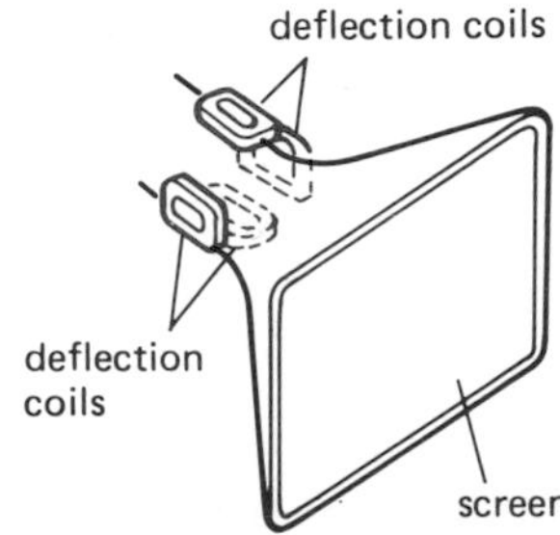

Fig. 21.2

Fluorescent screen

The inside of the wide end of the tube is coated with a substance called a *phosphor* which emits light where it is hit by the electrons. The colour of the light emitted and the 'afterglow' or 'persistence', i.e. the time for which the emission is visible after the electron bombardment has stopped, depends on the phosphor.

Phosphors emitting bluish light are popular for oscilloscopes. For radar tubes long persistence types producing orange light are common. For black-and-white TV tubes white phosphors of short persistence are used while for colour TV red, green and blue phosphors are employed.

Flat screen CRT

Small, flat screen CRTs are now available for use in pocket-sized, battery-operated black-and-white television receivers, Fig. 21.3. The electron beam producing the picture comes in from the side and is turned more or less through 90° by an electric field due to a pair of electrodes.

Fig. 21.3

The screen is viewed from the side on which the electrons strike the phosphor and, unlike a traditional CRT, no light is lost going through the phosphor. It is claimed that the picture is three times as bright as that in a normal tube of the same size with one-tenth of the power.

Questions

1. State the job done by each of the three main parts of a CRT.

2. In the CRT of Fig. 21.1 explain why the electrons travel with more or less constant speed between A_2 and S.

22 Pick-ups

Record player pick-ups are transducers which convert the sound stored on a disc into electrical energy.

Mono pick-ups

The pick-up for a conventional record player consists of a stylus and a cartridge. When the turntable revolves the stylus is made to vibrate as it follows the wavy shape of the sides of the groove in the disc. The vibrations are changed into electrical signals by the cartridge. The stylus tip is usually made of industrial diamond, with an average life of 1000 hours before the wear to it damages the groove.

There are two main types of conventional cartridge.

(a) Crystal. This type, shown in Fig. 22.1a, uses the piezoelectric effect (p. 43) like the crystal microphone and consists of a ceramic crystal of lead zirconium titanate. It has a good frequency response, an output of about 100 mV and the impedance of the system it supplies, e.g. an amplifier, should be 1 to 2 MΩ.

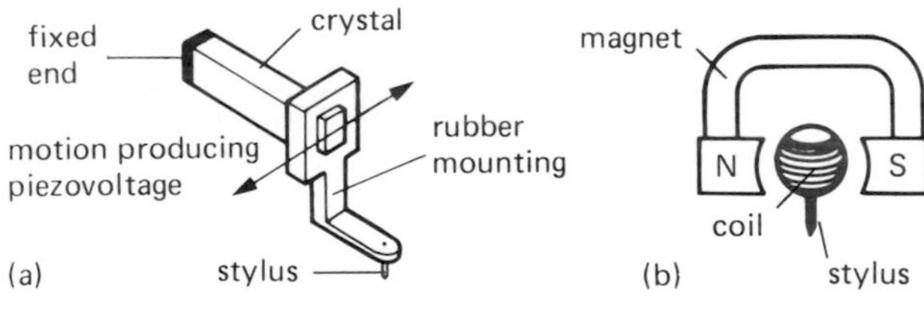

Fig. 22.1

(b) Magnetic. One version of this type, shown in Fig. 22.1b, depends on electromagnetic induction (p. 29) like the moving-coil microphone. The vibration of the stylus makes the coil move in a magnetic field and an output voltage in the range 5 to 10 mV is induced in the coil. A good quality magnetic cartridge gives excellent reproduction. It should supply a 'load' of impedance 50 to 100 kΩ.

Stereo pick-ups

A stereo record stores the sound as 'wavy shapes' on both walls of a right-angled groove in the disc. The groove is cut during recording by a stylus whose vibrations are controlled in the two directions shown in Fig. 22.2a by the signals from two microphones. The right-hand wall carries signal A and the left signal B, which may or may not be the same as A.

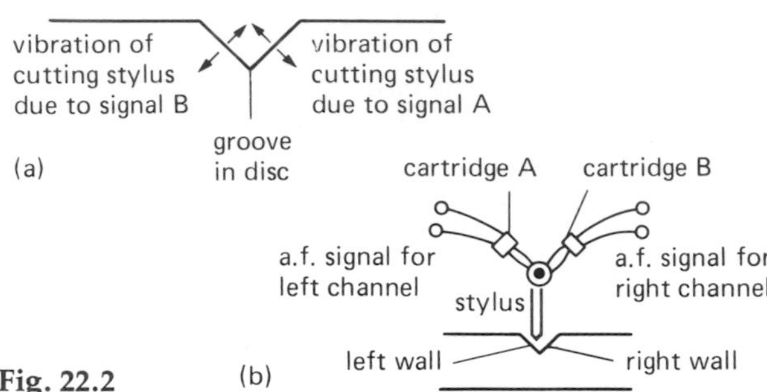

Fig. 22.2

A stereo pick-up, Fig. 22.2b, consists of two cartridges (crystal or magnetic) and the motion of the stylus, like that of the cutting stylus, is a combined horizontal and vertical one. Separate a.f. signals are produced in cartridges A and B and are amplified in two different electronic 'channels'.

Laser pick-up

This is the latest type of pick-up. It is used with digital audio disc (DAD) players (p. 172) in which disc wear is eliminated and distortion is one-hundredth of that in an ordinary player.

The pick-up shown in principle in Fig. 22.3 uses a disc with a reflecting surface in which the audio signal is recorded as a series of tiny pits of varying length and spacing. A very narrow intense beam of light from a laser passes through a partly-reflecting mirror and is focused by a lens system on the pits. The reflected beam is changed into an electrical signal by a photodiode (p. 61). The pick-up moves across the disc as it rotates.

Fig. 22.3

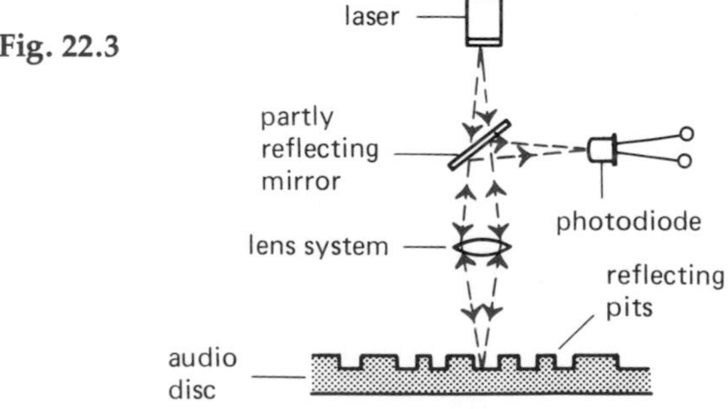

Digital video disc players (p. 172) also use it.

23 Relays and reed switches

Relays

A relay is a switch used to turn other circuits on or off. It is a transducer which changes an electrical signal to movement and then back to an electrical signal. It is useful if we want (i) a small current in one circuit to control another circuit containing a device such as a lamp or electric motor which needs a large current, (ii) several different switch contacts to be operated simultaneously.

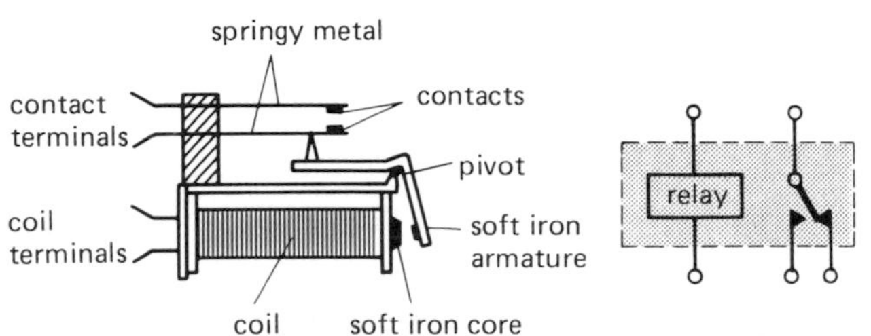

Fig. 23.1

The structure of a relay is shown in Fig. 23.1 with its symbol; the contacts may be normally open (as shown), normally closed or change-over. When the controlling current flows through the coil, the soft iron core is temporarily magnetized and attracts the iron armature. This rocks on its pivot and operates the contacts in the circuit being controlled. When the coil current stops, the armature is no longer attracted and the contacts return to their normal positions.

A relay with a nominal coil voltage of 12 V may, for example, work off any voltage between 9 V and 15 V. The coil resistance (e.g. 185 Ω) is also quoted as well as the current and voltage ratings of the contacts.

Relay circuits

(a) Simple intruder alarm. When light falls on the LDR in the circuit of Fig. 23.2a, the LDR resistance decreases, and the current through it and the relay coil is large enough to operate a relay. The normally open contacts close and the bell rings. If the light is

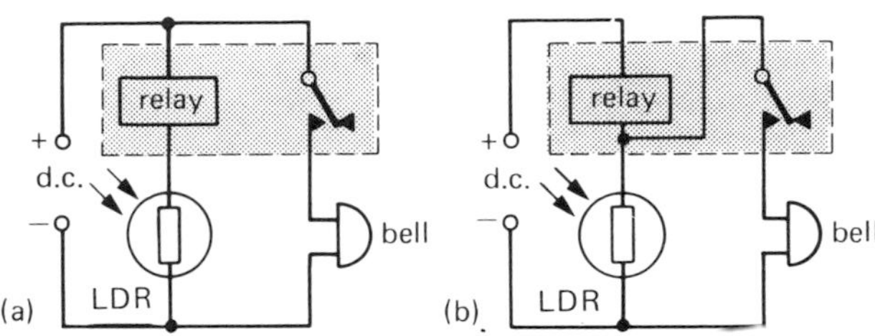

Fig. 23.2

removed the LDR current is too small to 'hold' the relay and the ringing stops.

(b) Latched intruder alarm. By connecting the relay contacts in series with the relay coil as in Fig. 23.2b, the bell keeps ringing when the LDR is no longer illuminated, i.e. the alarm is 'latched'.

Reed switches

Relays operate relatively slowly and for fast-switching, e.g. in a telephone exchange, reed switches are used. One with normally open contacts is shown in Fig. 23.3. The reeds are thin strips of easily magnetized and demagnetized nickel-iron sealed in a small glass tube.

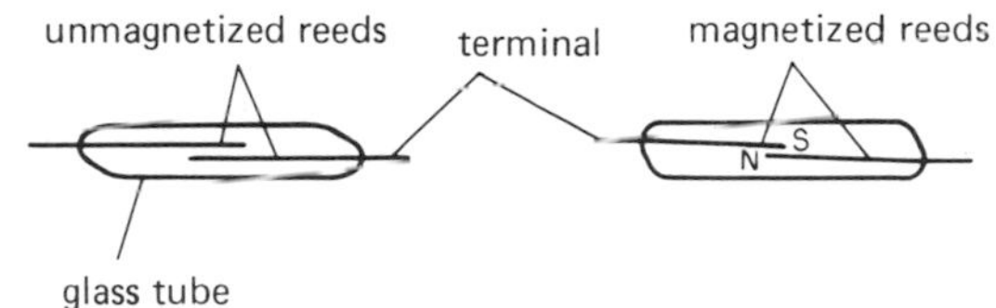

Fig. 23.3

The switch is operated either by passing a current through a coil surrounding it or by bringing a magnet near. In both cases the reeds become magnetized, attract and on touching complete the circuit connected to the terminals. The time for this may only be 1 millisecond and there is almost no *contact bounce*, i.e. no repeated making and breaking of the contacts when the reeds come together. When the current in the coil stops or the magnet is removed, the reeds separate.

Question

1. The circuit diagram in Fig. 23.4 consists of a 12 V electro-magnetic relay, a 12 V lamp operated from a 12 V power supply and a manually operated switch S.

(a) When the switch S is closed, (i) describe the behaviour of the relay, (ii) describe the behaviour of the lamp.

(b) A 1000 μF capacitor is now fitted in parallel with the relay coil. How would this change affect the behaviour of the lamp? (*A.E.B. 1982 Control Tech.*)

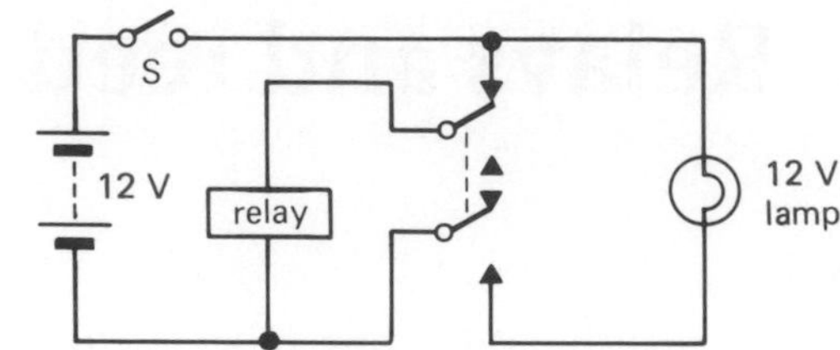

Fig. 23.4

24 Electric motors

Electric motors change electrical energy into mechanical energy. One of their newest uses is in robots, where 'stepper' motors, driven by a series of electrical pulses, turn the robot arm a small fixed amount for each pulse.

D.C. electric motors

(a) Principle. A simple d.c motor consists of a coil of wire fixed to an axle which can rotate between the poles of a permanent magnet, Fig. 24.1. Each end of the coil is connected to half of a split copper ring, called the *commutator*, which rotates with the coil. Two carbon blocks, the *brushes*, press lightly against the commutator and are connected to a d.c. supply.

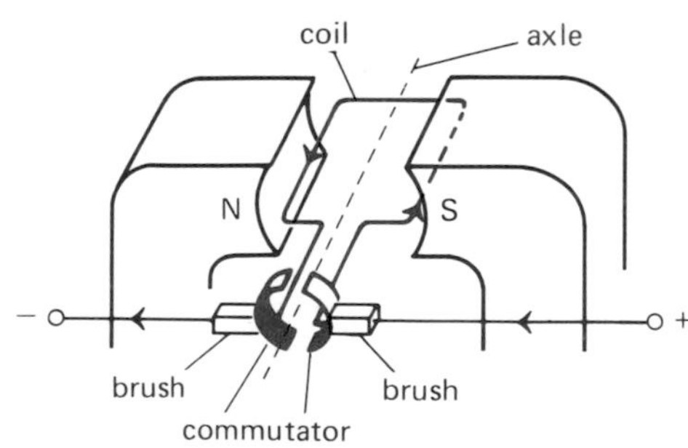

Fig. 24.1

When current passes through the coil, it becomes a 'flat' magnet and rotates so that its N pole is opposite the S pole of the permanent magnet and vice versa. The coil's inertia makes it overshoot the vertical position and the commutator halves change contact from one brush to the other. This reverses the current in the coil, reversing its magnetic polarity and making it carry on for another half-turn in the same direction to face opposite poles again, and so on.

(b) Practical motors. Small motors for driving models use permanent magnets. More powerful types have:

(i) an *armature* consisting of several multi-turn coils (each with its own pair of commutator segments) wound in slots in a soft-iron (laminated) core to give smoother running and self-starting. (The two-pole motor in Fig. 24.1 needs a push to start unless it is in the horizontal position, in which the force on it is a maximum.), and

(ii) an *electromagnet* instead of a permanent magnet, its windings being called the *field* coils.

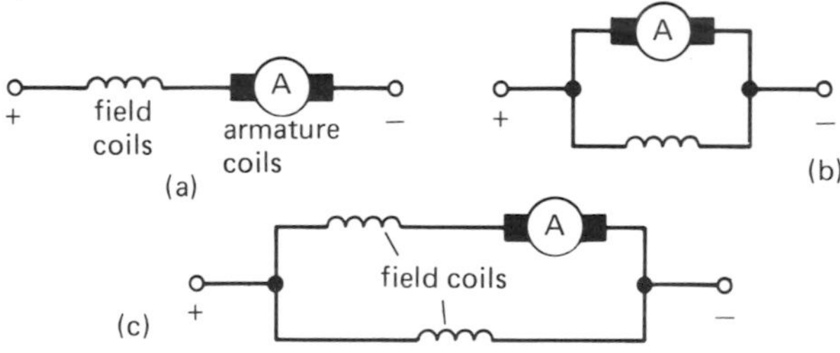

Fig. 24.2

A *series-wound* motor has the armature and field coils in series, Fig. 24.2a, and has a large turning effect (torque) when starting but its speed varies with the load. A *shunt-wound* motor has the armature and field coils in parallel, Fig. 24.2b, and runs at a steady speed when the load varies. A *compound-wound* motor, Fig. 24.2c, has some coils in series and some in parallel and gives the advantages of both types.

Small d.c. motors are only about 10% efficient, larger ones give higher values.

(c) Starting resistor. When a motor is running it behaves like a dynamo (p. 83) and produces a back e.m.f. which opposes and reduces the current in the armature (Lenz's law). On starting, the back e.m.f. is absent and a large, damaging current would flow unless there was a starting resistor in series with the armature to limit it. The resistance is reduced as the motor speeds up.

A.C. electric motors

The *induction* motor is the commonest a.c. type; its action depends on the production of eddy currents by a moving magnetic field. The large 'squirrel-cage' type is much used in industry; its rotor (armature) is an array of copper rods embedded in the surface of an iron cylinder (like a cage). The small 'shaded-pole' type, in which a copper plate covers part of the magnetic field, is used in record players.

The advantages of induction motors are:

(a) no brushes to wear or cause sparks that give radio interference,
(b) no starting resistor needed, and
(c) constant speed controlled by the mains frequency (not voltage).

Question

1. (a) What is the back e.m.f. of a motor?
(b) A d.c. motor has an armature resistance of 0.5 Ω and when working from a 12 V supply the current through it is 2 A. Comment on these figures.

25 Progress questions

1. Draw a labelled cross-sectional diagram of one type of microphone. Explain briefly how it works. (*L.*)

2. What is meant by the *sensitivity* of a *microphone*? Describe in detail how you would investigate the variation of sensitivity of a *crystal* microphone with *direction*. Show by means of a diagram the results you would expect to obtain.
Draw a labelled diagram showing the structure of the crystal microphone. (*C.*)

3. The resistance of a bead thermistor at different temperatures is found to be:

Temperature in °C	20	40	60	80	100
Resistance in kΩ	40	19	8.8	4.2	2.0

(a) Plot a graph of the resistance of the thermistor against temperature.
(b) Find from your graph the temperature of the thermistor at which it has a resistance of 15 kΩ.
(c) Explain why the resistance of the thermistor changes with temperature in this way.
(d) Draw a diagram of an electrical circuit suitable for measuring the resistance of the thermistor.
(e) Explain what happens to a thermistor when it is connected across the terminals of a battery so that power is converted in it. (*O.L.E.*)

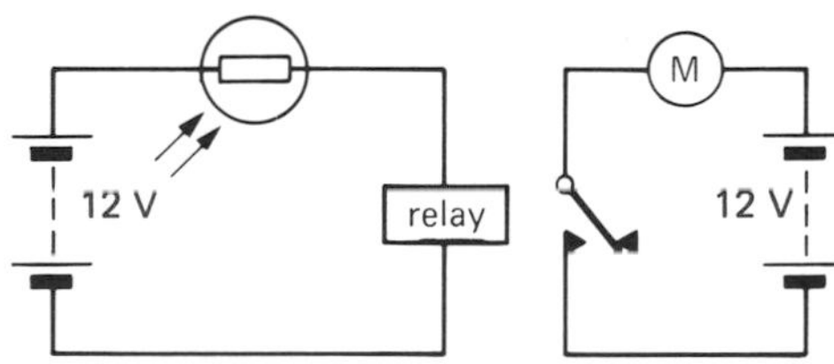

Fig. 25.1

4. The circuit in Fig. 25.1 shows a simple method of switching a motor on and off by making use of a light dependent resistor (LDR).
(i) Explain why the 12 V relay is likely to energize when light strikes the LDR.
(ii) If the relay coil has a resistance of 200 Ω and the LDR has a resistance of 40 Ω, calculate the current flowing in the relay coil.
(iii) Calculate the power dissipated in the LDR when it has a resistance of 40 Ω. (*A.E.B. 1982 Control Tech.*)

5. Draw a labelled diagram of a cathode ray tube. (Details of associated circuitry are not required.) (*L.*)

6. Explain briefly (a) the piezo effect, (b) how the piezo effect is used in the operation of a crystal pick-up cartridge. (*C.*)

7. Without the capacitor in the circuit of Fig. 25.2, the relay closes and opens normally when the switch S is closed and then opened.

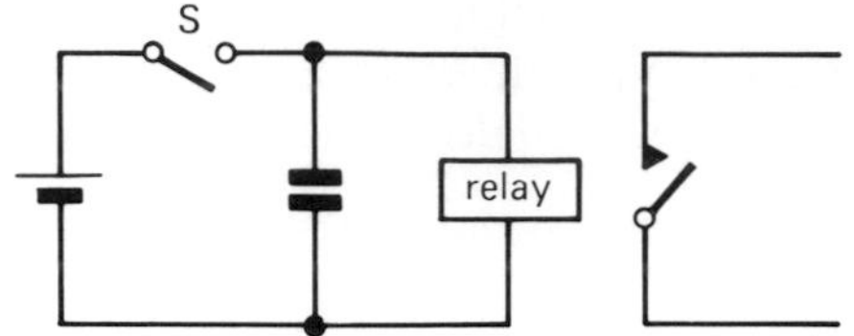

Fig. 25.2

When the switch S is closed the relay closes after a time delay. Explain.
State and explain what happens to the relay when the switch S is now opened. *(O.L.E. part qn.)*

8. Explain how a *back e.m.f.* is generated in the armature of a d.c. electric motor.
Why does this back e.m.f. (a) limit the speed of the motor, (b) necessitate a starting resistance in series with the armature? (C.)

Semiconductor diodes

26 Semiconductors

Semiconductors are materials whose electrical conductivity lies between that of good conductors like copper and good insulators like polythene. Examples are silicon, germanium, cadmium sulphide and gallium arsenide.

Intrinsic semiconduction

(a) Valence electrons. Many of the properties of materials can be explained if we assume that the electrons surrounding the nucleus of an atom are arranged in orbits or 'shells'. Those in the outermost shell are known as valence electrons because the chemical combining power or valency (as well as many physical properties) of the atom depend on them. The valence electrons form 'bonds' with the valence electrons of neighbouring atoms to produce in the case of most solids, a regularly repeating three dimensional pattern of atoms called a crystal lattice.

An atom of a semiconductor such as silicon or germanium has four valence electrons, Fig. 26.1a, and in the lattice each one is shared with a nearby atom to form four covalent bonds, Fig. 26.1b. Every

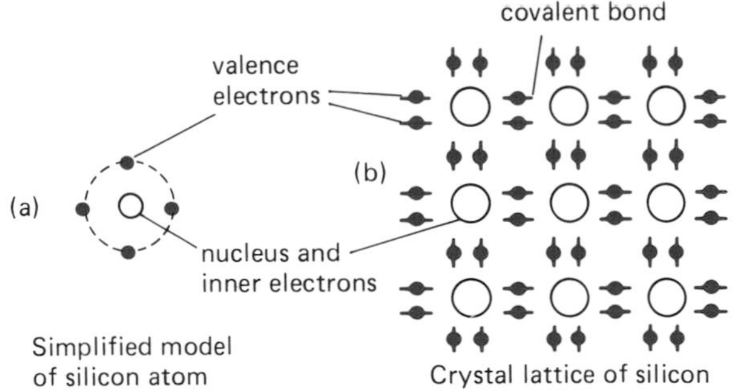

Fig. 26.1

atom has a half-share in eight valence electrons and it so happens that this number of electrons gives a very stable arrangement (most of the inert gases have it). A strong crystal lattice results in which it is difficult for electrons to escape from their atoms. Pure silicon and germanium are therefore very good insulators, being perfect at near absolute zero (−273 °C).

(b) Electrons and holes. At normal temperatures the atoms are vibrating sufficiently in the lattice for a few bonds to break, setting free some valence electrons. When this happens, a deficit of negative charge, called a vacancy or *hole* is left by each free electron in the outermost shell of the atom from which it came, Fig. 26.2a. The atom, now short of an electron, becomes a positive ion. We think of the hole as a positive charge equal in size to the negative charge on an electron.

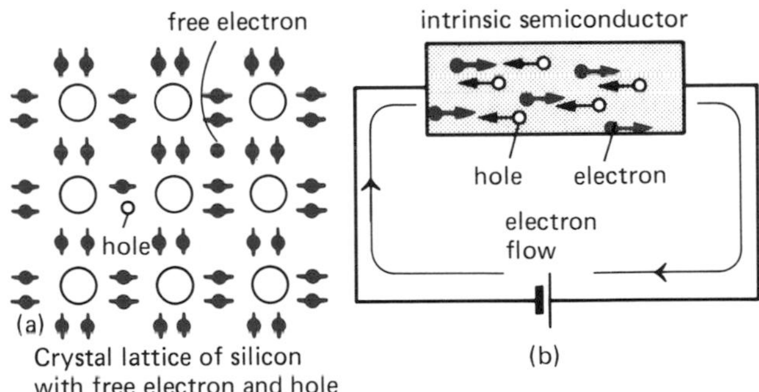

Fig. 26.2

When a battery is connected across a pure semiconductor, it attracts free electrons to the positive terminal and provides a supply of free electrons at the negative terminal. These free electrons drift through the semiconductor by 'hopping' from one hole to another nearer the positive terminal, making it *seem* that positive holes are moving to the negative terminal. The current in a *pure* semiconductor is very small and can be *thought* of as streams of free electrons and holes going in opposite directions as in Fig. 26.2b. It is called *intrinsic* semiconduction because the charge carriers come from inside the material.

If the temperature of a semiconductor rises, more bonds break and intrinsic conduction increases because more free electrons and holes are produced. The resistance of a semiconductor decreases with temperature rise.

Extrinsic semiconduction

The use of semiconductors in diodes, transistors and integrated circuits, depends on increasing the conductivity of very pure (i.e. intrinsic) semiconducting materials by adding to them tiny, but controlled amounts of certain 'impurities'. The process is called *doping* and the material obtained is known as an *extrinsic* semiconductor because the impurity supplies the charge carriers (which add to the intrinsic ones). The impurity atoms must have about the same size as the semiconductor atoms so that they fit into its crystal lattice without causing distortion. There are two kinds of extrinsic semiconductor.

(a) n-type. This type is made by doping pure semiconductor with, for example, phosphorus. A phosphorus atom has five valence electrons and Fig. 26.3a shows what occurs when one is introduced into the lattice of a silicon crystal. Four of its valence electrons form covalent bonds with four neighbouring silicon atoms but the fifth is spare and, being loosely held, it can take part in conduction. The impurity (phosphorus) atom is called a *donor*.

donor impurity atom (phosphorus)
silicon atom (nucleus and inner electrons)
spare electron
covalent bond
(a)
Crystal lattice of n-type silicon

majority charge carrier (electron)
n-type semiconductor
minority charge carrier (hole)
electron flow
(b)

Fig. 26.3

The 'impure' silicon is an n-type semiconductor since the *majority* charge carriers are negative electrons. (The overall charge in the crystal is zero since every atom present is electrically neutral.) Impurity atoms are added in the amount which produces the required conductivity.

A few positive holes are also present in n-type material (as intrinsic charge carriers formed by broken bonds between silicon atoms); they are *minority* carriers. Fig. 26.3b shows conduction in an n-type semiconductor.

(b) p-type. To make this type, silicon or germanium is doped with, for example, boron. A boron atom has three valence electrons and Fig. 26.4a shows what happens when one is introduced into the lattice of a silicon crystal. Its three valence electrons each share an electron with three of the four silicon atoms surrounding it. One bond is incomplete and the position of the missing electron, i.e. the hole, behaves like a positive charge since it can attract an electron from a nearby silicon atom, so forming another hole. For that reason the impurity (boron) atom is called an *acceptor*.

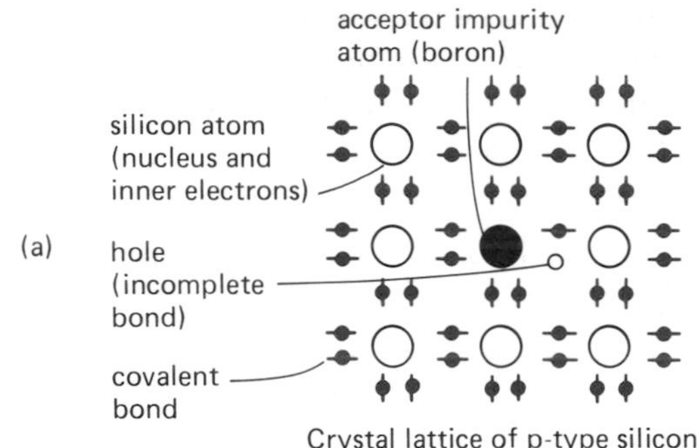

Crystal lattice of p-type silicon

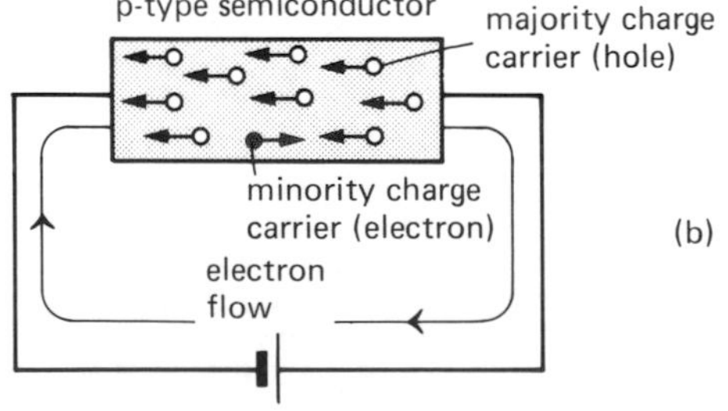

Fig. 26.4

The 'impure' silicon is a p-type semiconductor since the *majority* charge carriers causing conduction are positive holes. (Note again that the semiconductor as a whole is electrically neutral.) In p-type material a few electrons are present as *minority* carriers. Fig. 26.4b shows conduction in such material.

In both n- and p-types of semiconductor a temperature rise increases the proportion of minority carriers and upsets the semiconductor device.

Questions

1. Explain the terms (i) intrinsic, (ii) extrinsic, semiconductor.

2. What are the majority charge carriers in (i) n-type, (ii) p-type, semiconductors?

27 Junction diode

The p-n junction

The operation of many semiconductor devices depends on effects which occur at the boundary (junction) between p- and n-type materials formed in the same continuous crystal lattice.

(a) Unbiased junction. A p-n junction is represented in Fig. 27.1a. As soon as the junction is produced, free electrons near it in the n-type material are attracted across into the p-type material where they fill holes. As a result the n-type material becomes positively charged and the p-type material becomes negatively charged (both previously being neutral). At the same time holes pass across the junction from p-type to n-type, capturing electrons there.

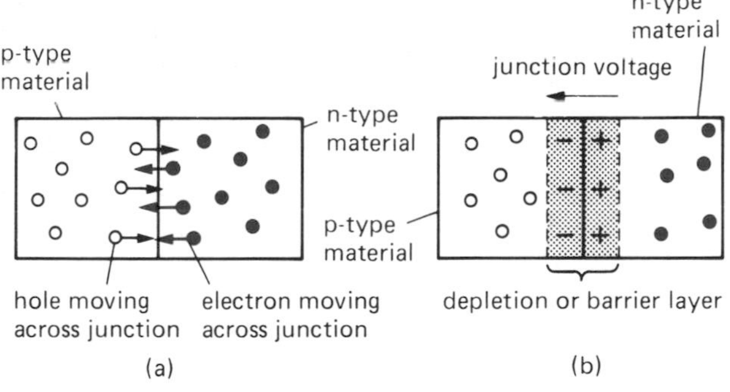

Fig. 27.1

The exchange of charge soon stops because the negative charge on the p-type material opposes the further flow of electrons and the positive charge on the n-type opposes the further flow of holes. The region on either side of the junction becomes fairly free of majority charge carriers, Fig. 27.1b, and is called the *depletion* or *barrier layer*. It is less than 10^{-6} m wide and is, in effect, an insulator. A p.d. known as the *junction voltage*, acts across the depletion layer from n- to p-type, being about 0.6 V for silicon and 0.1 V for germanium.

(b) Reverse biased junction. If a battery is connected across a p-n junction with its positive terminal joined to the n-type side and its negative terminal to the p-type side, it helps the junction voltage. Electrons and holes are repelled farther from the junction and the depletion layer widens, Fig. 27.2a. Only a few minority carriers cross the junction and a tiny current, called the *leakage* or *reverse* current, flows. The resistance of the junction is very high in reverse bias.

(c) Forward biased junction. If a battery is connected so as to oppose the junction voltage, the depletion layer narrows. When the battery voltage

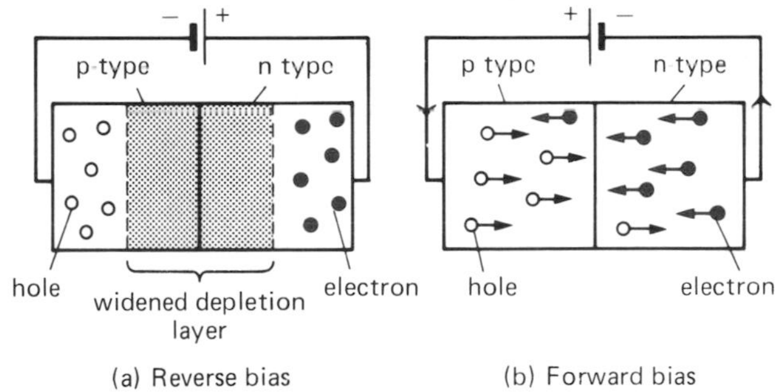

Fig. 27.2

exceeds the junction voltage, appreciable current flows because majority carriers are able to cross the junction. Electrons travel from the n- to the p-side and holes in the opposite direction, Fig. 27.2b. The junction is then forward biased, i.e. the *p-type* side is connected to the *positive* terminal of the battery and the *n-type* side to the *negative* terminal. A small leakage current again flows due to minority carriers. The resistance of the junction is very low in forward bias.

Junction diode

(a) Construction. A junction diode consists of a p-n junction with one connection to the p-side, the *anode* A, and another to the n-side, the *cathode* K. Its symbol is shown in Fig. 27.3a. In actual diodes the cathode end is often marked by a band, Fig. 27.3b; it is the end from which conventional current leaves the diode when forward biased. A simplified, much enlarged section of a silicon diode is shown in Fig. 27.3c but in fact the boundaries are neither straight nor well-defined.

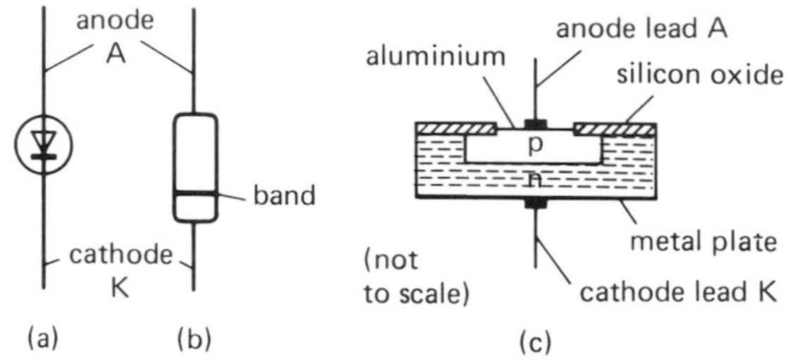

Fig. 27.3

(b) Characteristics. Typical characteristic curves for silicon and germanium diodes at 25°C are shown in Fig. 27.4. Conduction does not start until the forward voltage V_F is about 0.6 V for silicon and 0.1 V for germanium; thereafter, a very small change in V_F causes a sudden, large increase in the forward current I_F. When I_F is limited to a value within the power rating of the diode (by a resistor in the circuit), the p.d. across a silicon diode is about 1 V.

The reverse currents I_R are negligible (note the change of scales on the negative axes of the graph) and remain so until the reverse voltage V_R is large enough (from 5 V up to 1000 V depending on the level of doping) to break down the insulation of the depletion layer. I_R then increases suddenly and rapidly and permanent damage to the diode occurs. I_R is smaller and more constant for silicon than for germanium.

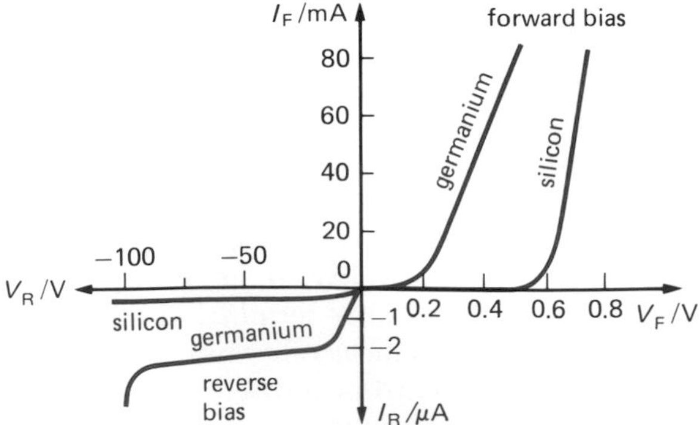

Fig. 27.4a

The *average forward current* I_F (av) and the *maximum reverse voltage* V_{RRM} (previously called the peak inverse voltage) are usually quoted for a diode. For example, for the 1N4001, I_F (av) = 1 A and V_{RRM} = 50 V.

The characteristic for an ideal diode is shown in Fig. 27.4b.

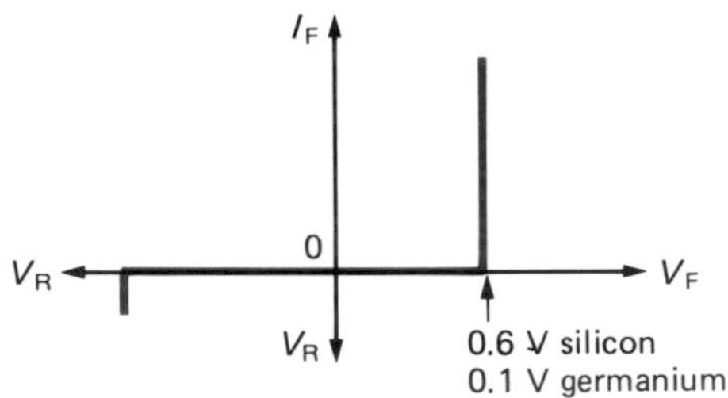

Fig. 27.4b

(c) Uses. Junction diodes are used

(i) as *rectifiers* to change a.c. to d.c. in power supplies (p. 89). Silicon is preferred to germanium because its much lower reverse current makes it a more efficient rectifier, i.e. it provides a more complete conversion of a.c. to d.c. Silicon also has a higher breakdown voltage and can work at higher temperatures, and
(ii) to *prevent damage to a circuit by a reversed power supply*. In Fig. 27.5a the battery is correctly connected and forward biases the diode. In Fig. 27.5b it is incorrectly connected but since it reverse biases the diode, no current passes and no damage occurs to the circuit.

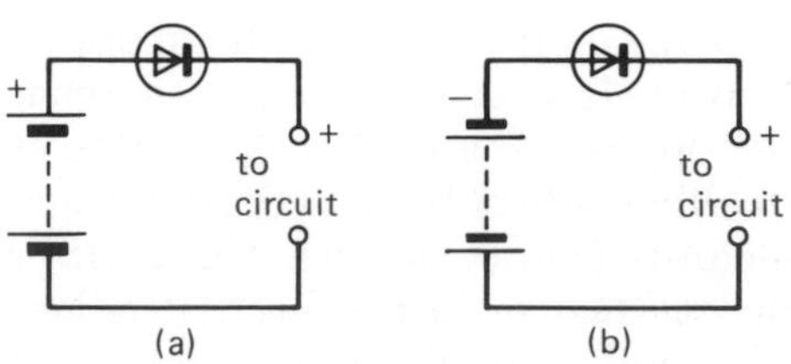

Fig. 27.5

Questions

1. In the circuits of Fig. 27.6 say which of the 6 V 60 mA lamps are bright, dim or off.

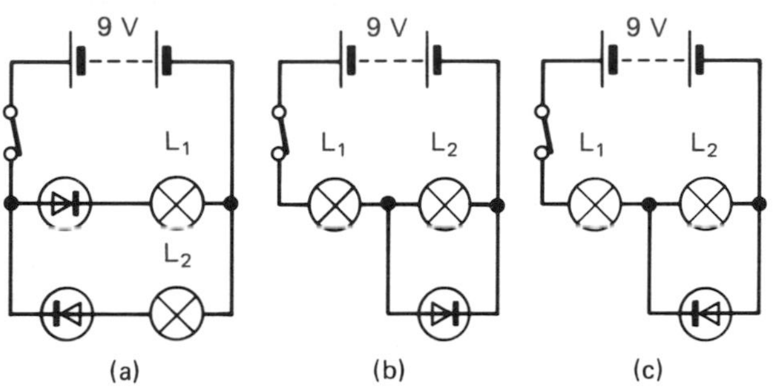

Fig. 27.6

2. The voltage drop across a 1N4001 diode is 1 V when the forward current is 1 A. If it is used on a 6 V supply as in Fig. 27.7, calculate

(a) the value of the safety resistor R, and
(b) the power dissipated in the diode.

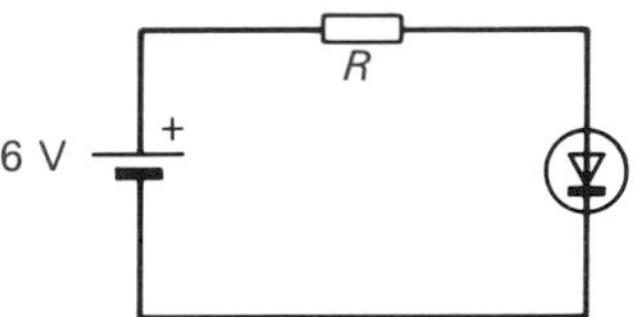

Fig. 27.7

28 Other diodes

Zener diode

A Zener diode is a silicon junction diode designed for stabilizing, i.e. keeping steady, the output voltage of a power supply (p. 90). One is shown in Fig. 28.1a,b with its symbol, the cathode end often being marked by a band.

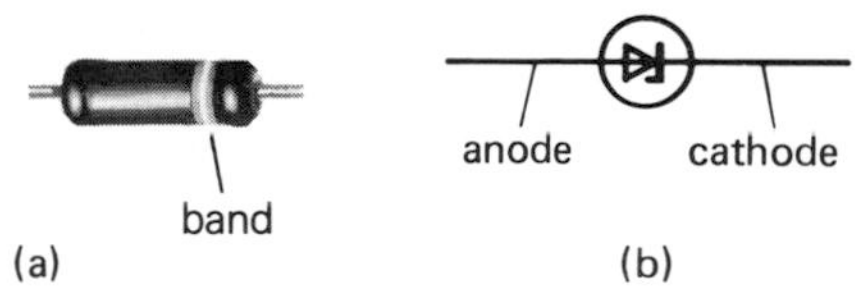

Fig. 28.1

(a) Characteristic. The characteristic in Fig. 28.2 shows that when forward biased, a Zener conducts at about 0.6 V, like an ordinary silicon diode. However it is normally used in reverse bias. Then, the reverse current I_R is negligible until the reverse p.d. V_R reaches a certain value V_Z, called the *Zener* or *reference voltage*, when it increases suddenly and rapidly as the graph indicates. Damage may occur to the diode by overheating unless I_R is limited by a series resistor. If this is connected, the p.d. across the diode remains constant at V_Z over a wide range of values of I_R, i.e. part AB of the characteristic is at right angles to the V_R axis. It is this property of a Zener diode which makes it useful for stabilizing power supplies.

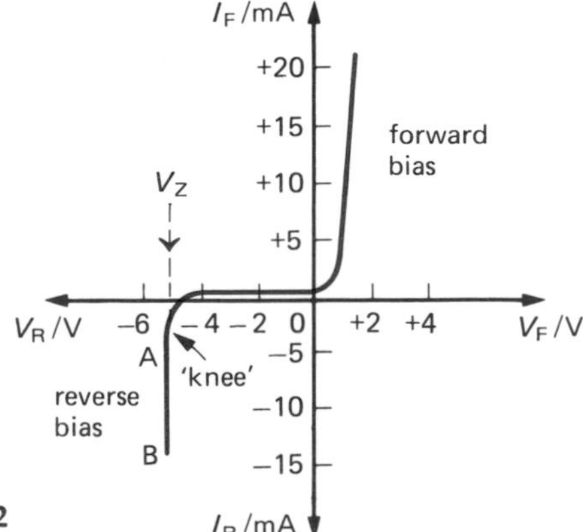

Fig. 28.2

The value of the current-limiting resistor should ensure that the power rating of the diode is not exceeded. For example, a 400 mW (0.4 W) Zener diode for which V_Z = 10 V can pass a maximum current I_{max} given by

$$I_{max} = \frac{\text{power}}{\text{voltage}} = \frac{0.4}{10} = 0.04\text{ A} = 40\text{ mA}$$

Zener diodes with specified Zener voltages are made, e.g. 3.0 V, 3.9 V, 5.1 V, 6.2 V, 9.1 V, 10 V, up to 200 V by controlling the doping.

(b) Demonstration. When the input p.d. V_i in Fig. 28.3 is gradually increased we see that:

(i) if $V_i < 6.2$ V (the Zener voltage), $I_R = 0$ and the output p.d. $V_o = V_i$ since there is no current through R and so no p.d. across it, and

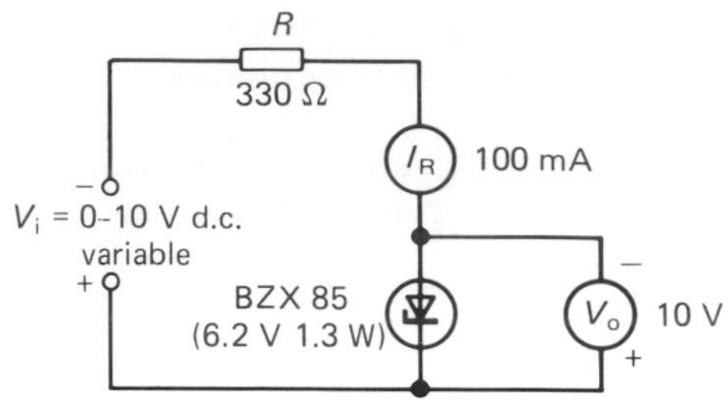

Fig. 28.3

(ii) if $V_i \geq 6.2$ V, the diode conducts, even although it is reverse biased, and $V_o = 6.2$ V, any excess p.d. appearing across R.

The Zener diode thus regulates V_o so long as V_i is not less than V_Z (here 6.2 V), i.e. there must be current through it.

Light-emitting diode (LED)

(a) Action. A LED, shown in Fig. 28.4 with its symbol, is a junction diode made from the semiconductor gallium arsenide phosphide. When forward biased it conducts and emits red, yellow, or green light depending on its composition. No light emission occurs on reverse bias which, if it exceeds 5 V, may damage the LED.

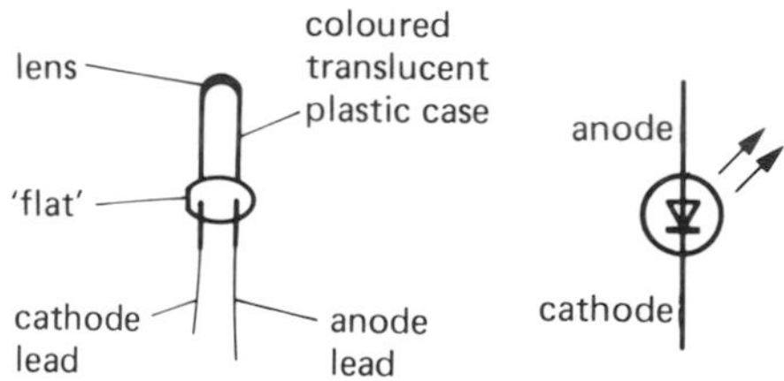

Fig. 28.4

When a p-n junction diode is forward biased, electrons move across the junction from the n-type side to the p-type side where they recombine with holes near the junction. The same occurs with holes crossing from the p-type side. Every recombination results in the release of a certain amount of energy, causing in most semiconductors, a temperature rise. In gallium arsenide phosphide some of the energy is emitted as light which escapes from the LED because the junction is very close to the surface of the material.

(b) External resistor. Unless a LED is of the 'constant-current' type (incorporating an integrated circuit regulator, p. 91) it must have an external resistor R connected in series to limit the forward current which, typically may be 10 mA (0.01 A). The voltage drop V_F is greater across a conducting LED than across an ordinary diode and is about 2 V, therefore R can be calculated from:

$$R = \frac{(\text{supply voltage} - 2.0)\,\text{V}}{0.01\,\text{A}}$$

For example, on a 5.0 V supply, $R = 3.0/0.01 = 300\,\Omega$.

(c) Uses. LEDs are used as indicator lamps, especially in digital electronic circuits to show whether outputs are 'high' or 'low'. One way of using a LED to test for a 'high' output (9 V in this case) is shown in Fig. 28.5a and for a 'low' output (0 V) in Fig. 28.5b. In the first case the output acts as the 'source' of the LED current, in the second case the output has to be able to accept or 'sink' the current (but also see p. 137).

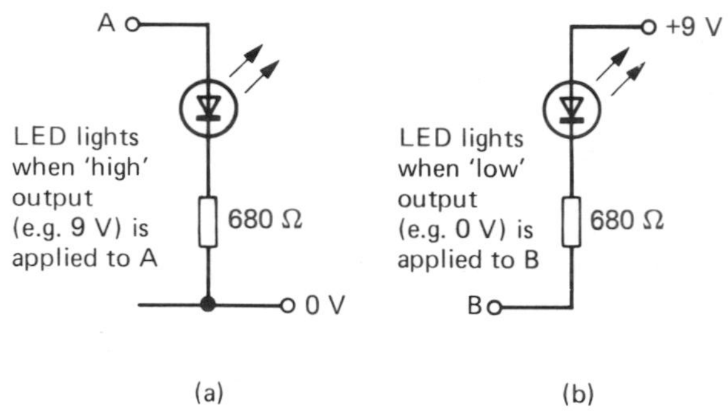

Fig. 28.5

The use of LEDs in seven-segment displays was considered earlier (p. 47).

The advantages of LEDs are small size, reliability, long life and high operating speed.

Point-contact diode

(a) Construction. The construction of a germanium point-contact diode is shown in Fig. 28.6 The tip of a

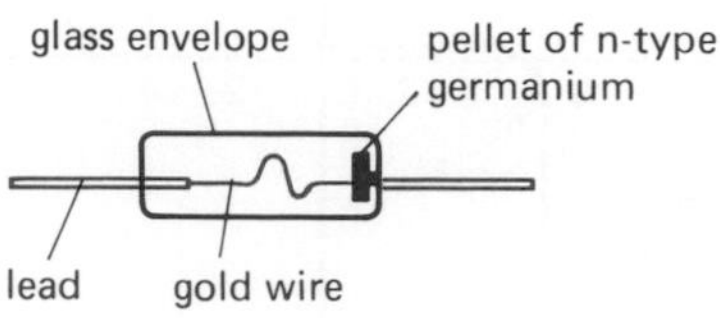

Fig. 28.6

gold wire presses on a pellet of n-type germanium. During manufacture a brief current is passed through the diode and produces a tiny p-type region in the pellet around the tip, so forming a p-n junction of very small area.

(b) Use. Point-contact diodes are used as signal diodes to detect radio signals (a process similar to rectification in which radio frequency a.c. is converted to d.c., p. 182) because of their very low capacitance. When reverse biased, the depletion layer in a junction diode acts as an insulator sandwiched between two conducting 'plates' (the p- and n-regions). It therefore behaves as a capacitor and the larger its junction area and the thinner the depletion layer, the greater is its capacitance (p. 57).

A capacitor 'passes' a.c. and the higher the frequency and the greater the capacitance, the less opposition does it offer ($X_C = 1/(2\pi fC)$). At radio frequencies therefore a normal junction diode would not be a very efficient detector (rectifier) because of the comparatively large junction area; its opposition in the reverse direction would not be large enough. A point-contact diode on the other hand is more suited to high frequency signal detection because of its tiny junction area.

Germanium is used for signal diodes because it has a lower 'turn-on' voltage than silicon (about 0.1 V compared with 0.6 V) and so lower signal voltages start it conducting in the forward direction. For the OA91 point-contact diode I_F (av) = 20 mA, $V_F \approx 1$ V and V_{RRM} = 100 V.

Photodiode

A photodiode consists of a normal p-n junction in a case with a transparent 'window' through which light can enter. One is shown in Fig. 28.7a,b with its symbol. It is operated in reverse bias and the leakage (minority carrier) current increases in proportion to the amount of light falling on the junction. The effect is due to the light energy breaking bonds in the crystal lattice of the semiconductor to produce electrons and holes.

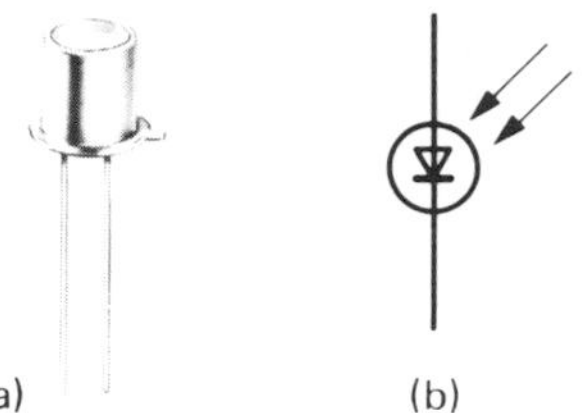

Fig. 28.7

Photodiodes are used as 'fast' counters which generate a current pulse every time a beam of light is interrupted.

Questions

1. In the circuit of Fig. 28.8 a 6 V 1 W Zener diode D in parallel with a 6 V 60 mA lamp L is connected in reverse bias to a battery via a protective resistor R.
Explain how the lamp behaves as the battery voltage is increased gradually from 3 V to 9 V.

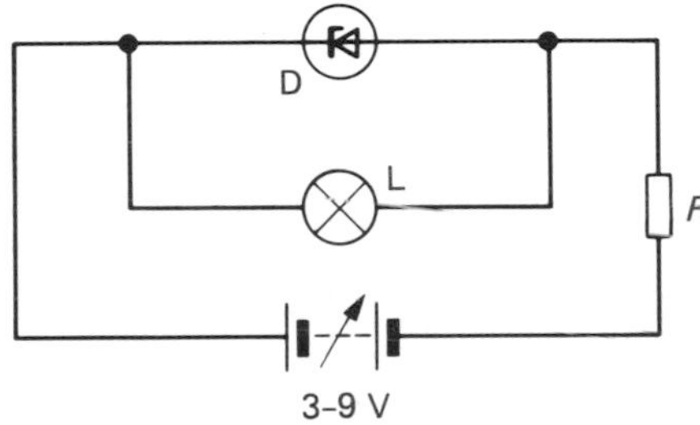

Fig. 28.8

2. Calculate the maximum current a Zener diode can pass without damage if its breakdown voltage is 10 V and its maximum power rating is 5 W.

3. In Fig. 28.9 the LED is bright when a current of 10 mA flows and the forward voltage (V_F) across it is 2 V. Calculate the value (E12 preferred) of R if a 9 V supply is used.

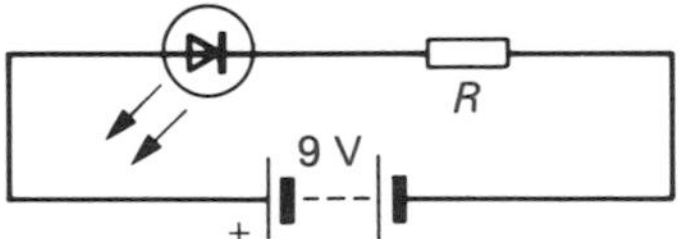

Fig. 28.9

4. State one use for (a) a Zener diode, (b) a LED, (c) a point-contact diode, (d) a photodiode.

29 Progress questions

1. (a) Explain how an electric current is conducted in pure silicon when a p.d. is applied to it.
(b) Why does the resistance of pure silicon fall as its temperature rises?
(c) If silicon is doped with phosphorus (a pentavalent element) it becomes an n-type semiconductor. Explain what this means.
(d) How can silicon be made a p-type semiconductor?
(e) What happens to the depletion layer in a p-n junction under (i) reverse bias, (ii) forward bias?

2. In Fig. 29.1 L_1, L_2, L_3 and L_4 are 12 V lamps.

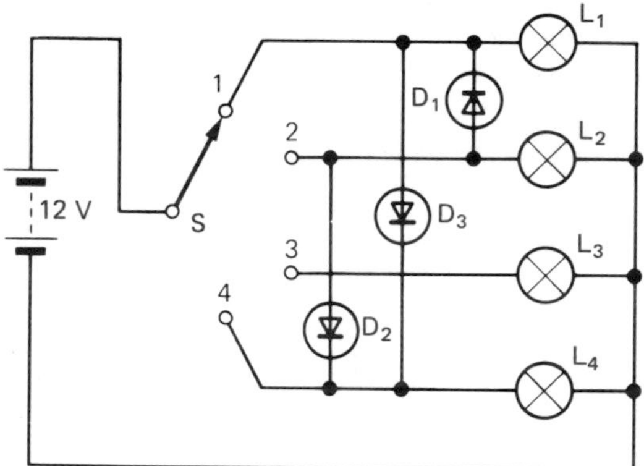

Fig. 29.1

(a) How would you describe the contact arrangement of the rotary switch S?
(b) Which lamps will illuminate with the switch in position (i) 1, (ii) 2, (iii) 3, (iv) 4?

(A.E.B. 1981 Control Tech.)

3. (a) Draw the reverse characteristics of a Zener diode, giving typical values on your axes. In what way does its behaviour differ from a normal p-n diode in reverse bias?
(b) The Zener diode Z shown in Fig. 29.2 has a breakdown voltage of 6 V. What is the current through it? Discuss what happens to the current through the diode, and the p.d. across it when the positive supply is altered to (i) 12 V, (ii) 4 V. *(C.)*

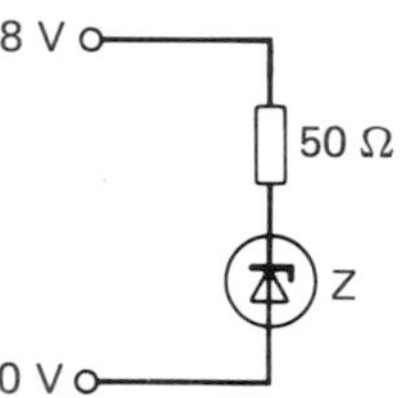

Fig. 29.2

4. Fig. 29.3 shows a circuit diagram of a stabilized voltage supply together with the characteristic of the diode in the circuit.
(i) What type of diode is represented in the circuit?
(ii) What type of bias is applied to the diode?
(iii) What will be the output voltage?
(iv) Explain why the output voltage is constant for a range of input voltages. *(O.L.E.)*

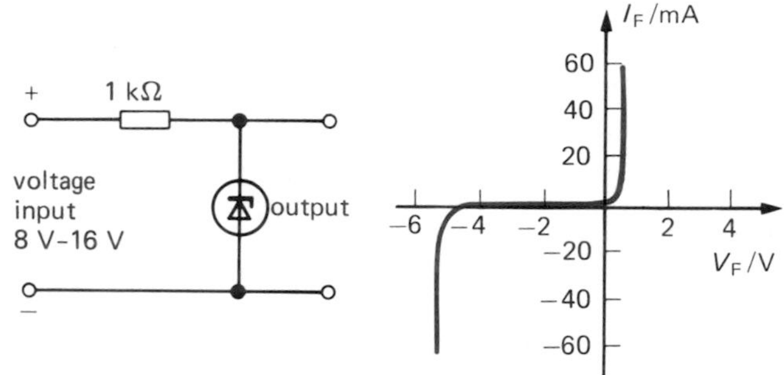

Fig. 29.3

5. (a) In which of the circuits in Fig. 29.4 would the LED light when the point Q is raised from 0 V to +9 V?
(b) Why is the 680 Ω resistor included in the circuits?
(c) Calculate the current through the LED which lights if the forward voltage dropped across the LED is 2 V.

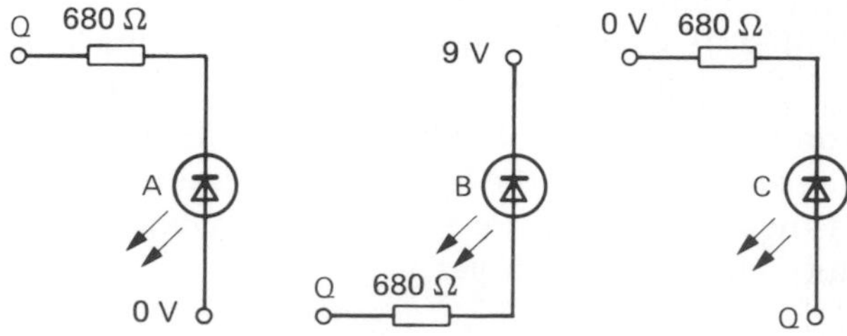

Fig. 29.4

Transistors

30 Transistor as a current amplifier

About transistors

Transistors are the tiny semiconductor devices which have revolutionized electronics. They have three connections and are made as separate (discrete) components, like those in Fig. 30.1 in their cases, and also as parts of integrated circuits (ICs) where many thousands may be packed on a 'chip' of silicon. Transistors are used as current, voltage and power *amplifiers* and as high-speed *switches*.

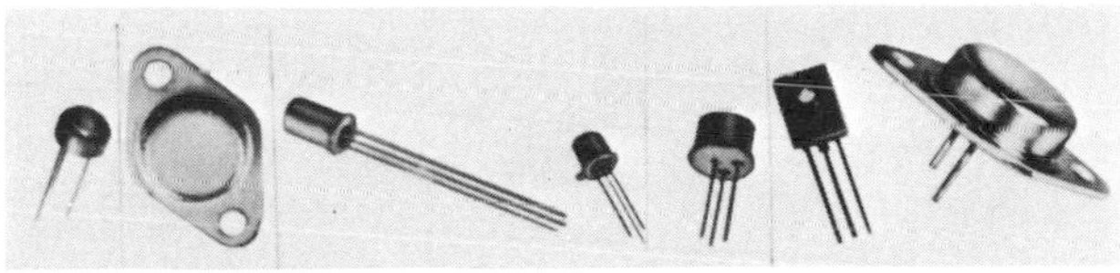

Fig. 30.1

There are two basic types, the common *bipolar* or *junction* transistor, to be considered in this chapter and the next two, and the less common *unipolar* or *field effect* transistor (FET), to be discussed later.

A junction transistor consists of two p-n junctions (in effect two diodes back-to-back) in the same semiconductor crystal. A very thin slice of lightly doped p- or n-type material called the *base* B, is sandwiched between two thicker, more heavily doped slices of the opposite type of material, called the *collector* C and the *emitter* E, the latter being smaller. The two possible arrangements are shown simplified in Fig. 30.2 along with the symbols for n-p-n and p-n-p transistors. The arrows on the symbols indicate the direction in which conventional (positive) current would flow; electron flow is in the opposite direction.

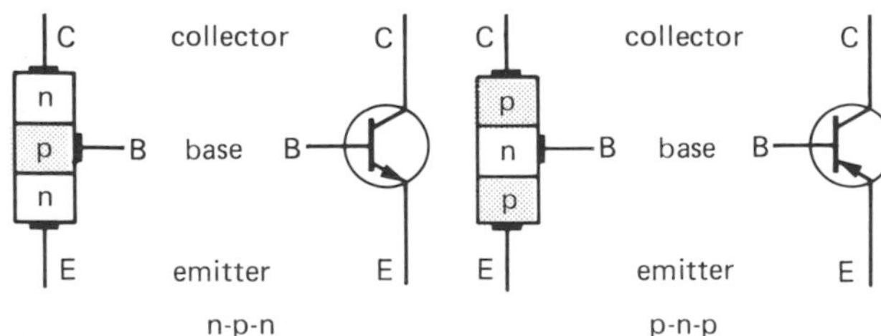

Fig. 30.2

In general, silicon is preferred to germanium for making transistors because it can work at higher temperatures (175 °C compared with 75 °C) and has a lower leakage current (p. 57). Silicon n-p-n types are easier to mass-produce than p-n-p types and so are commoner, although it is sometimes useful to have both types in a circuit.

What a transistor does

(a) Action. There are two current paths through a transistor. One is the base-emitter path and the other is the collector-emitter (via base) path. The transistor's value arises from the fact that it can link circuits connected to each path so that the current in one controls that in the other.

If a p.d. (e.g. +6 V) is connected across an n-p-n silicon transistor so that the collector becomes positive with respect to the emitter, the base being unconnected, the base-collector junction is reverse biased (since the positive of the supply goes to the n-type collector). Current cannot flow through the transistor.

If the base-emitter junction is now *forward biased* by applying a p.d. V_{BE} of about +0.6 V (+0.1 V for a germanium transistor), Fig. 30.3, electrons flow from the n-type emitter across the junction (as they would in a junction diode) into the p-type base. Their loss is made good by electrons entering the emitter from the external circuit to form the *emitter current* I_E.

In the base, only a small proportion (about 1%) of the electrons from the emitter combine with holes because the base is very thin (less than 10^{-3} mm) and is lightly doped. Most of the electrons pass through the base under the strong attraction of the

...tor as a current amplifier

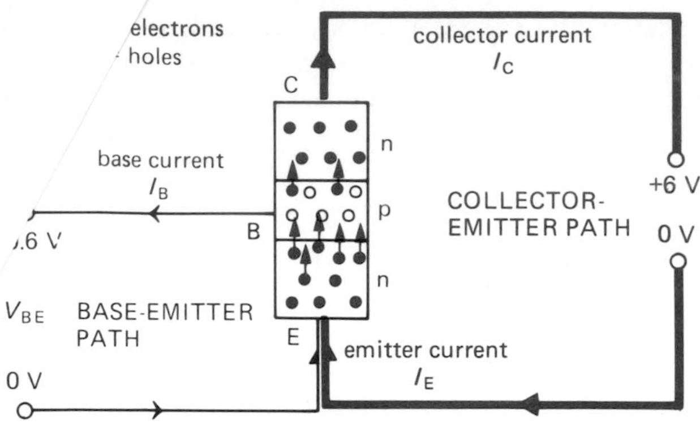

Fig. 30.3

positive voltage on the collector. They cross the base-collector junction and become the *collector current* I_C in the external circuit.

The small amount of electron-hole recombination occurring in the base gives it a momentary negative charge which is immediately compensated by a flow of positive holes from the base power supply. This flow of holes to the base from the external circuit creates a small *base current* I_B and enables the transistor to maintain the much larger collector current.

If we regard I_B as the *input* current to the transistor and I_C as the *output* current from it, then the transistor acts

(i) as a *switch* in which I_B turns on and controls I_C, i.e. $I_C = 0$ if $I_B = 0$ ($I_B = 0$ until $V_{BE} \approx +0.6$ V), and increases when I_B does, and

(ii) as a *current amplifier* since I_C is greater than I_B.

(b) Further points. Note that for an n-p-n transistor the collector and base must be positive with respect to the emitter; for a p-n-p type they must be negative.

The input (base-emitter) and output (collector-emitter) circuits in Fig. 30.3 have a common connection at the emitter. The transistor is said to be in *common-emitter* mode. Two other less usual modes are *common-collector* and *common-base*.

(c) d.c. current gain. Typically I_C is 10 to 1000 times greater than I_B depending on the type of transistor. The *d.c. current gain* h_{FE} is an important property of a transistor and is defined by

$$h_{FE} = \frac{I_C}{I_B}$$

For example, if $I_C = 5$ mA and $I_B = 0.05$ mA (50 μA), $h_{FE} = 5/0.05 = 100$. Although h_{FE} is approximately constant for one transistor over a limited range of I_C values, it varies between transistors of the same type due to manufacturing tolerances.

Also note that since the current leaving a transistor equals that entering, we have

$$I_E = I_B + I_C$$

In the above example $I_E = 5.05$ mA.

Demonstration

(a) Current amplification and control. In the circuit of Fig. 30.4 only one power supply is used to provide the positive voltages to the collector and base of the n-p-n transistor and rheostat R controls I_B.

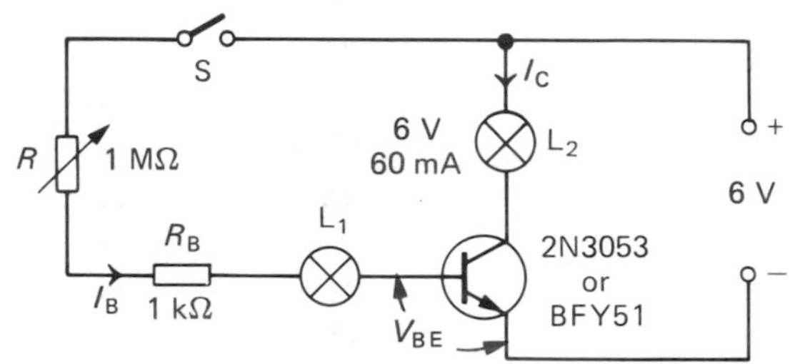

Fig. 30.4

With switch S open, $I_B = 0$, therefore $I_C = 0$ and lamps L_1 and L_2 are not lit.

If S is closed and I_B is gradually increased by reducing R, I_C increases and when it is large enough, L_2 lights up but not L_1. In this way a small current I_B (too small to light L_1) produces and controls smoothly the much larger current I_C required to light L_2. The rheostat need only have a low power rating but the transistor must be able to handle I_C (60 mA).

(b) Base resistor R_B. The base-emitter junction is, in effect, a forward biased p-n diode and so the p.d. across it (V_{BE}) cannot rise much above 0.6 V (p. 58). If input p.ds greater than 0.6 V or so were applied *directly* to the base (e.g. by adjusting R to have zero resistance in Fig. 30.4 and omitting R_B), I_B (and I_C) would become excessive and the transistor would be destroyed by over-heating. By limiting I_B, R_B prevents this when the input exceeds 0.6 V and *must always be present*.

Darlington pair

Greater current gains can be obtained by connecting two transistors as in Fig. 30.5, called a *Darlington pair*. The emitter current I_E of Tr_1 is nearly the same as its collector current and equals h_{FE1} times its base

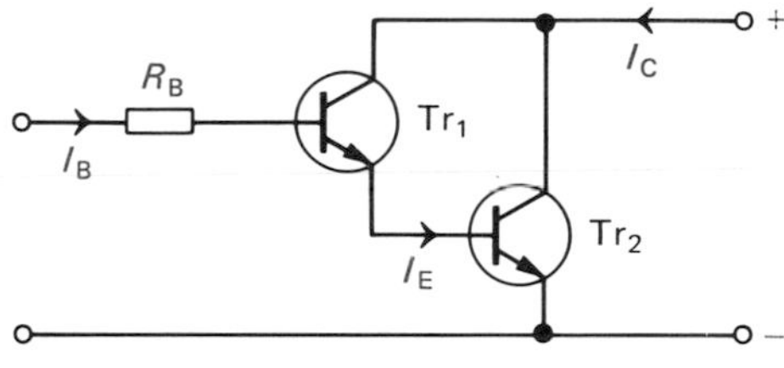

Fig. 30.5

(input) current I_B. This emitter current forms the base current of Tr_2. The collector (output) current I_C of Tr_2 equals h_{FE2} times its base current. The overall current gain is $h_{FE} = I_C/I_B = h_{FE1} \times h_{FE2}$ (typically 10^4).

Questions

1. In the labelled circuit of Fig. 30.6 the lamp L lights when the switch S is closed. Which is (a) the base-emitter current path, (b) the collector-emitter current path?

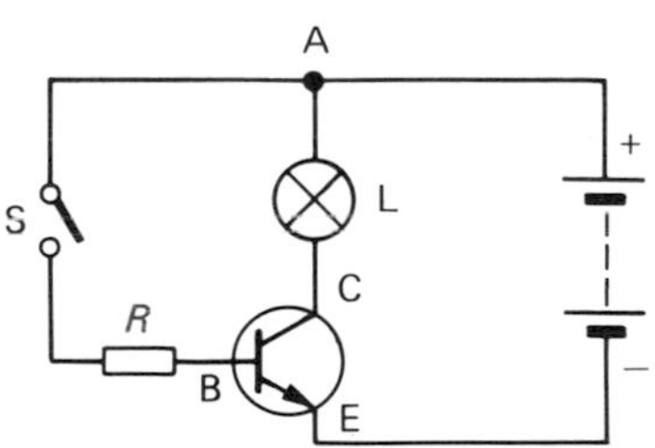

Fig. 30.6

2. In which circuits of Fig. 30.7 will lamp L light? (B_1 = 3 V, B_2 = 6 V, R = 1 kΩ, L = 6 V 60 mA.)

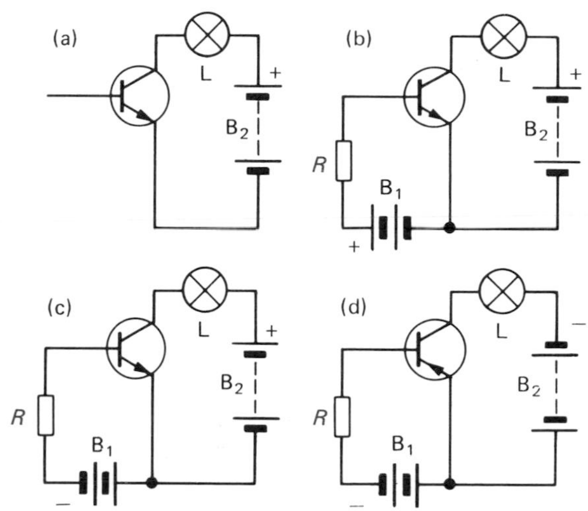

Fig. 30.7

3. For the transistor in Fig. 30.8 what is the value of (a) h_{FE} and (b) I_E?

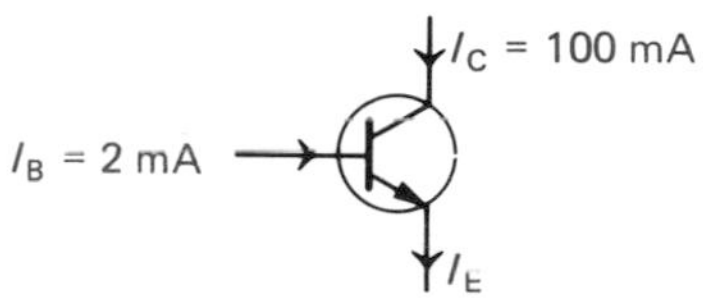

Fig. 30.8

4. A transistor has a d.c. current gain of 200 when the base current is 50 μA. Calculate the collector current in mA.

5. Two transistors have d.c. current gains of 80 and 100. What is their equivalent gain as a Darlington pair?

31 Transistor as a switch

Transistors have many advantages over other electrically operated switches such as relays. They are small, cheap, reliable, have no moving parts, their life is almost indefinite (in well-designed circuits) and they can switch millions of times a second.

Transistor switching circuit

The basic common-emitter switching circuit is shown in Fig. 31.1. It contains a protective resistor R_B in the base circuit and a 'load' resistor R_L in the

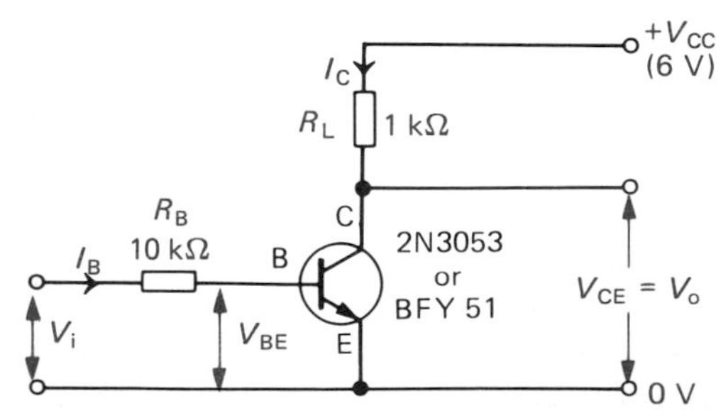

Fig. 31.1

collector circuit. R_L and the transistor form a potential divider across the power supply voltage V_{CC} to the collector (note the double suffix).

V_i is the input p.d. (usually d.c.) applied to the base-emitter circuit. Depending on its value, V_i causes switching of the output p.d. V_o (= V_{CE} and which is taken off across the collector and emitter) between a 'high' value (e.g. near V_{CC}) and a 'low' value (e.g. near 0 V), as we will now see.

(a) Action. If V_i is gradually increased from 0 to 6 V and corresponding values of V_i, V_o, I_C and I_B measured with appropriate meters, voltage and current graphs like those in Fig. 31.2a, b can be plotted.

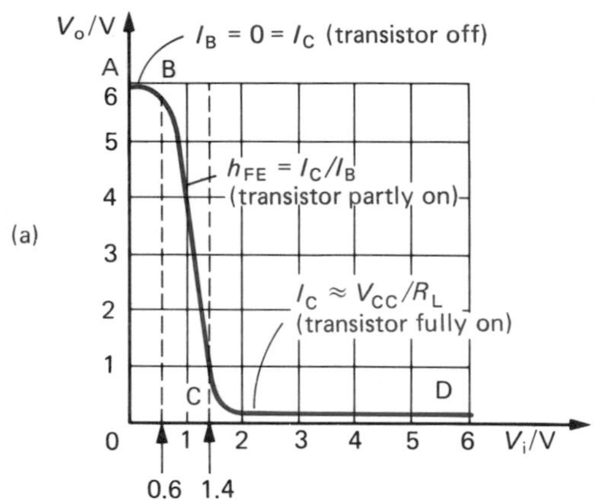

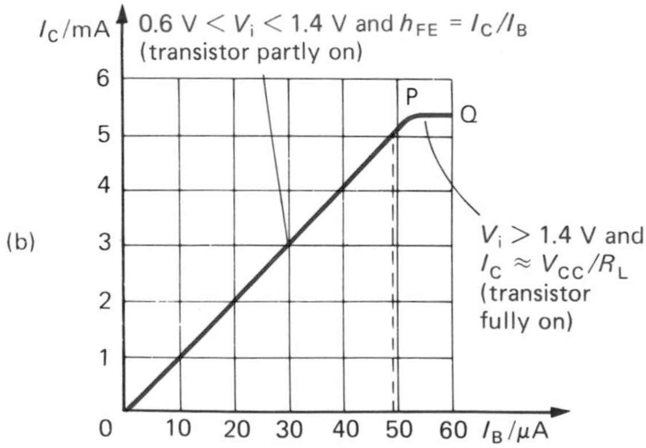

Fig. 31.2

The graphs show that

(i) When $V_i < 0.6$ V, $V_o = V_{CC} = 6$ V, i.e. AB on graph (a), and $I_B = I_C = 0$, i.e. point 0 on graph (b).
(ii) when $V_i > 0.6$ V but < 1.4 V, V_o falls rapidly as V_i increases, i.e. BC on graph (a), and I_C increases linearly with I_B, i.e. 0P on graph (b), and
(iii) when $V_i > 1.4$ V, $V_o \approx 0$ V, i.e. CD on graph (a), and I_C reaches a maximum even although I_B is increased further, i.e. PQ on graph (b).

(b) Explanation. The transistor and R_L are in series across V_{CC}; therefore since d.c. voltages can be added directly, we can say for the collector-emitter circuit:

$$V_{CC} = I_C R_L + V_{CE}$$

Rearranging the equation and putting $V_{CE} = V_o$ gives

$$V_o = V_{CC} - I_C R_L \qquad (1)$$

That is, V_o is always less than V_{CC} by the voltage drop across R_L.

When $V_i < 0.6$ V, $V_o = V_{CC} = 6$ V because $I_B = 0 = I_C$, therefore from (1), $I_C R_L = 0$. The transistor is *off* and behaves like a very high resistor, i.e. an open switch, Fig. 31.3a, with none of V_{CC} dropped across R_L since there is no current through it.

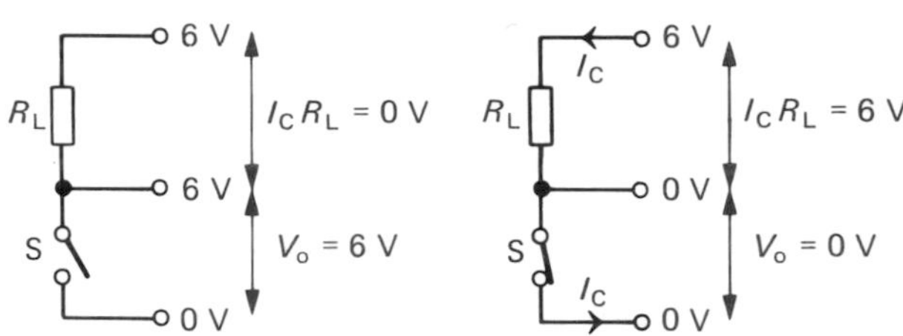

Fig. 31.3

When $V_i > 0.6$ V but < 1.4 V, the transistor is *partly on*, i.e. its resistance is decreasing, I_B and I_C flow and part of V_{CC} is dropped across R_L and part across the transistor, according to (1). In this region I_B and I_C are related by $h_{FE} = I_C/I_B$.

When $V_i > 1.4$ V, $V_o \approx 0$ V, therefore from (1), $V_{CC} \approx I_C R_L \approx 6$ V. Nearly all V_{CC} is now dropped across R_L and very little across the transistor which is behaving as if it had almost zero resistance, i.e. like a closed switch, Fig. 31.3b. The resistance of the collector-emitter circuit is now more or less constant and equal to R_L. Therefore although I_B increases when V_i rises from 1.4 V to 6 V, I_C does not but approaches its maximum value given by V_{CC}/R_L (i.e. 6 V/1 kΩ = 6 mA), as can be seen from (1) by putting $V_o = 0$. In this region the transistor is *fully on* and is said to be *saturated* (or *bottomed*) because increasing I_B does not increase I_C, i.e. $h_{FE} = I_C/I_B$ no longer applies.

(c) Summary. Depending on whether V_i is 'low' (0 to 0.6 V) or 'high' (> 1.4 V), V_o switches between the two voltage levels V_{CC} ('high') and 0 V ('low') respectively.

V_i	V_o
'low' (< 0.6 V)	'high' (6 V)
'high' (> 1.4 V)	'low' (0 V)

Further points

(a) Base-emitter circuit. In the circuit of Fig. 31.1, R_B and the base-emitter junction form a potential divider across V_i. Hence

$$V_i = I_B R_B + V_{BE} \qquad (2)$$

For a silicon transistor $V_{BE} \approx 0.6$ V *whatever the value of* V_i. When $V_i > 0.6$ V, the 'extra' voltage ($V_i - 0.6$ V) is dropped across the protective resistor R_B.

Note that I_B can be calculated from (2) if V_i is known and we assume $V_{BE} = 0.6$ V. It is obtained by rearranging (2), to give

$$I_B = \frac{V_i - 0.6}{R_B}$$

(b) Power considerations. The power P used by a transistor as a switch should be as small as possible. It is given by $P = V_{CE} \times I_C$. When the transistor is off, $I_C = 0$ and so $P = 0$. When the transistor is fully on, $V_{CE} \approx 0$ (typically it is 0.1 V, ideally it should be 0 V), therefore $P \approx 0$.

Power is used only when the transistor is partly on, i.e. along BC in Fig. 31.2a. It is therefore important to ensure (i) it is either fully off or fully on and (ii) it switches rapidly between these two states.

A two-transistor circuit called a *Schmitt trigger* meets these fast-switching requirements and is considered on p. 153.

Worked example

(a) *In the circuit of Fig. 31.4, $h_{FE} = 100$ for the transistor. Calculate the value of R_B for the lamp current I_C to be 60 mA. (Assume the transistor is not saturated and that $V_{BE} = 0.6$ V.)*

(b) *If the lamp has resistance $R_L = 100\,\Omega$ would the transistor be saturated?*

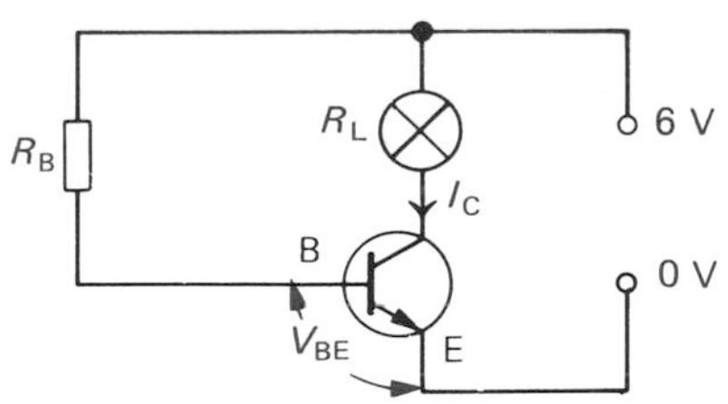

Fig. 31.4

(a) If the transistor is not saturated $h_{FE} = I_C/I_B$, therefore

$I_B = I_C/h_{FE} = 60/100 = 0.6$ mA.

Rearranging equation (2) we get

$R_B = (V_i - V_{BE})/I_B$

where $V_i = V_{CC} = 6$ V (since the input is connected to the collector power supply), $V_{BE} = 0.6$ V and $I_B = 0.6$ mA.

Substitution gives

$R_B = (6 - 0.6)\text{ V}/0.6\text{ mA} = 5.4/0.6 = 9\text{ k}\Omega$

(b) When the transistor is saturated I_C is a maximum, given by

$I_C = V_{CC}/R_L = 6\text{ V}/100\,\Omega = 0.06\text{ A} = 60\text{ mA}$

The transistor would *just* be saturated.

Alarm circuits

In many alarm circuits a transducer in a potential divider circuit is used to switch on a transistor which then activates the alarm.

(a) Light-operated. A simple circuit which switches on a lamp L when it gets dark is shown in Fig. 31.5. R and the light dependent resistor (LDR) form a potential divider across the 6 V supply. The input is the p.d. V_i across the LDR and in bright light is small because the resistance of the LDR is low (e.g. 1 kΩ) compared with that of R (10 kΩ). V_{BE} is less than the 0.6 V or so required to turn on the transistor.

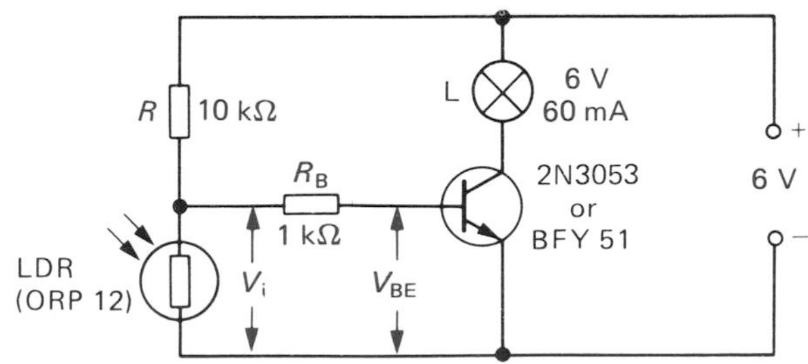

Fig. 31.5

In the dark more of the 6 V supply is dropped across LDR, due to its greater resistance (e.g. 1 MΩ), and less across R. V_i is then large enough for V_{BE} to reach 0.6 V and switch on the transistor which creates a collector current sufficient to light L. If R is replaced by a variable resistor, the light level at which L comes on can be adjusted.

When R and the LDR are interchanged, L is on in the light and off in the dark and the circuit could act as an intruder alarm.

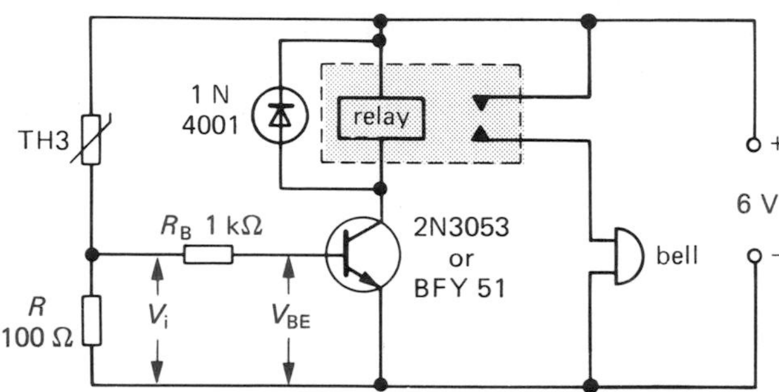

Fig. 31.6

(b) Temperature-operated. In the high-temperature alarm circuit of Fig. 31.6 an n.t.c. thermistor (p. 46) and resistor R form a potential divider across the 6 V supply. When the temperature of the thermistor rises, its resistance decreases, so increasing V_i and V_{BE}. When $V_{BE} \approx 0.6$ V, the transistor switches on and collector current (too small to ring the bell directly) flows through the relay coil. The relay contacts close, enabling the bell to obtain directly from the 6 V supply, the larger current it needs.

The diode protects the transistor from damage by the large e.m.f. induced in the relay coil (due to its inductance) when the collector current falls to zero (at switch-off). The diode is forward biased to the induced e.m.f. and, because of its low resistance, offers an easy path to it. To the power supply the diode is reverse biased and its high resistance does not short-circuit the relay coil when the transistor is on.

(c) Time-operated. In the circuit of Fig. 31.7 when S_1 and S_2 are closed, L is on and the transistor is off because $V_{BE} = 0$ (due to S_2 short-circuiting C and stopping it charging up). If S_2 is opened, C starts to charge through R and, *after a certain time*, $V_{BE} = 0.6$ V causing the transistor to switch on. This operates the relay whose contacts open and switch off L. The *time delay* between opening S_2 and L going off depends on the time constant CR.

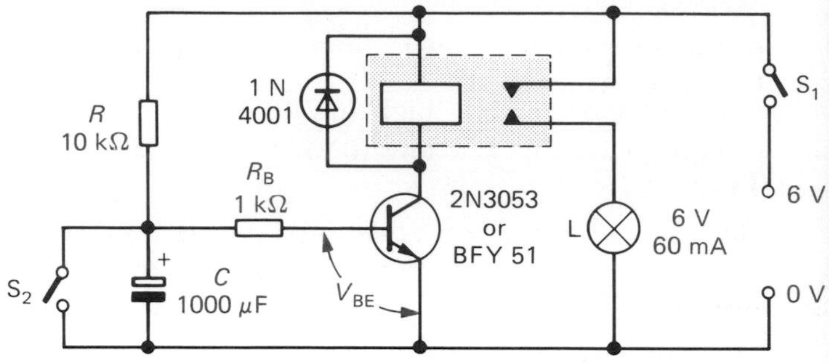

Fig. 31.7

The circuit could be used as a *timer* to control a lamp in a photographic dark room. It is reset by opening S_1 and closing S_2 to let C discharge.

Questions

1. A transistor circuit is shown in Fig. 31.8. (a) What is I_B assuming $V_{BE} = 0$? (b) Calculate I_C if $h_{FE} = 80$. (c) Is the transistor saturated? Justify your answer.

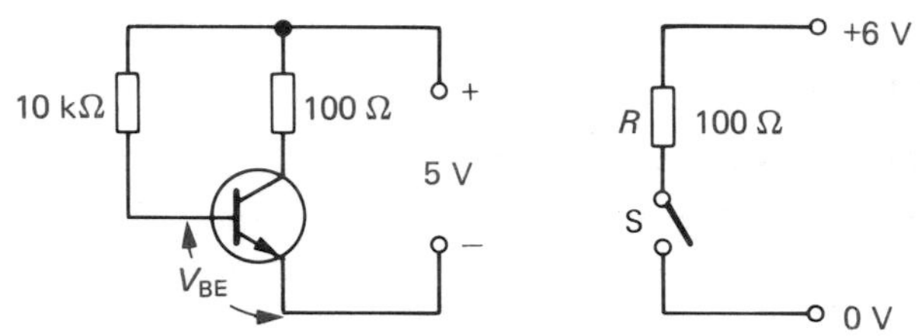

Fig. 31.8 Fig. 31.9

2. In the circuit of Fig. 31.9, what is the p.d. across R when (a) S is open, (b) S is closed?

3. The circuit in Fig. 31.10 is for a silicon transistor Tr.
(a) When Tr is off what is the p.d. across (i) R_L (ii) Tr?
(b) When Tr is saturated what is the p.d. across (i) R_L (ii) Tr?
(c) Will R_1/R_2 be large or small when Tr is (i) off, (ii) saturated?
(d) What is the value (or range of values) of V_{BE} when Tr is (i) off, (ii) saturated?
(e) Why is (i) R_B, (ii) R_L necessary?

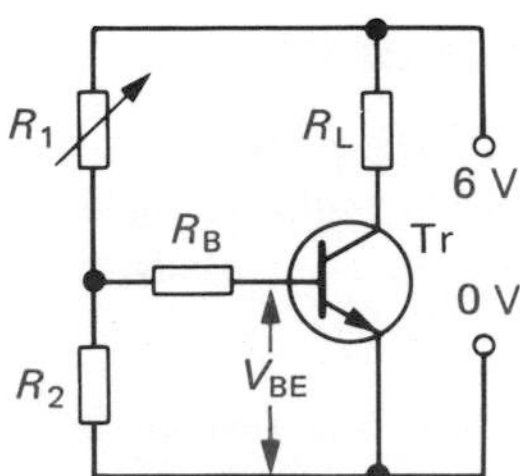

Fig. 31.10

32 More about transistors

In this chapter the junction (bipolar) transistor is treated in more detail.

Transistor characteristics

These are graphs, found by experiment, which show the relationships between various currents and voltages and enable us to see how best to use a transistor. A circuit for investigating an n-p-n transistor in common-emitter connection is given in Fig. 32.1. Three characteristics are important.

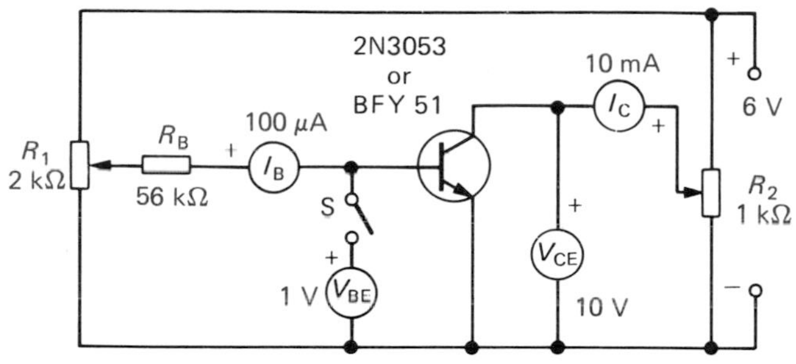

Fig. 32.1

(a) Transfer characteristic ($I_C - I_B$). V_{CE} is kept constant (e.g. at the supply voltage) and R_1 varied to give several pairs of values of I_B and I_C. If the results are plotted on a graph, a more or less straight line is obtained, like that in Fig. 32.2a, showing that I_C is directly proportional to I_B.

The *a.c. current gain*, h_{fe} (important when the transistor is handling changing currents) is defined by

$$h_{fe} = \frac{\Delta I_C}{\Delta I_B}$$

where ΔI_C (delta I_C) is the change in I_C produced by a change of ΔI_B in I_B. Since the $I_C - I_B$ graph is almost linear, $h_{fe} \approx h_{FE}$ (the d.c. current gain $= I_C/I_B$).

(b) Input characteristic ($I_B - V_{BE}$). Again keeping V_{CE} constant, corresponding values of I_B and V_{BE} are obtained by varying R_1 in the circuit of Fig. 32.1. A typical graph for a silicon transistor is given in Fig. 32.2b. It shows that $I_B = 0$ until $V_{BE} \approx 0.6\,V$ and thereafter small changes in V_{BE} cause large changes in I_B (but V_{BE} is always near 0.6 V whatever the value of I_B).

The *a.c. input resistance* r_i of the transistor is defined by

$$r_i = \frac{\Delta V_{BE}}{\Delta I_B}$$

where ΔI_B is the change in I_B due to a change of ΔV_{BE} in V_{BE}. Since the input characteristic is non-linear, r_i varies but is of the order of 1 to 5 kΩ.

Note. The voltmeter for measuring V_{BE} should be an electronic type with a very high resistance (e.g. 10 MΩ). If a moving coil type is used allowance must be made for the current it takes due to its 'low' resistance. To do this, the voltmeter is disconnected and R_1 is adjusted to make $I_B = 10\,\mu A$ (say). I_C is noted. The voltmeter is connected and R_1 adjusted until I_C has the noted value. The corresponding value of V_{BE} is recorded. The process is repeated for other values of I_B.

(c) Output characteristic ($I_C - V_{CE}$). In the circuit of Fig. 32.1, I_B is set to a low value (e.g. 10 μA) and I_C is measured as V_{CE} is increased in stages by R_2. This is

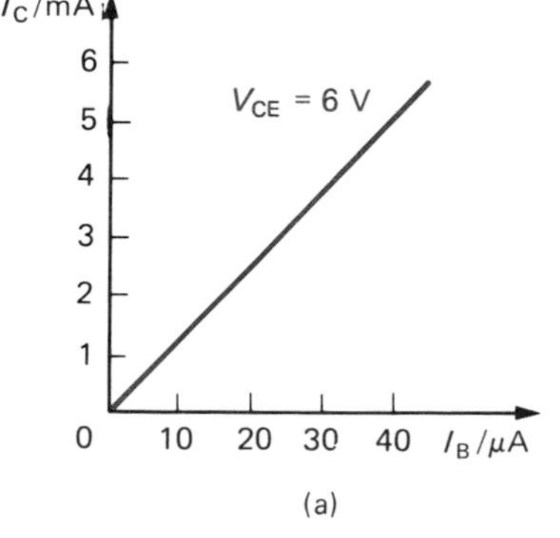

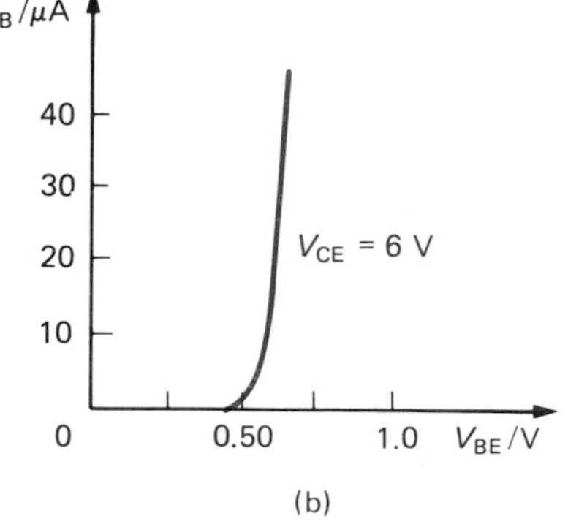

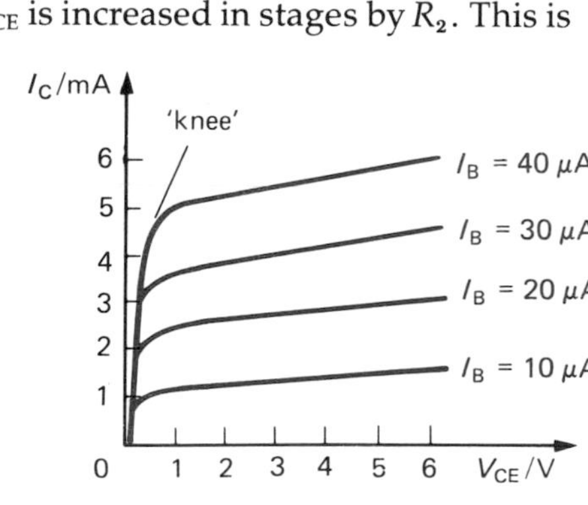

Fig. 32.2

Table 32.1

	BC108	ZTX300	2N3705	BFY51	2N3053
I_C *max*/mA	100	500	500	1000	1000
h_{FE} at I_C mA	110–800 2	50–300 10	50–150 50	40 (min) 150	50–250 150
P_T/mW	360	300	360	800	800
V_{CEO} *max*/V	20	25	30	30	40
V_{EBO} *max*/V	5	5	5	6	5
f_T/MHz	250	150	100	50	100
Outline	TO18	E line	TO92 (A)	TO39	TO39

repeated for different values of I_B, enabling a family of curves like those in Fig. 32.2c to be plotted. They show that I_C hardly changes when V_{CE} does except below the 'knee' of the graphs where V_{CE} is less than a few tenths of a volt.

The *a.c. output resistance* r_o of the transistor is defined by

$$r_o = \frac{\Delta V_{CE}}{\Delta I_C}$$

where ΔI_C is the change in I_C caused by a change ΔV_{CE} in V_{CE} on the part of the characteristic to the right (i.e. beyond) the 'knee'. In this region the slope ($\Delta I_C/\Delta V_{CE}$) is small, making r_o fairly high, typically 50 kΩ.

Transistor data

Transistors are identified by one of several codes. In the American system they start with 2N followed by a number, e.g. 2N3053. In the Continental system the first letter gives the semiconductor material (A = germanium, B = silicon) and the second letter gives the use (C = audio frequency amplifier, F = radio frequency amplifier). For example, the BC108 is a silicon a.f. amplifier. Some manufacturers have their own code.

While one type of transistor may replace another in many circuits, it is often useful to study the published data when making a choice. Table 32.1 lists the main ratings, called *parameters*, for five popular n-p-n silicon transistors. All are general purpose types, suitable for use in amplifying or switching circuits. The BC108 is a low current, high gain device, the others are medium current, medium gain types.

(a) Current, voltage and power ratings. The symbols used have the following meanings.

I_C *max* is the maximum collector current.

V_{CEO} *max* is the maximum collector-emitter p.d. when the base is open-circuited.

V_{EBO} *max* is the maximum emitter-base p.d. when the collector is open-circuited.

P_T *max* is the maximum power rating at 25°C and equals $V_{CE} \times I_C$ approximately.

(b) d.c. current gain h_{FE}. Owing to manufacturing spreads, h_{FE} is not the same for all transistors of the same type. Usually minimum and maximum values are quoted but sometimes only the former. Also, since h_{FE} decreases at high and low collector currents, the I_C at which it is measured is stated.

In general, when selecting a transistor type we have to ensure that its minimum h_{FE} gives the current gain required by the circuit.

(c) Transition frequency f_T. This is the frequency at which h_{FE} = 1 and is important in high-frequency circuits.

(d) Outlines. These are given in Fig. 32.3 for the transistors in Table 32.1.

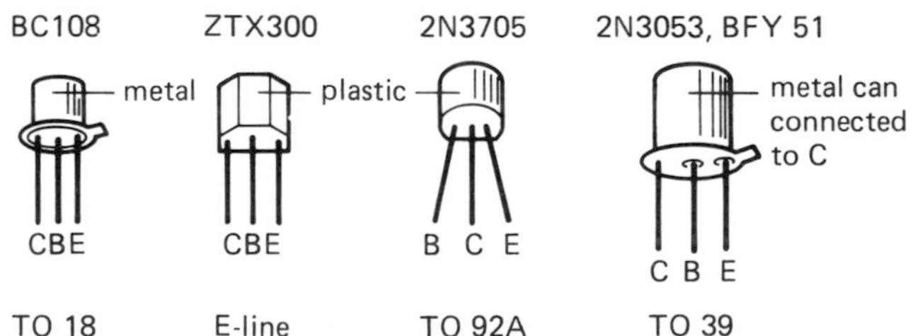

Fig. 32.3

Questions

1. State typical values of (i) h_{fe} (ii) r_i (iii) r_o (iv) V_{EBO} *max*, for a junction transistor.

2. Manufacturers sometimes quote 'V_{CE} *sat max* at I_C/I_B mA' for a transistor, e.g. for BFY51 it is 0.35 V at 150/15. What does this mean? In what type of circuit is it important and why?

33 Field effect transistor

Introduction

In a junction transistor, the small input (base) current controls the larger output (collector) current; it is a *current*-controlled amplifier. In a field effect transistor (FET), the input *voltage* controls the output current; the input current is usually negligible (less than 1 pA = 1 picoampere = 10^{-12} A). This is very useful when the input is from a device such as a crystal pick-up, which cannot supply much current.

A FET consists of a bar or 'channel' of n- or p- type semiconductor having two metal contacts at its ends, called the *drain* D and the *source* S. A third contact, the *gate* G, is connected to a small p- or n-type region between D and S which forms a p-n junction. A simplified diagram of the commoner n-channel FET is shown in Fig. 33.1a.

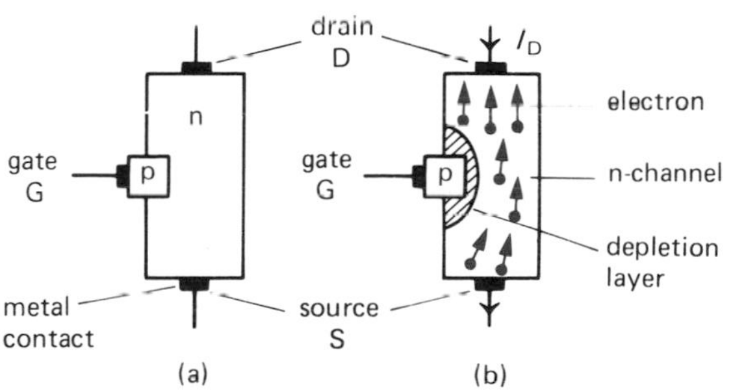

Fig. 33.1

Action of a FET

The channel acts as a conductor which is narrower in the middle due to the depletion layer at the p-n junction behaving as an insulator (p. 57). When the drain is made *positive* with respect to the source, electrons flow from source to drain (as the names suggest). Usually the gate is *negative* relative to the source, thus reverse biasing the p-n junction and widening the depletion layer, Fig. 33.1b. This narrows the channel further and reduces the electron flow, i.e. the drain current I_D.

For a given drain-source p.d. V_{DS}, I_D is controlled by the gate-source p.d. V_{GS} (more correctly by the electric field produced by V_{GS}) and decreases as V_{GS} goes more negative.

The operation of a junction transistor depends on the flow of both majority and minority carriers (p. 56), hence *bipolar*. In a FET or *unipolar* transistor only majority carriers are involved, these being electrons in an n-channel type. FETs are therefore less upset than bipolar types by a temperature rise since this increases minority carriers.

Characteristics of a FET

Normally only the transfer and output characteristics are plotted because the gate (input) current is negligible. Those for an n-channel FET, such as the general purpose 2N3819, may be found using the circuit of Fig. 33.2 which also shows the FET symbol.

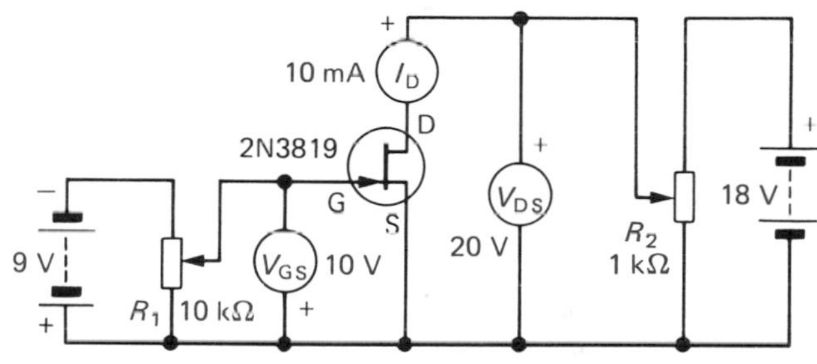

Fig. 33.2

(a) Transfer characteristic ($I_D - V_{GS}$). Since a FET is voltage-operated, a transfer characteristic gives the relation between the input (gate) p.d. V_{GS} and the output (drain) current I_D (for fixed V_{DS}). A typical graph is given in Fig. 33.3; it is nearly linear.

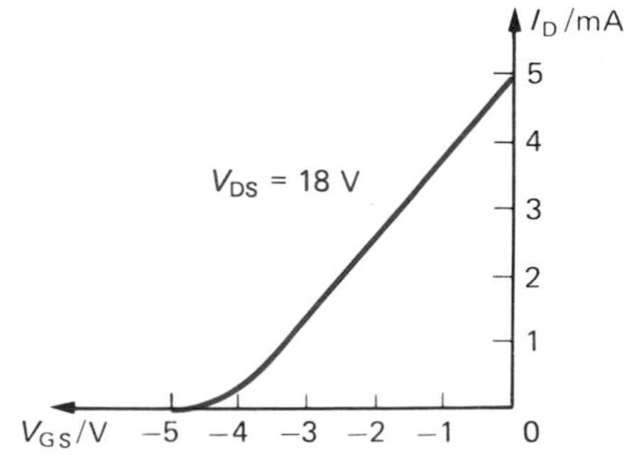

Fig. 33.3

The gain of a FET is measured by its *transconductance*, g_m, defined by

$$g_m = \frac{\Delta I_D}{\Delta V_{GS}}$$

where ΔI_D is the change in I_D caused by a change ΔV_{GS} in V_{GS}. Its value can be found from a transfer characteristic and is in the range 1 to 10 mA/V.

(b) Output characteristics ($I_D - V_{DS}$), Fig. 33.4. They are similar to those of the bipolar transistor, I_D rising sharply initially and then remaining almost constant as V_{DS} increases over a wide range of values. Their slope to the right of the 'knee' however is less, indicating a higher a.c. output resistance r_o (50 kΩ to 1 MΩ).

(c) Input resistance. The a.c. input resistance r_i of a FET is very high, greater than $10^9\,\Omega$. It arises from the fact that the p-n junction is always reverse biased and the only input current flowing in or out of the gate is the tiny one needed to change its potential.

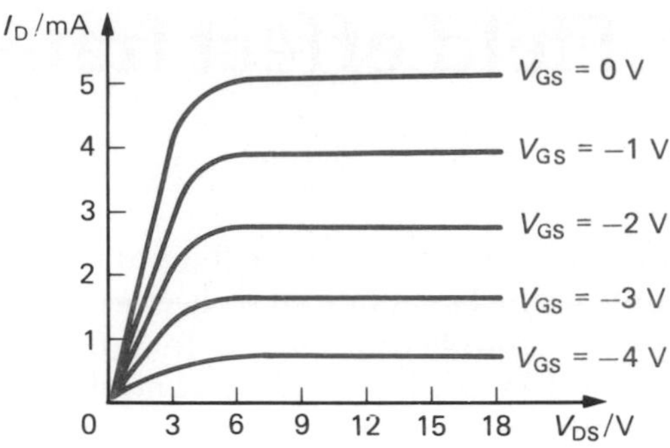

Fig. 33.4

Questions

1. Compare the electrical characteristics of a FET with those of a junction transistor.

2. In Fig. 33.3 the gate-source *cut-off voltage* is −5 V and the *transconductance* is 1 mA/V. Explain these statements.

3. Why is a FET less affected by temperature changes than a bipolar type?

34 Integrated circuits

Microelectronics

One important aspect of microelectronics is concerned with building electronic systems using integrated circuits (ICs). It has been responsible for many of the exciting developments in electronics in recent years.

An IC is a 'densely populated' miniature electronic circuit. It contains transistors and often diodes, resistors and small capacitors, all made from, and connected together on, a 'chip' of silicon no more than 5 mm square and 0.5 mm thick. A small part of one is shown in Fig. 34.1, much magnified.

The IC in Fig. 34.2 is in its protective plastic case (its 'package') which has been partly removed to show the leads radiating from it to the pins that enable it to communicate with the outside world. This type of package is the popular dual-in-line (d.i.l.) arrangement with the pins (from 6 to 40 in number but usually 14 or 16) 0.1 inch apart, in two lines on either side of the case.

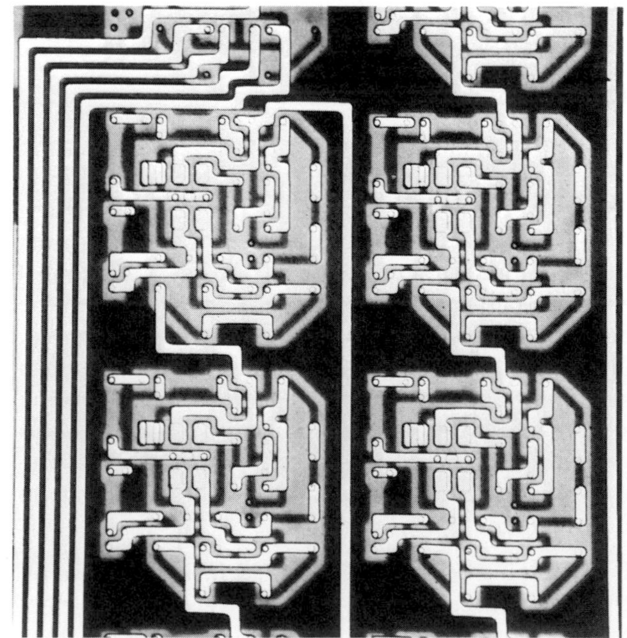

Fig. 34.1

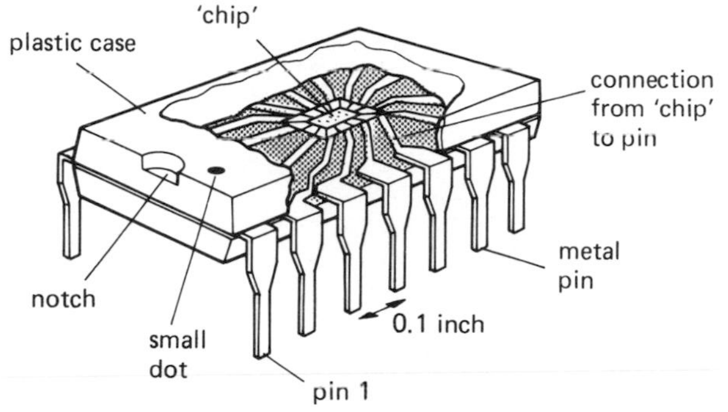

Fig. 34.2

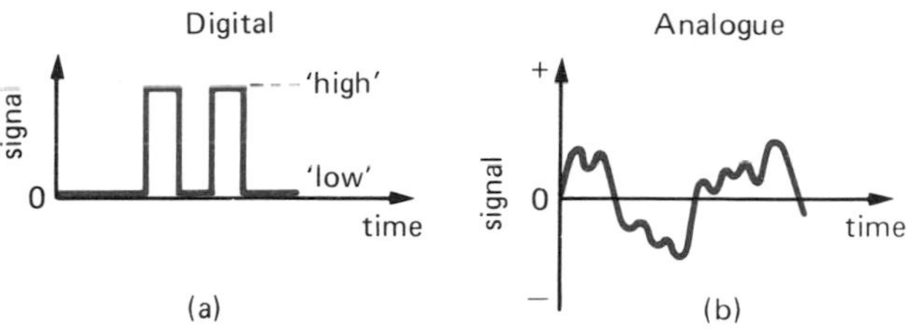

Fig. 34.3

The first ICs were made in the early 1960s and were fairly simple circuits with just a few components per chip; they were small-scale integrated (SSI) circuits. The complexity has increased rapidly through medium-scale integration (MSI) and large-scale integration (LSI) until today very-large-scale integrated (VLSI) circuits may have thousands of components.

Types of IC

There are two broad groups.

(a) Digital ICs. These contain *switching-type* circuits handling electrical signals which have only one of *two values*, Fig. 34.3a, i.e. their inputs and outputs are either 'high' (e.g. near the supply voltage) or 'low' (e.g. near 0 V). They were the earliest ICs because they were easier to make and large markets were available for them (e.g. as the 'brain' in pocket calculators).

(b) Linear ICs. These include *amplifier-type* circuits of many kinds, for both audio and radio frequencies. They handle signals that are often electrical representations, i.e. analogues, of physical quantities such as sound, which change smoothly and continuously over a *range of values*, Fig. 34.3b. One of the most versatile linear ICs and the first of this type (1964), is the operational amplifier (op amp).

Manufacture of ICs

Silicon which is 99.999 999 9 % pure is used and is produced chemically from silicon dioxide, the main constituent of sand. It is first melted and a single, near perfect crystal grown from it in the form of a cylindrical bar (up to 10 cm in diameter and 1 m long). The bar is cut into $\frac{1}{4}$ to $\frac{1}{2}$ mm thick wafers, Fig. 34.4a.

Depending on their size, several hundred identical circuits (the 'chips') are formed simultaneously on each wafer by the series of operations shown in Fig. 34.4b. This creates 'windows' in an insulating layer of silicon oxide first laid down on the surface of the wafer. Doping then occurs, in one method, by exposing the wafer at high temperature to the vapour of either boron or phosphorus, so that their atoms diffuse through the 'windows' into the silicon. The p- and n-type regions so obtained for the various components are then interconnected to give the required circuit by depositing thin 'strips' of aluminium, Fig. 34.4c.

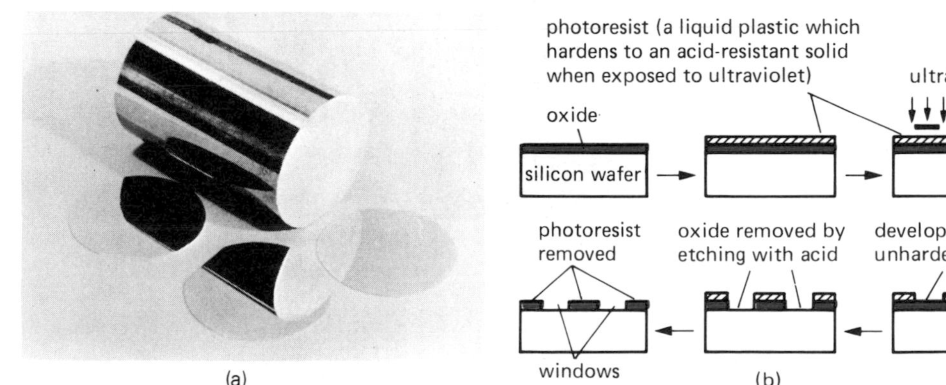

Fig. 34.4

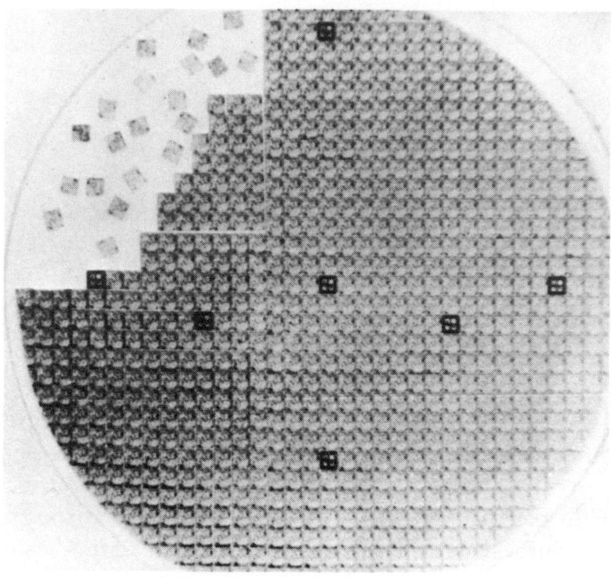

Fig. 34.5

Integrated diodes and transistors have the same construction as their discrete versions. Integrated resistors are thin layers of p- or n-type silicon of different lengths, cross-section areas and degrees of doping. One form of integrated capacitor consists of two conducting areas (of aluminium or doped silicon) separated by a layer of silicon oxide as dielectric.

Every chip is tested, Fig. 34.5a, and faulty ones discarded—up to 70% may fail. The wafers are next cut into separate chips, Fig. 34.5b, and each one is then packaged and connected automatically by gold wires to the pins on the case.

The complete process, which can require up to three months, must be done in a controlled, absolutely clean environment, Fig. 34.6. Although design and development costs are high, volume production makes the whole operation economically viable.

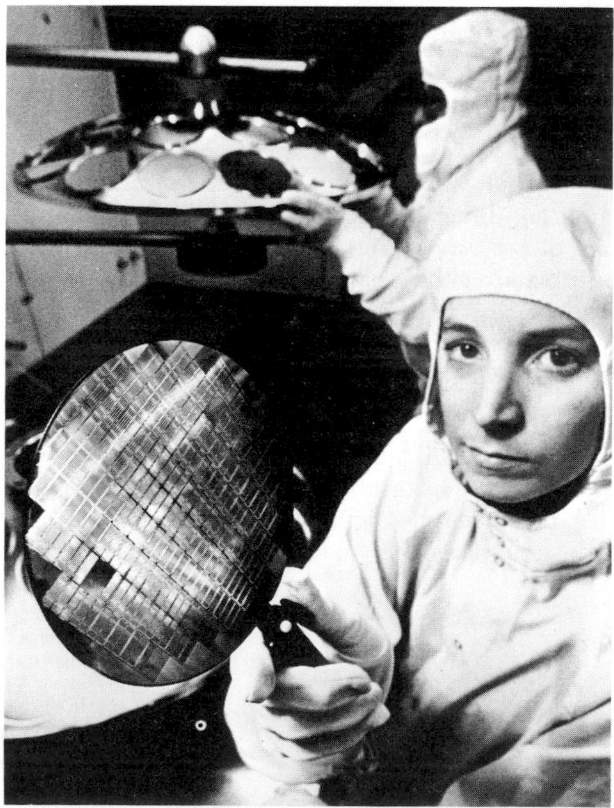

Fig. 34.6

Question

1. State some (a) advantages, (b) limitations, of building circuits from ICs compared with those built from discrete components.

35 Progress questions

1. (a) If the variable resistor R_1 in Fig. 35.1 is adjusted so as to gradually reduce its resistance, describe the effect upon the 12 V lamp.
(b) (i) If A_1 reads 10 mA and A_2 reads 1 A, calculate a value for the current gain of the transistor.
(ii) What reading would you expect to be indicated on ammeter A_3? *(A.E.B. 1981 Control Tech.)*

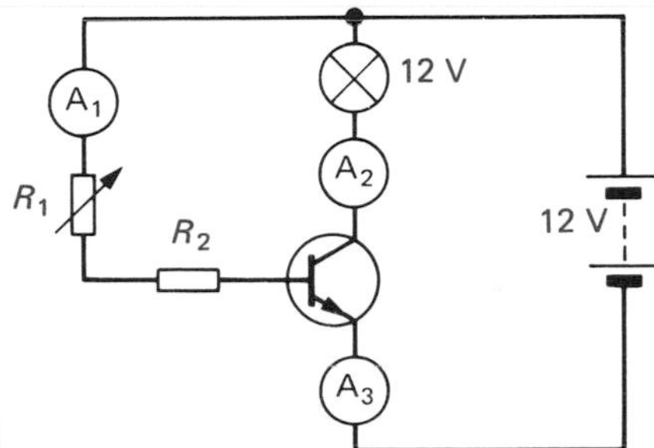

Fig. 35.1

2. Given that the h_{FE} of the transistor in the circuit in Fig. 35.2 is 100, calculate the collector current and the potential difference between the collector and the emitter. (For the purposes of the calculation you should assume that $V_{BE} = 0$.) *(L.)*

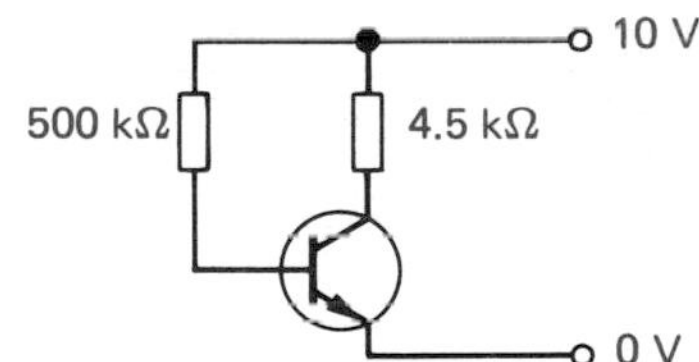

Fig. 35.2

3. An npn transistor has a collector load of 1 kΩ. It is used in a circuit with supply rails of 0 V and 9 V. What is the maximum current that the transistor can conduct?

What is the power dissipation (a) in the resistor, (b) in the transistor, when the collector current is adjusted to half its maximum value by altering the base current? *(O. and C.)*

4. (a) The input voltage shown is applied to the circuit in Fig. 35.3. Copy the graph and sketch the output voltage immediately below the input. Indicate values of the output voltage on your output graph.
(b) Suggest a value for R_B. *(L.)*

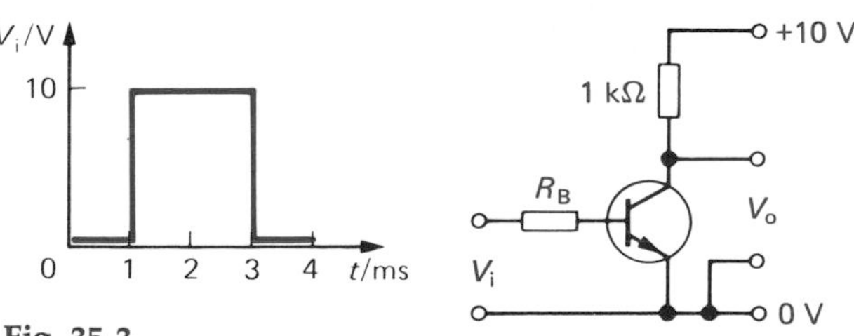

Fig. 35.3

5. Fig. 35.4 shows a relay circuit. The coil of the relay needs a current of 20 mA to close the contacts.
(a) If the current gain h_{FE} of the transistor is 50, calculate the base current needed for the relay to be just switched on.
(b) Neglecting the base-emitter voltage, calculate the value of R for this to happen.
(c) What purpose is served by the diode?
(d) This circuit may be used as the basis of a frost-warning alarm in which a lamp is switched on when the temperature falls below a certain value (which can be altered). Draw a labelled diagram of such a circuit. *(O.L.E.)*

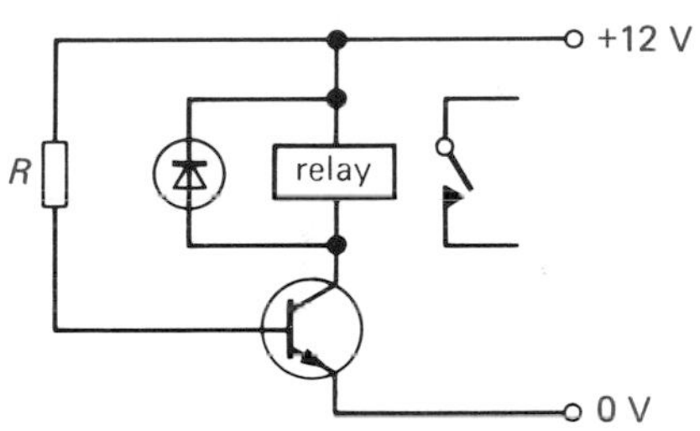

Fig. 35.4

6. For the circuit shown in Fig. 35.5
(a) state what conditions, light or dark, cause the bulb B to be lit,
(b) explain how, and why, the potential at the point X varies as conditions change from light to dark,
(c) explain how the bulb B becomes lit. *(C.)*

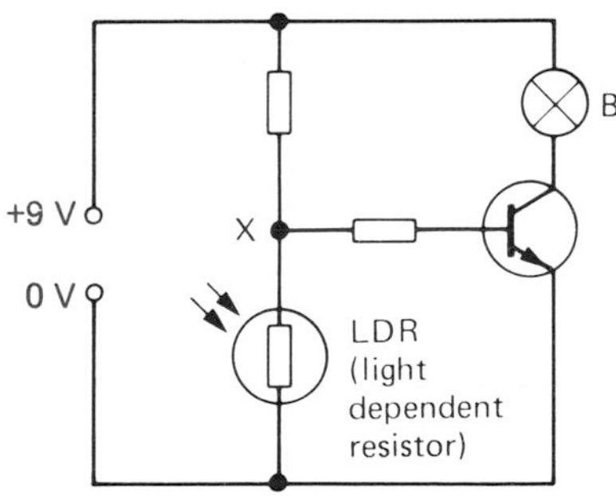

Fig. 35.5

7. What effect does a change in p.d. between the collector and the emitter of a transistor have on the collector current?
What effect does a change in the p.d. between the base and the emitter of a transistor have on the collector current?
Use a sketch graph to illustrate your answer in each case. *(L.)*

8. A manufacturer's description of the performance of a bipolar n-p-n transistor includes the following information:

V_{CE} volts (max)	V_{EB} volts (max)	I_C mA (max)	*Power* mW	h_{fe} at I_C(mA)
45	5	100	300	125–500 at 2

Explain what these entries mean. Describe how you would determine experimentally the value of h_{fe} for a given transistor.

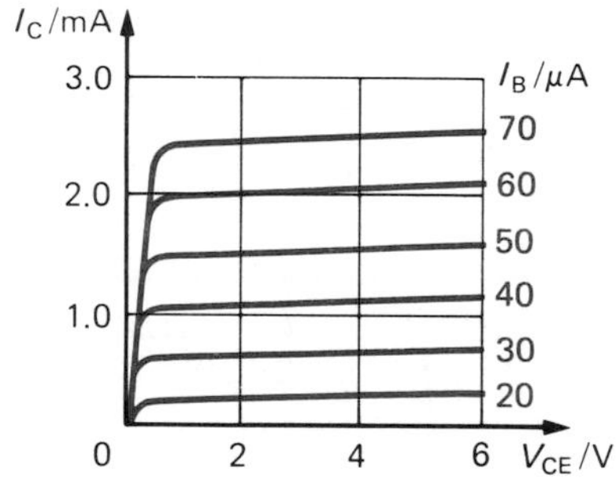

Fig. 35.6

Fig. 35.6 gives an alternative method for quoting some of the data for a transistor. Use these data to estimate a value for h_{fe} for this transistor at (i) $I_C = 0.5$ mA, (ii) $I_C = 2.0$ mA. Explain how your estimate is made. (*O. and C.*)

9. (a) Explain why the circuit in Fig. 35.7a switches the transistor *off* a certain time *after* switch S is opened.
(b) Explain why the circuit in Fig. 35.7b switches the transistor *on* a certain time *after* switch S is opened.

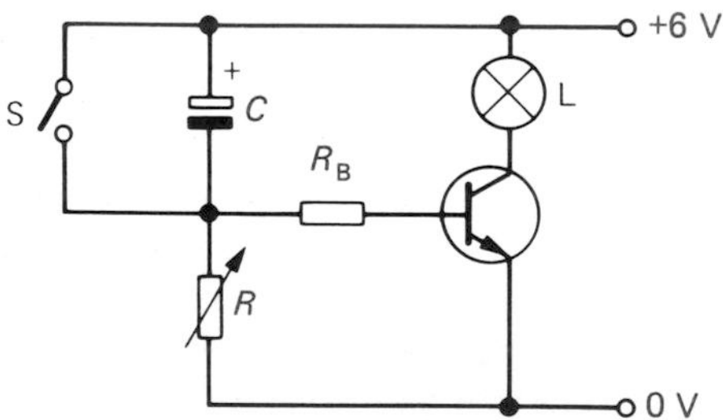

Fig. 35.7a

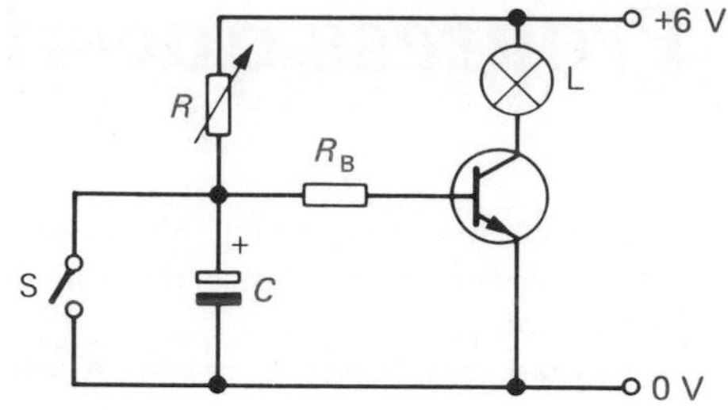

Fig. 35.7b

10. What are the important properties of the field effect transistor (FET)?
Figs. 35.8a and b show two sets of data for a FET. Describe how you would obtain the data in Fig. 35.8b experimentally.
What parameters for the FET may be obtained from the figures? How is this information useful in applications of the FET? (*O. and C.*)

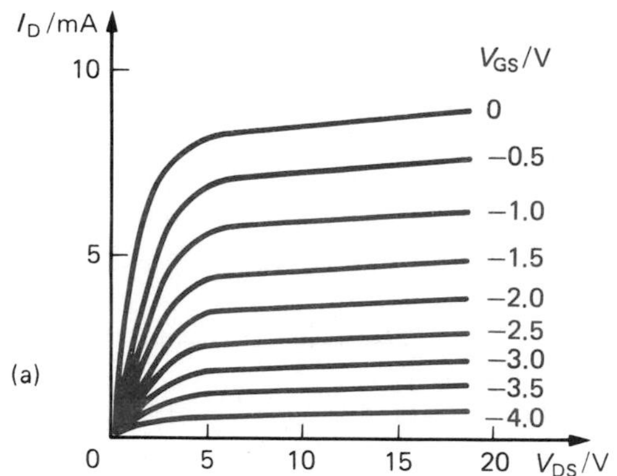

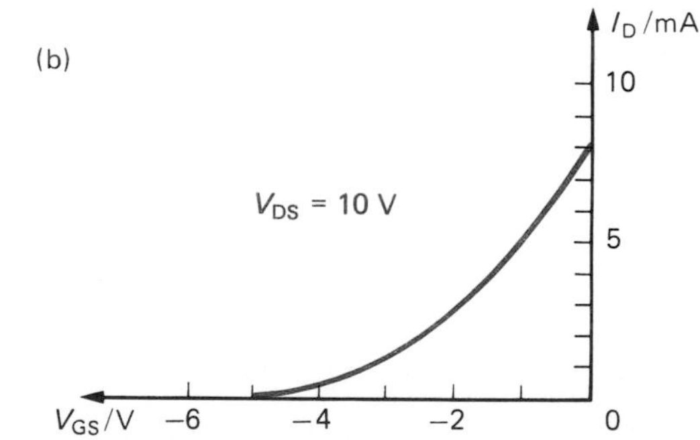

Fig. 35.8

Power supplies and safety

36 Electricity in the home

Electric lighting

(a) Filament lamps, Fig. 36.1a. The filament is a small coiled coil of tungsten wire, Fig. 36.1b, which, because of its high melting point (3400 °C), becomes and stays white hot (at about 2300 °C) when a suitable current passes through it. Most filament lamps contain nitrogen or argon gas. This reduces evaporation of the tungsten which in a vacuum would condense as a black deposit on the cool glass bulb and in air would burn away. Convection currents occur in the gas but their cooling effect on the filament is small since it is compact.

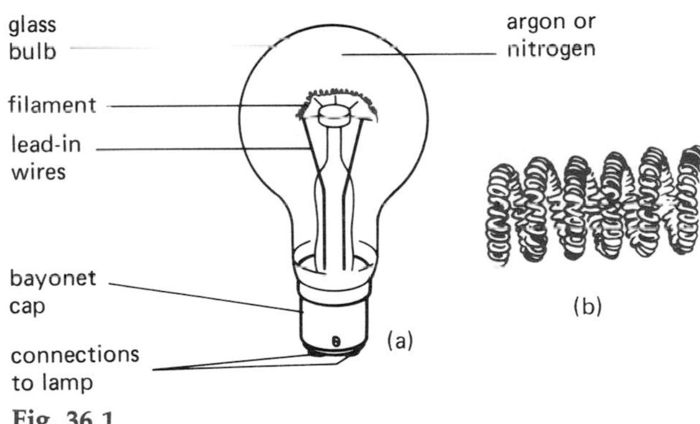

Fig. 36.1

(b) Fluorescent lamps, Fig. 36.2. When operating, electrons are emitted by both electrodes and accelerated to the one that is positive (each one is positive for half of every cycle of the a.c. supply). The high-speed electrons so produced collide with mercury vapour atoms in the tube and either *excite* them, by supplying energy which they then re-emit as ultraviolet radiation, or they *ionize* them, by knocking out electrons to form positive mercury vapour ions (which are accelerated to the negative electrode). The flow of electrons and ions is called a *gas discharge*. The ultraviolet makes the phosphor on the inside of the tube fluoresce, i.e. emit (visible) light.

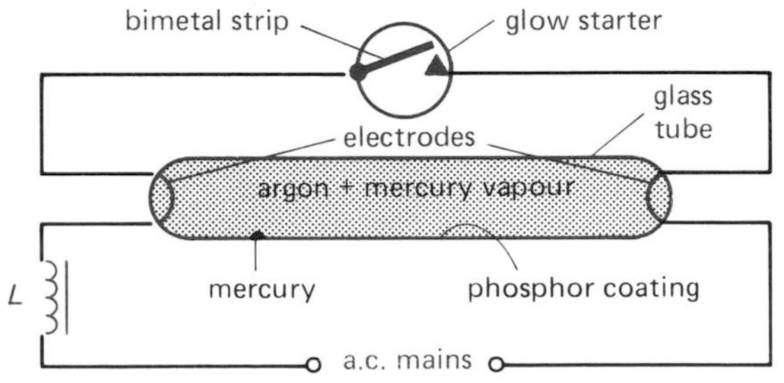

Fig. 36.2

Since the discharge creates more electrons and ions, the process could 'run away' and is controlled by an inductor L (rather than a resistor) in series with the tube.

The discharge is begun by a *glow starter* consisting of a glass tube containing rare gases and a switch, that is normally open, controlled by a bimetal strip. At switch-on, a discharge occurs in the starter and warms the bimetal strip. This bends and closes the contacts. The discharge in the starter stops (why?), the current through the electrodes increases due to the smaller resistance of the starter, the bimetal strip cools and the starter contacts open. This cuts off the current suddenly and an e.m.f. of 800 to 1500 V is induced in L (p. 30), causing a large p.d. between the tube electrodes. These, being hot, emit electrons by thermionic emission (p. 48) and set up the discharge in the tube. (When the tube is working the p.d. across it is not enough to start a discharge in the starter.)

(c) Efficiency. This tells how good a lamp is at changing electrical energy into light. It is expressed in *lumen per watt*, a lumen being a measure of the light energy emitted per second. If the efficiency of a filament lamp is 12.5 lumen per watt, then since 1 watt = 625 lumen, the percentage efficiency is 12.5 × 100/625 = 2%, i.e. most of the electrical energy becomes heat. A fluorescent lamp not only lasts 3–4 times as long as a filament lamp but its efficiency is 3–4 times greater.

Electric heating

(a) Heating elements. In domestic appliances such as electric fires, cookers, kettles and irons the 'elements', Fig. 36.3, are made from nichrome wire. This is an alloy of nickel and chromium which does not oxidize (and become brittle) when the current makes it red hot.

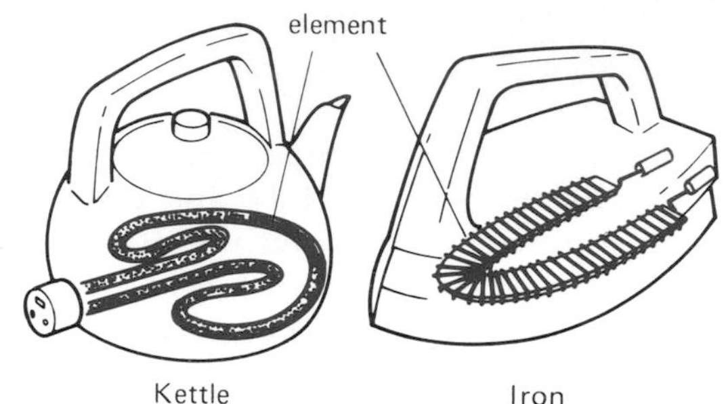

Fig. 36.3

The elements in *radiant* electric fires are at red heat (about 900°C) and the radiation emitted is directed into the room by polished reflectors. In *convector* types the element is below red heat (about 450°C) and is designed to warm air which is drawn through the heater by natural or forced convection. In *storage* heaters the elements heat fire-clay bricks during the night using off-peak electricity. Next day these cool down, giving off the stored heat.

(b) Fuses. A fuse is a short length of wire (often tinned copper) which melts and breaks the circuit

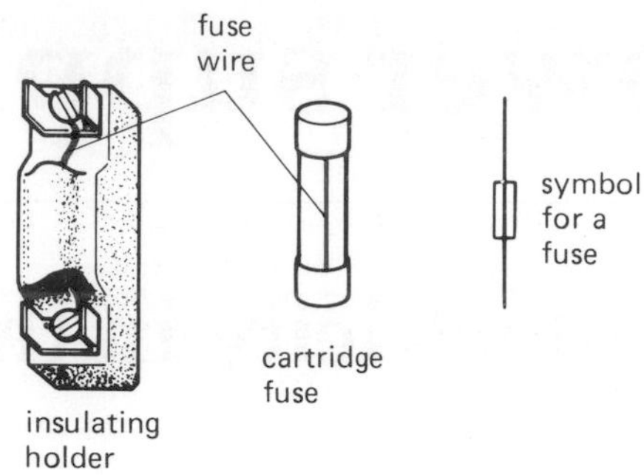

Fig. 36.4

when the current in it exceeds its rated value. A 3 A fuse may need as much as 5 A to blow at once: it will carry up to 3 A indefinitely. Two types of fuse are shown in Fig. 36.4.

House circuits

Electricity usually enters our homes by an underground cable containing two wires, the *live* (L) and the *neutral* (N). The neutral is earthed at the local substation by connection to a metal plate buried in the ground and so is at 0 V. The supply is a.c. and the live wire is alternately positive and negative. A modern house circuit is shown in Fig. 36.5.

(a) Circuits in parallel. Every circuit is connected in parallel with the supply, i.e. across the live and neutral and receives the full mains p.d. (240 V in U.K.).

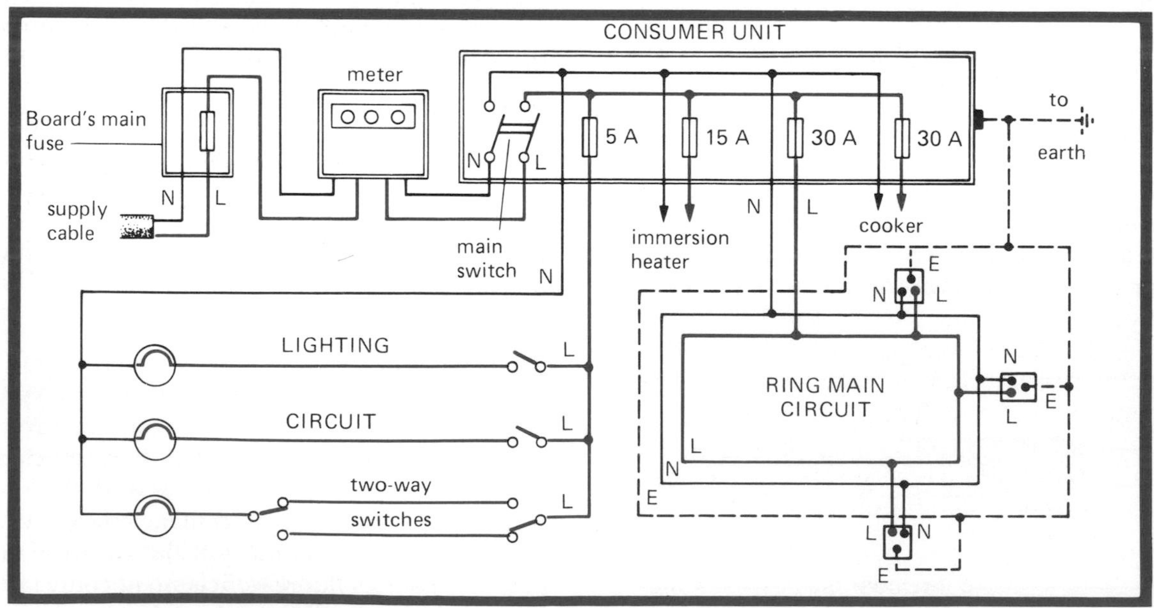

Fig. 36.5

(b) Switches and fuses. These are always in the *live* wire. If they were in the neutral, lamp and power sockets would be 'live' when switches were 'off' or fuses 'blown'. A shock (fatal) could then be obtained by, for example, touching the element of an electric fire when it was switched off.

(c) Ring main circuit. The live and neutral wires run in two complete rings round the house and the power sockets, each rated at 13 A, are tapped off across them. Thinner wires can be used since the current to each socket goes by two paths, i.e. in the whole ring. The ring has a 30 A fuse and if it has ten sockets all can be used so long as the total current does not exceed 30 A.

(d) Fused plug. Only one type of plug is used in a ring main circuit. It is wired as in Fig. 36.6 and has its own cartridge fuse, 3 A (blue) for appliances with powers up to 720 W and 13 A (brown) for those between 720 W and 3 kW.

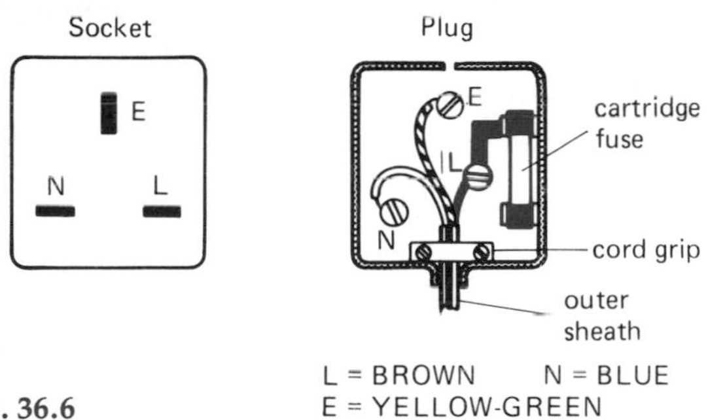

Fig. 36.6

(e) Earth wire. A ring main has a third wire which goes to the top socket on all power points and is earthed by connection either to a metal water pipe in the house or to an earth connection on the supply cable. This third wire is a safety precaution to prevent electric shock should an appliance develop a fault.

The earth pin on a three-pin plug is connected to the metal case of the appliance which is thus joined to earth by a path of almost zero resistance. If then, for example, the element of an electric fire breaks or sags and touches the case, a large current passes to earth and 'blows' the fuse. Otherwise the case becomes 'live' and anyone touching it receives a shock which *might be fatal*, especially if they were 'earthed' by, say, standing on a concrete floor or holding a water tap.

(f) Miniature circuit breaker (m.c.b.). This is a modern alternative to a fuse. It has the advantages of being instantly resettable by a switch or button

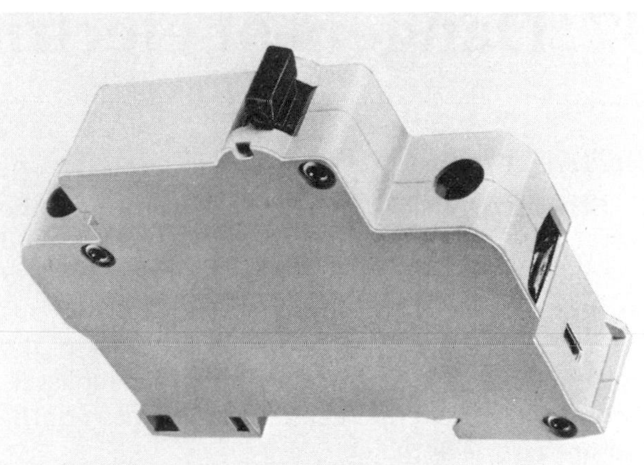

Fig. 36.7

and of breaking an overloaded circuit in less than one-hundredth of a second—a fuse takes much longer. As with fuses there are various ratings, e.g. 6 A, 10 A, 16 A, 32 A. It contains an electromagnet which becomes strong enough when the current through it exceeds the rated value to separate contacts that break the circuit. One is shown in Fig. 36.7.

Paying for electricity

Electricity boards charge for the *electrical energy* they supply. A joule is a very small unit of energy and a larger unit, the *kilowatt-hour* (kWh) is used. It is the electrical energy used by a 1 kW appliance in 1 hour.

A 3 kW electric fire working for 2 hours uses 6 kWh of electrical energy—usually called 6 'units'. At present (1985) a 'unit' costs about 5p.

Typical powers of some appliances are:

Lamps	60, 100 W	Fire	1, 2, 3 kW
Fridge	150 W	Kettle	2–3 kW
TV set	150 W	Immersion	3 kW
Iron	750 W	Cooker	8 kW

Questions

1. Why is an inductor rather than a resistor used to prevent 'run-away' in a fluorescent lamp?

2. What steps should be taken before replacing a blown fuse in a plug?

3. What size fuse (3 A or 13 A) should be used in a plug connected to (a) a 150 W TV set, (b) a 750 W electric iron and (c) a 2 kW electric kettle?

4. What is the cost of using a 3 kW immersion heater for 90 minutes if electricity costs 5p per kWh?

37 Dangers of electricity: safety precautions

Electric shock

(a) Effects. An electric current passing through the body causes electric shock. Effects are produced especially on the heart (upsetting its rhythm and the flow of blood), the muscles and the nervous system. Their severity depends on the *value* of the current, Table 37.1, and the *time* for which it passes, although age, general health and moistness of the skin are also factors.

Table 37.1

Current in mA	Effect
1	Maximum safe current
2–5	Begins to be felt by most people
10	Muscular spasm, i.e. unable to let go and could become fatal
100	Probably fatal if through heart.

The value of the current depends on the voltage applied to the body and its resistance. The latter may exceed 10 kΩ or it might be less than 1.5 kΩ if the skin is moist. If the voltage is 240 V and the body resistance 1.5 kΩ, the current through someone directly in contact with the earth is 240 V/1.5 kΩ = 160 mA (which would be fatal), Fig. 37.1a. In Fig. 37.1b the current is less (e.g. 50 mA) because the resistance of the tiles and the wood floor has to be added to that of the body but a severe shock would still be obtained.

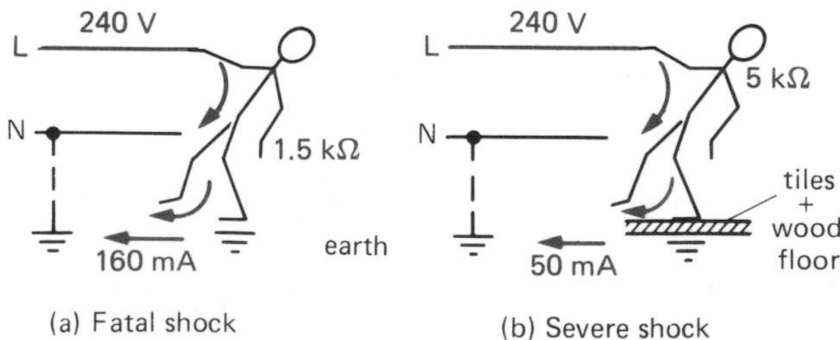

(a) Fatal shock (b) Severe shock

Fig. 37.1

(b) Treatment. If the shocked person is still touching live equipment, *switch off the supply at once.* If this is impossible, pull the victim away using a dry non-conducting article such as a stick, loop of rope or coat and preferably stand on a good insulator (e.g. a rubber mat). *Do not use your bare hands.* Send for qualified medical assistance immediately.

If the heart has stopped, try to restart it by striking the chest smartly three times over the heart.

If breathing has stopped, apply the 'kiss of life' by

(i) laying the patient on his back and clearing any obstruction from his mouth,
(ii) tilting his head well back, lifting his chin up to open the air passage and pinching his nostrils closed, Fig. 37.2a,

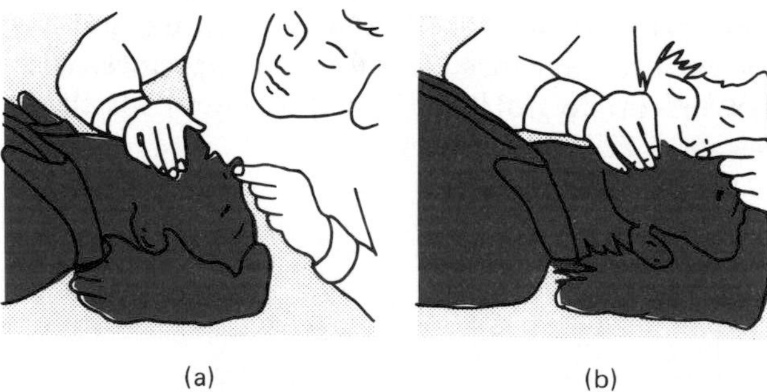

(a) (b)

Fig. 37.2

(iii) taking a deep breath, opening your mouth wide, sealing your lips round the patient's mouth, blowing into his lungs, Fig. 37.2b, watching for his chest to rise, then removing your mouth, and
(iv) taking a deep breath while his chest falls and repeating the process rapidly six times, then ten times a minute until the patient starts breathing or medical help arrives.

If the patient is shocked but still breathing, keep him warm and lying still. Should he be unconscious, lie him on his side and do not give drinks since there is a risk of choking.

Other dangers

(a) Burns. These may be external, for example on the hands, due to an electric shock or to touching a hot soldering iron. The treatment is to cool the affected part with cold water till the pain goes. Extensive burns require medical care.

Internal burns can result, with little feeling of electric shock, if high frequency currents pass through the body.

(b) **Fires.** Each year in the U.K. about 30 000 house fires and 8000 industrial/commercial fires have an electrical origin. The main causes include defective insulation on wiring producing excessive currents when conductors touch (i.e. short circuit), overloaded conductors, poor connections, sparking at switch contacts and overheating of electric motors, cooking, lighting and heating appliances.

Electrical fires are best fought with carbon dioxide or powder extinguishers. Water or foam types may cause short circuits.

(c) **Explosions.** Care is needed when electrical equipment is used in hazardous atmospheres. These are created not only by flammable gases and vapours but by carbon-based dusts (produced for example, when grain, sugar or coal are ground) since they form explosive concentrations if mixed with air in certain proportions. An explosion can be started by an electric spark or by an electrical appliance with a surface temperature high enough to ignite any vapour or dust present.

Safety precautions

The following help to prevent accidents.

1. *Mains plugs* must be correctly connected, the outer sheath of the cable secured by the cord grip and the correct fuse fitted (p. 79).

2. *Cables to equipment* should have a rubber grommet where they enter the metal case, to prevent the insulation wearing.

3. *Fuses, circuit breakers and single-pole switches* must be in the live (brown) lead of mains-operated equipment (p. 79).

4. *Earthing regulations* must be satisfied. In most accidents someone receives a 'shock to earth'. To prevent this, there is a legal requirement that one point of every supply system shall be earthed. As was stated before (p. 79), this is usually done at the Electricity Board substation by connecting the neutral wire to a metal plate buried deep in the ground.

Also, if the non-statutory IEE (Institution of Electrical Engineers) regulations are adopted, each consumer has an earth connection (e.g. to a metal water pipe) and the 'fault current' then passes from the consumer's faulty circuit or appliance back to the substation via the earth, Fig. 37.3. The resistance of

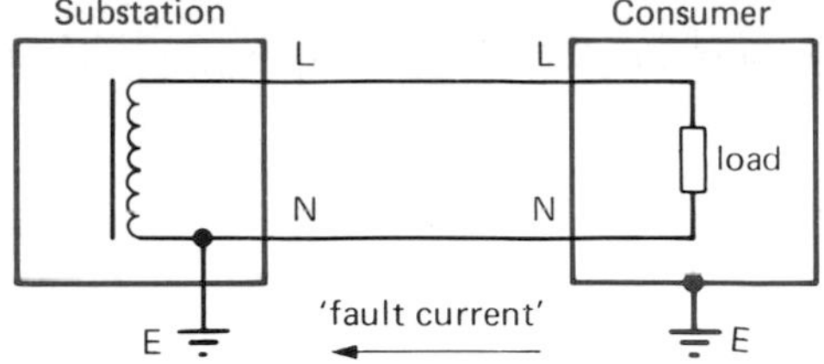

Fig. 37.3

this 'fault circuit' must be very low to enable the 'fault current' to be large enough to blow the circuit fuse, i.e. good earth connections are essential. In practice the 'fault current' should be at least three times the fuse current. Therefore for a 30 A fuse on 240 V mains, the total resistance of the 'fault circuit' must not be more than $240\,V/90\,A = 2.7\,\Omega$.

5. *Capacitors* may hold a lethal charge for a long time even though the apparatus is disconnected from the supply. They can be discharged by holding a metal bar with a good insulating handle across the terminals.

6. *Laboratory discipline.* Never work alone in a laboratory or without proper supervision and know how to summon help in an emergency.

More protective devices

(a) **Earth-leakage circuit breaker (e.l.c.b.).** This is an adaptation of the miniature circuit breaker (p. 79) and is used when the resistance of the earth path (between the consumer and the substation) is not small enough for the fault-current to blow the fuse. In the type shown in Fig. 37.4, current passes to

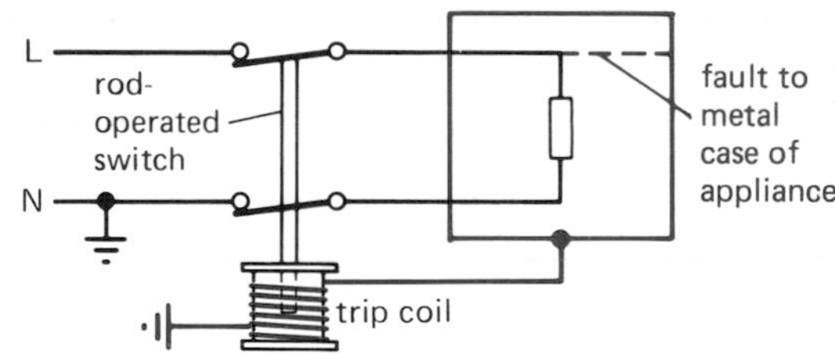

Fig. 37.4

earth through a relay-type 'trip coil' when, for example, the metal case of the appliance becomes 'live' due to a fault. As a result, the rod is pulled into the coil by magnetic attraction so opening the switch which can be set to break the circuit before the case voltage exceeds 30 V.

Earth-leakage circuit breakers are available for plugging into sockets supplying power to portable appliances such as electric lawnmowers and hedge trimmers. In such cases the risk of electrocution is greater because the user is generally making a good earth connection to ground via his feet and so additional protection is advisable.

(b) Thermal trips. To stop an appliance overheating a thermal trip can be placed in contact with it so that it opens the circuit at a certain temperature. The principle of such a device is shown in Fig. 37.5 using

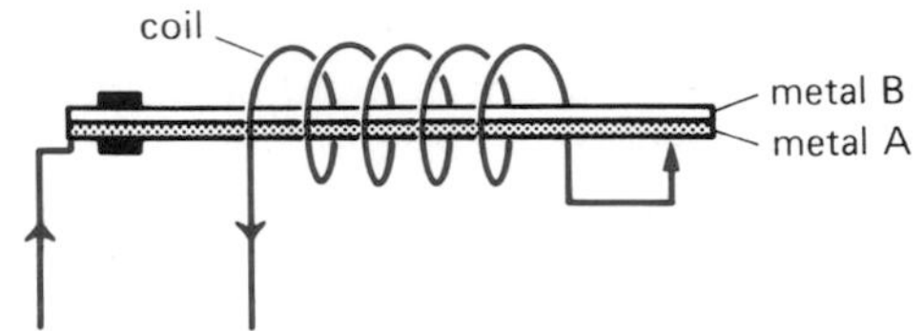

Fig. 37.5

a bimetal strip in which metal A expands more than metal B when heated. The strip bends upwards and breaks the contact.

(c) Isolating transformer. It separates a piece of equipment from the mains supply. Its floating secondary winding reduces the risk of a shock since one connection from the live wire to earth does not give a complete circuit (but two connections will). An earthed metal screen between the windings (turns ratio usually 1:1) offers further protection.

(d) Double insulation. Appliances such as vacuum cleaners, hair dryers and food mixers are usually double insulated, shown by the sign ⧈ on their specification plate.

Connection to the supply is by a 2-core insulated cable, with no earth wire, and the appliance is enclosed in an insulating plastic case. Any metal attachments which the user might touch are fitted into this case so that they do not make a direct connection with the internal electrical parts, e.g. a motor. There is then no risk of a shock should a fault develop.

Questions

1. The circuits of Fig. 37.6a,b, show 'short circuits' between the live (L) and neutral (N) wires. In both the fuse has blown but whereas (a) is now safe, (b) is still dangerous even though the lamp is out and suggests the circuit is safe. Explain.

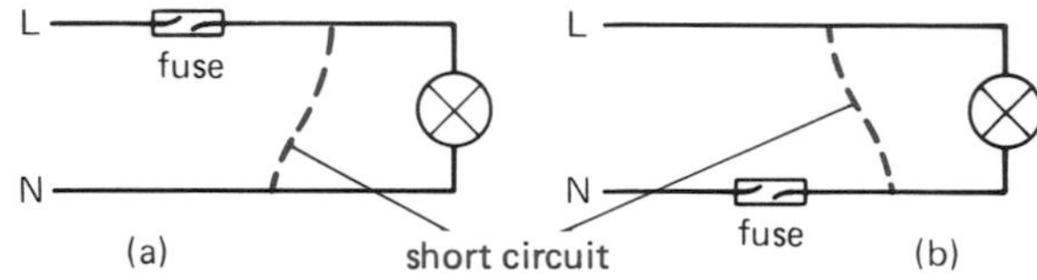

Fig. 37.6

2. The circuits in Fig. 37.7a,b, show 'earth faults' on an appliance. Why does the fuse blow in (a) but not in (b)? Why is the circuit and the appliance in (b) liable to overheat and catch fire?

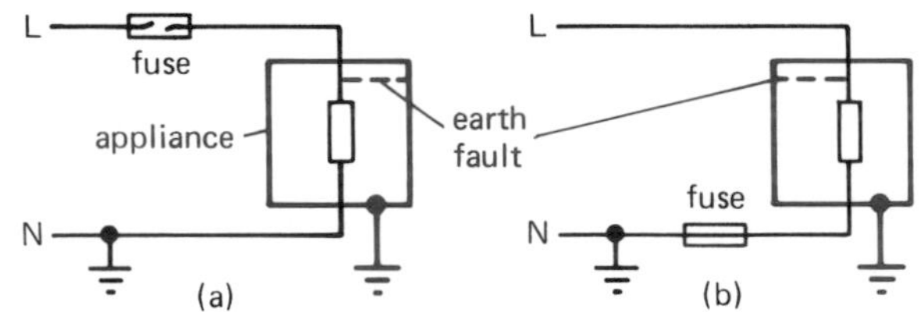

Fig. 37.7

3. The 1:1 isolating transformer in Fig. 37.8 has its secondary centre-tapped to earth. How does this reduce the *severity* of a shock to earth?

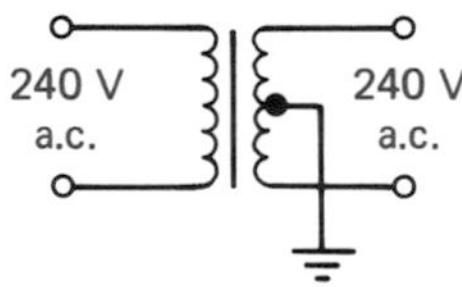

Fig. 37.8

38 Generating electricity

Simple generators

(a) a.c. generator (alternator). The basic construction is shown by the simple arrangement in Fig. 38.1a. If the coil is rotated by an external force, an alternating e.m.f. is induced in it. One complete rotation produces one cycle of a.c. The e.m.f. has its maximum value when the coil is passing the horizontal position, Fig. 38.1b, since the magnetic field through it is then *changing* most rapidly (p. 29).

(b) d.c. generator (dynamo). An a.c. generator becomes a d.c. one if the slip rings are replaced by a commutator like that in a d.c. motor, Fig. 38.2a. The brushes are arranged so that as the coil goes through the vertical, they change over from one half of the commutator to the other. In this position the e.m.f. induced in the coil reverses, making one brush always positive and the other negative. Fig. 38.2b shows the e.m.f. at the brushes; it is a varying d.c.

Practical generators

In actual generators several coils are wound on a soft iron cylinder (the *armature*) and electromagnets (with *field* windings) usually replace permanent magnets.

(a) Cars. They were previously fitted with dynamos but these have now been replaced by alternators because of their greater output current at low engine speeds. The a.c. is rectified (changed to d.c.), part is used to energize the field coils (there is always enough residual magnetism in their iron cores to start the process) and the rest fed out via the slip rings to charge the battery.

(b) Power stations. In power station alternators, the electromagnets rotate (the *rotor*) while the field coils and their iron core are at rest (the *stator*). The large p.ds and currents (25 kV at several hundred amperes) induced in the stator are then led away

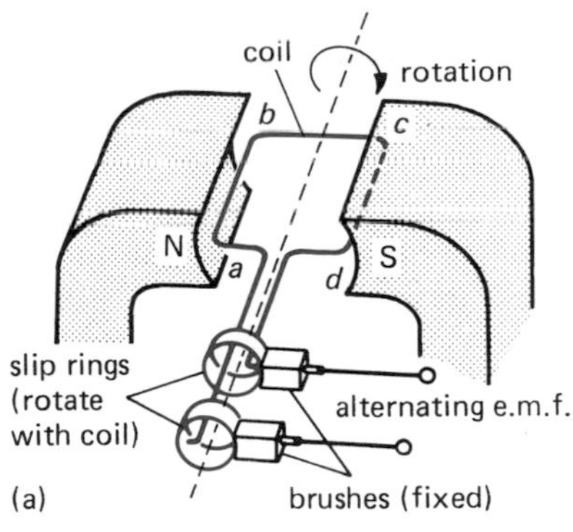

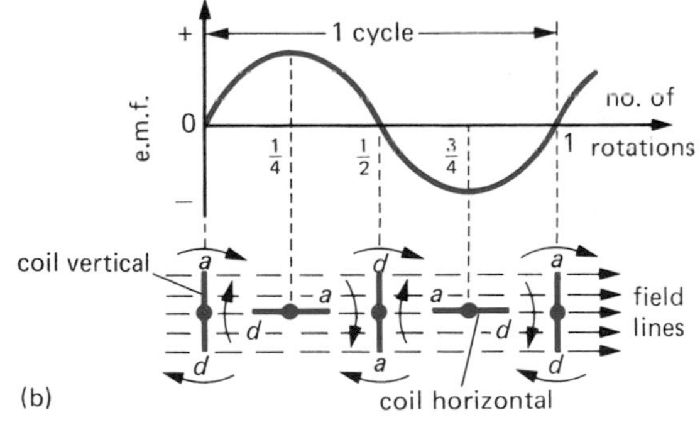

Fig. 38.1

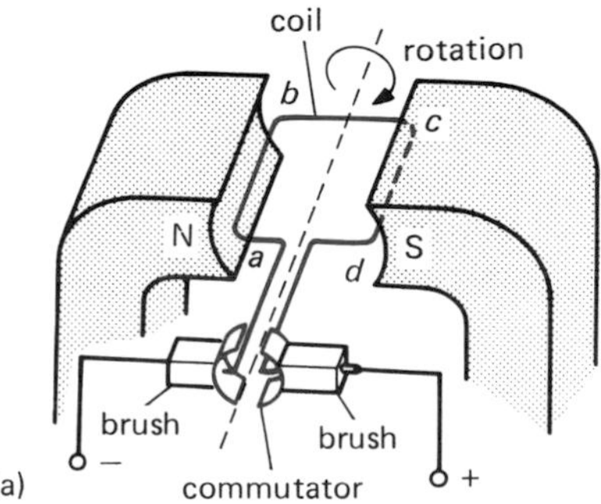

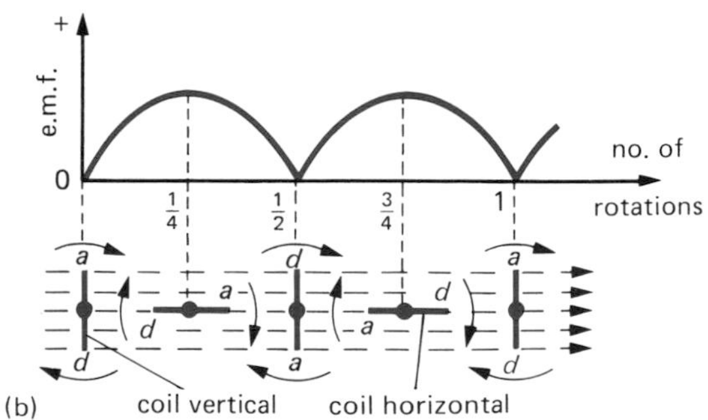

Fig. 38.2

through stationary cables, otherwise they would quickly destroy the slip rings by sparking. Instead, the relatively small d.c. required by the rotor is fed via the slip rings from a small dynamo (the *exciter*), which is driven by the same turbine (usually steam) as the rotor, Fig. 38.3.

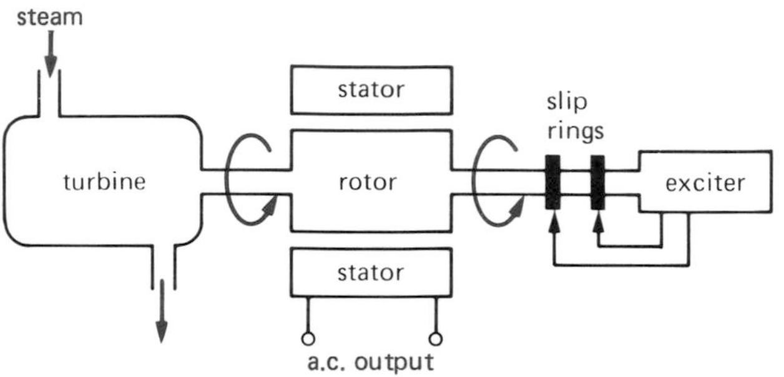

Fig. 38.3

(c) Bicycles. In this case too a.c. is induced in a stationary coil but the rotor is a permanent magnet, Fig. 38.4.

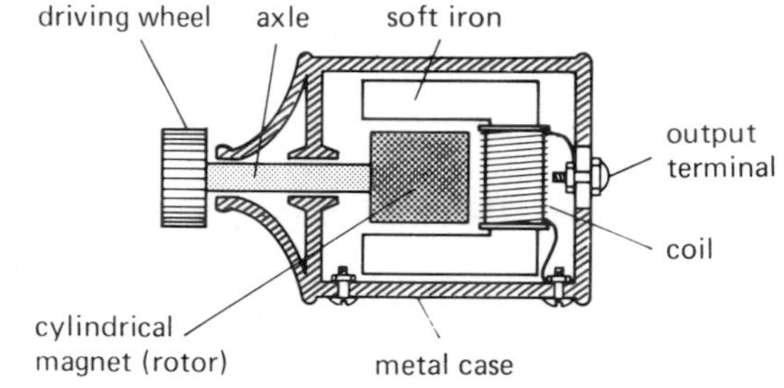

Fig. 38.4

Electricity supply

The Grid System is a network of cables, most supported on pylons, which connect about 200 power stations in the U.K. (via seven Area Control Centres) to consumers. The different voltages involved are shown in Fig. 38.5.

The power loss (I^2R) in a cable is less if electrical energy is transmitted at high voltage and low current since the amount of unwanted heat produced (due to the resistance R of the cable) is proportional to the square of the current I. The use of a.c. and not d.c. is due to the high efficiency of transformers.

Energy sources

(a) Fuels. Most of Britain's power stations use some form of fuel (80% coal, 8% oil, 10% nuclear) to turn water into steam for driving the turbine that rotates the alternator. Present-day stations are only about 30% efficient. Much of the energy supplied by the fuel is lost as heat to the atmosphere when steam from the turbine is condensed in cooling towers. In the future, some of the heat may be used to warm nearby buildings and increase the overall efficiency.

Coal is bulky, generates waste but is readily available in the U.K. Estimates suggest it will run out in about 300 years. *Oil*, the other main fossil fuel, is more convenient but is no longer cheap and reserves may only last for a few decades. *Nuclear* fuels (uranium and plutonium) provide cheaper

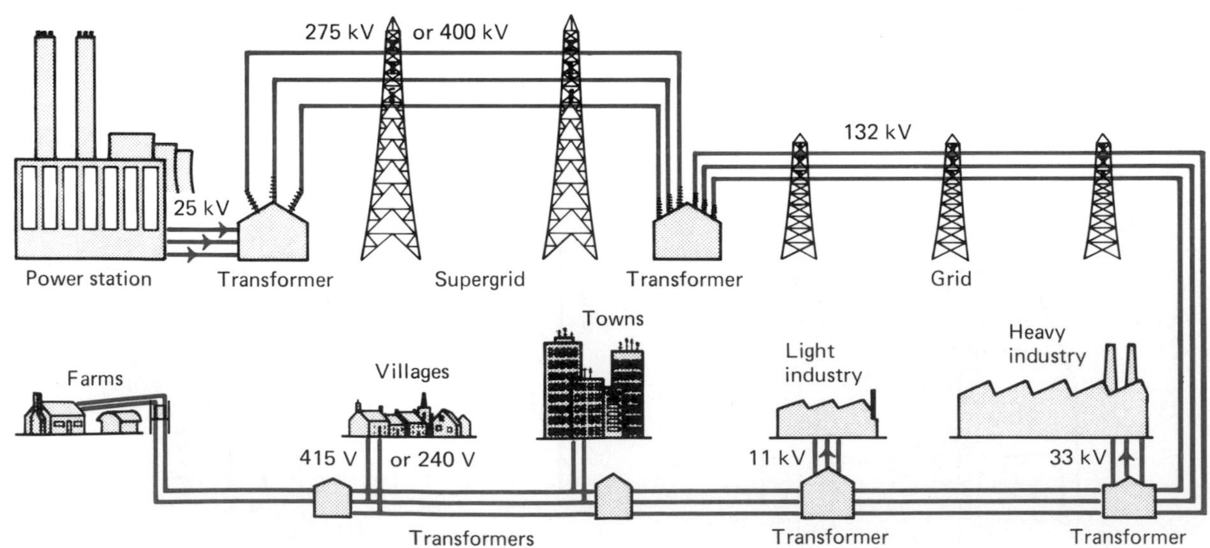

Fig. 38.5

electricity but involve high capital costs and produce radioactive waste.

(b) Other sources. The potential energy of water stored in the mountains of Scotland and Wales is harnessed for low running-cost *hydroelectric* power stations. *Wind, tidal* and *solar* energy are also being exploited and *geothermal* energy is under investigation in Cornwall where the heat stored in granite rocks deep in the earth has been used to warm cold water.

Questions

1. Draw a block diagram to show the forms of energy involved in the generation of electricity in a fossil-fuel power station.

2. Why does the Grid System use (i) a.c., (ii) high p.ds?

39 Sources of e.m.f.

Primary cells

In an electric cell an *electrolyte* reacts chemically with two *electrodes*, making one positive and the other negative, to produce electrical energy. Primary or 'dry' cells are those which are discarded when the chemicals are used up.

(a) Zinc-carbon cell, Fig. 39.1. It has a zinc negative electrode, a manganese dioxide positive electrode and the electrolyte is a solution of ammonium chloride. The carbon rod is in contact with the positive electrode (but is not involved in the chemical reaction) and is called the *current collector*. The e.m.f. is 1.5 V and the internal resistance about 0.5 Ω. This is the most popular cell for low current or occasional use, e.g. in torches. The high power (HP) version uses specially prepared manganese dioxide rather than the natural ore.

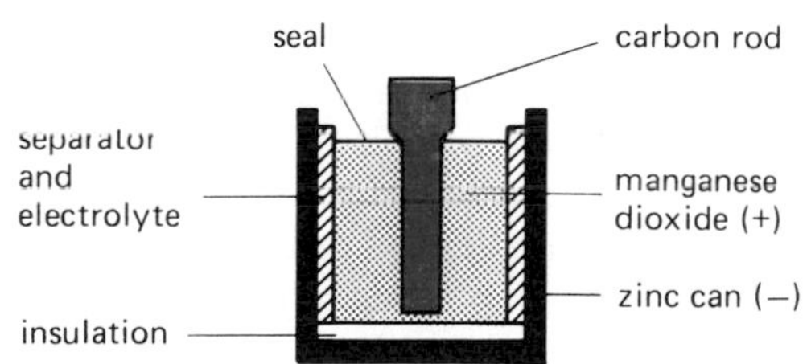

Fig. 39.1

(b) Alkaline-manganese cell. The same electrode materials are used as in the zinc-carbon cell. The electrolyte is a strong solution of the alkali potassium hydroxide which has more ions (to carry current) than an ammonium chloride solution of the same strength. This allows the cell to give larger continuous currents than the zinc-carbon type, for the HP version of which it is a longer lasting (but dearer) replacement in for example electric shavers and tape recorders. It is also leak-proof since the steel case has no part in the reaction; the zinc is in powder form. The e.m.f. is 1.5 V.

(c) Silver oxide cell. The positive electrode is silver oxide, the negative is zinc and the electrolyte is potassium hydroxide. It has a long life for a small cell and its e.m.f. is almost constant at 1.5 V. It is made as a 'button' cell and used in watches, calculators and cameras where small, occasional currents are required. Mercury and zinc-air cells have similar properties and uses.

Secondary cells

Secondary cells or accumulators, can be recharged by passing a current through them in the opposite direction to that in which they supply one.

(a) Lead-acid accumulator. The positive electrode is lead dioxide (brown) and the negative one lead (grey), the materials being held in lead-alloy plates. The electrolyte is dilute sulphuric acid, Fig. 39.2a. During discharge both electrodes change to lead sulphate (white) and the acid becomes more dilute.

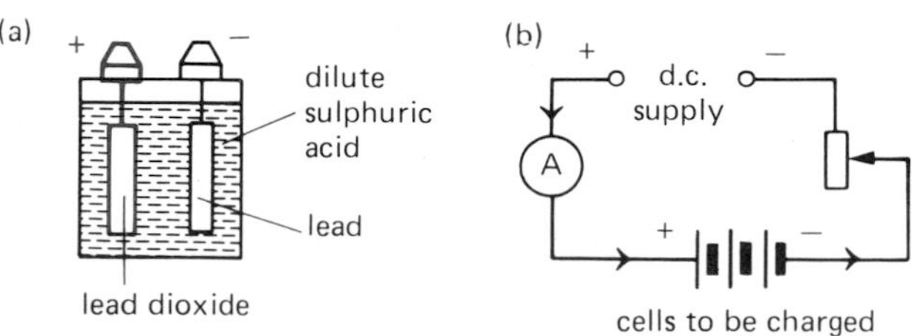

Fig. 39.2

The e.m.f. is steady at 2 V and the low internal resistance of 0.01 Ω allows quite large, continuous currents to be maintained.

The state of charge of a cell can be found by measuring the relative density of the acid with a hydrometer. When fully charged it is 1.25 and falls to 1.18 at full discharge. A 'flat' cell should not be left, otherwise the lead sulphate hardens and cannot be changed back to lead dioxide and lead. Recharging once a month is advisable using a circuit like that in Fig. 39.2b. The supply must be d.c. of greater e.m.f. than that of the cells to be recharged. Note that the + of the supply goes to the + of the cells and that the current is adjusted by the rheostat to the value recommended on the cells.

Overcharging splits water in the acid solution into hydrogen and oxygen gas which bubbles up round the electrodes. Any water lost must be replaced by topping up with distilled water to keep the electrodes covered. Maintenance-free sealed types are now available with gas escape safety vents.

A 12 V car battery consists of six lead-acid cells in series.

(b) Nickel-cadmium cell (Nicad). The electrodes are of nickel (+) and cadmium (−) and the electrolyte is potassium hydroxide. It has an e.m.f. of 1.2 V and is made in the same sizes as primary cells e.g. HP2, PP3; button types are also available, Fig. 39.3.

Fig. 39.3

High currents can be supplied. Recharging must be by a *constant current* power supply (p. 91) because of the very low internal resistance.

(c) Capacity. This states how long a cell can supply current and is measured in *ampere-hours* (A h) for a 10-hour discharge time. A 30 A h cells would sustain 3 A for 10 hours, but while 1 A would be supplied for more than 30 hours, 6 A would not flow for 5 hours.

Fuel cells

In one type, hydrogen (the fuel) and oxygen combine to form water and generate an e.m.f. between two porous, metal electrodes in an electrolyte, Fig. 39.4. The electrodes also act as catalysts to speed up the reaction.

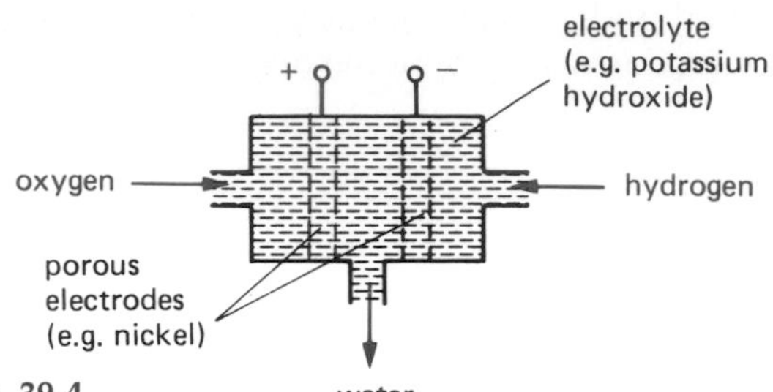

Fig. 39.4

Fuel cells are used in spacecraft where small, light power supplies are essential; the water produced provides a drinking supply.

Photovoltaic cells

Fig. 39.5

These changes light into electrical energy directly. Types designed as solar cells, Fig. 39.5, are made of silicon. They produce about 0.5 V per cell in full sunlight with a maximum current of about 35 mA per cm^2 of cell and an efficiency of 10%. Panels of solar cells are used in artificial satellites to power electronic equipment.

Thermocouples

If two different metals, e.g. copper and iron, are joined as in Fig. 39.6a, an e.m.f. is produced when

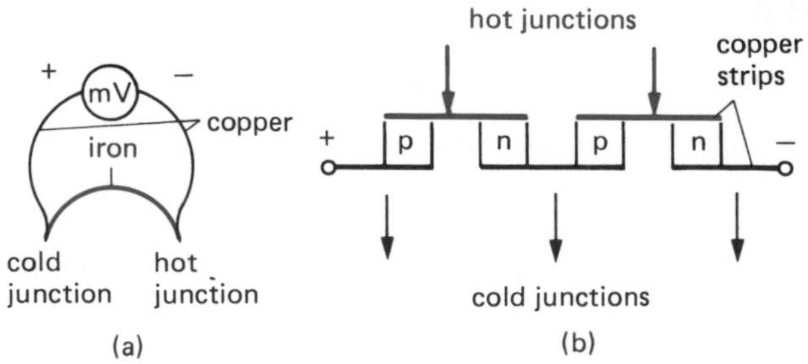

Fig. 39.6

there is a temperature difference between the junctions. Its value is of the order of millivolts and depends on the metals and the temperature difference. The arrangement, called a *thermocouple*, is used to measure high temperatures.

Semiconductor thermocouples are now available, Fig. 39.6b, which are suitable as small power supply units in devices such as weather buoys, the necessary heating of the hot junctions being done by a radioactive source.

Piezoelectric effect

The generation of an e.m.f. by certain crystals when bent or struck was considered previously (p. 43).

Questions

1. What kind of cell would you use for (a) a torch, (b) a cassette recorder, (c) an LCD watch, (d) an electric shaver, (e) a calculator with LED display and (f) a hearing aid?

2. What current should be supplied by a 20 A h accumulator discharging at the 10-hour rate?

40 Rectifier circuits

Power supplies

The voltages required for circuits containing semiconductor devices seldom exceed 30 V and may be as low as 1.5 V. Current demands vary from a few microamperes to many amperes in large systems. Usually the supply must be d.c.

Batteries are suitable for portable equipment but in general power supply units operated from the a.c. mains are employed. In these, a.c. has to be converted to d.c., the process being called *rectification*. Apart from not needing frequent replacement, as batteries may, power supply units are more economical and reliable and can provide more power. They can also supply very steady p.ds when this is required.

In most power supply units a transformer steps down the a.c. mains from 240 V to a much lower voltage. This is then converted to d.c. using one or more junction diodes in a rectifier circuit.

Half-wave rectifier

In the circuit of Fig. 40.1, the load is represented by a resistor R, but in practice it will be a piece of electronic equipment. Positive half-cycles of the alternating input p.d. V_i from the transformer secondary, forward bias diode D which conducts, creating a pulse of current. This produces a p.d. across R having almost the same peak value as V_i, if the small p.d. (about 1 V) across D is ignored.

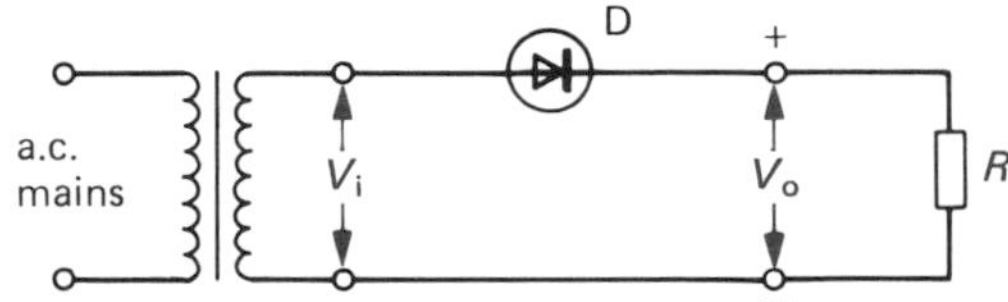

Fig. 40.1

The negative half-cycles of V_i reverse bias D, there is little or no current in the circuit and V_o is zero. Fig. 40.2 shows the V_i and V_o waveforms. V_o varies but is unidirectional, i.e. it is d.c.

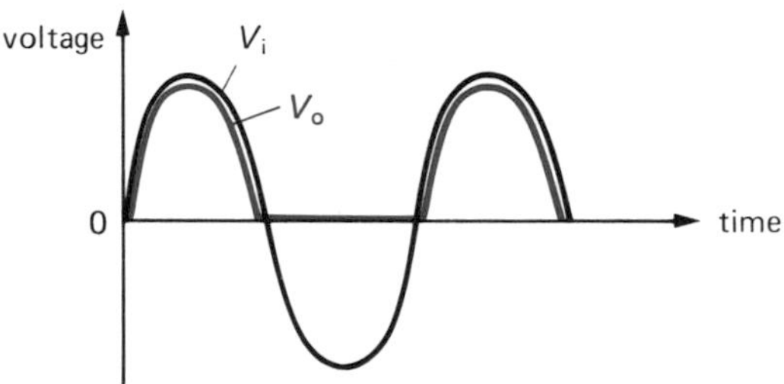

Fig. 40.2

Centre-tap full-wave rectifier

In full-wave rectification, both halves of every cycle of input p.d. produce current pulses.

In the circuit of Fig. 40.3, two diodes D_1 and D_2 and a transformer with a centre-tapped secondary are used. Suppose that at a certain time during the first half-cycle the input p.d. V_i is 12 V. If we take the centre-tap O as the reference point at 0 V, then

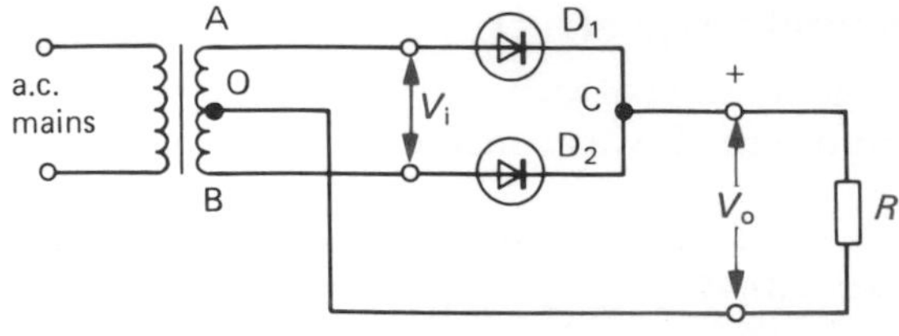

Fig. 40.3

when A is at +6 V, B will be at −6 V. D_1 is forward biased and conducts, giving a current pulse via the path AD_1CROA. V_o is again almost equal to V_i. During this half-cycle D_2 is reverse biased by the p.d. across OB (since B is negative with respect to O).

On the other half of the first cycle, B becomes positive relative to O and A negative. D_2 conducts to give current via the path BD_2CROB and D_1 is now reverse biased. In effect, the circuit consists of two half-wave rectifiers working into the same load R on alternate half-cycles of V_i. The current through R is in the same direction during both half-cycles and V_o is a fluctuating direct voltage with a waveform as in Fig. 40.4. It is more continuous than V_o in the half-wave circuit and has a higher average d.c. value.

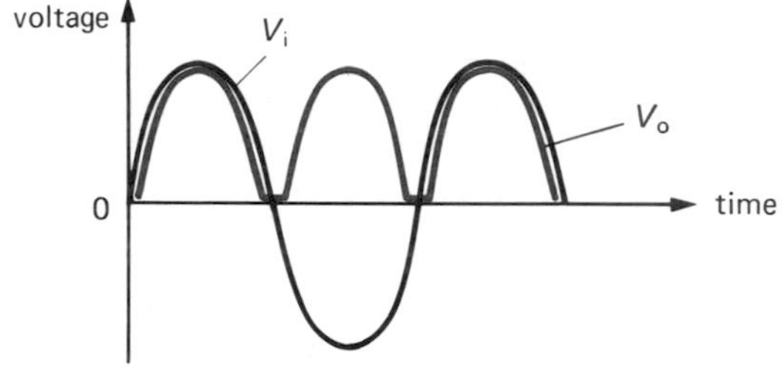

Fig. 40.4

Bridge full-wave rectifier

Only half of the secondary winding is used at any time in the circuit of Fig. 40.3 and the transformer has to produce twice the voltage required. A more popular arrangement using four diodes in a bridge network is shown in Fig. 40.5.

If A is positive with respect to B during the first half-cycle, D_2 and D_4 conduct and current takes the path AD_2RD_4B. On the next half-cycle when B is positive, D_1 and D_3 are forward biased and current follows the path BD_3RD_1A. Once again current through R is unidirectional during both half-cycles of the input V_i. The output V_o is a varying d.c. voltage with a waveform like that in Fig. 40.4.

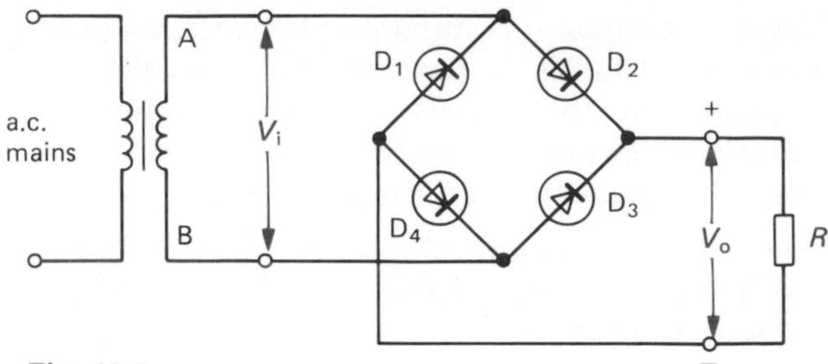

Fig. 40.5

All four diodes are available in one package, with two a.c. input connections and two output connections.

Heat sinks

High current diode rectifiers are mounted on *heat sinks* with cooling fins (made of aluminium sheet and painted black), as shown in Fig. 40.6, to stop them overheating. Manufacturers state what the

Fig. 40.6

thermal resistance of the heat sink should be for a particular rectifier. For example, if it is 2 °C W^{-1}, the temperature of the heat sink rises by 2 °C above its surroundings for every watt of power it has to get rid of. To dissipate 10 W, the rise will be 20 °C.

Questions

1. What is meant by rectification?

2. Strictly speaking the peak value of V_o in the circuit of Fig. 40.5 is less than the peak value of V_i by the voltage drop across *two* forward biased diodes. Explain why. Under what circumstances will the peak values of V_i and V_o be equal?

3. The V_o graphs in Figs. 40.2 and 40.4 are varying direct voltages having a steady d.c. component and an a.c. one. What is the frequency of the latter in each case if that of V_i is 50 Hz?

4. Explain the statement that a heat sink has a *thermal resistance* of 4 °C W^{-1}.

41 Smoothing circuits

The varying d.c. output voltage from a rectifier circuit can be used to charge a battery but must be 'smoothed' to obtain the steady d.c. required by electronic equipment.

Reservoir capacitor

The simplest way to smooth an output is to connect a large capacitor, called a *reservoir capacitor*, across it as in Fig. 41.1 for half-wave rectification. Its value on a 50 Hz supply may range from 100 μF to 10 000 μF depending on the current and smoothing needed.

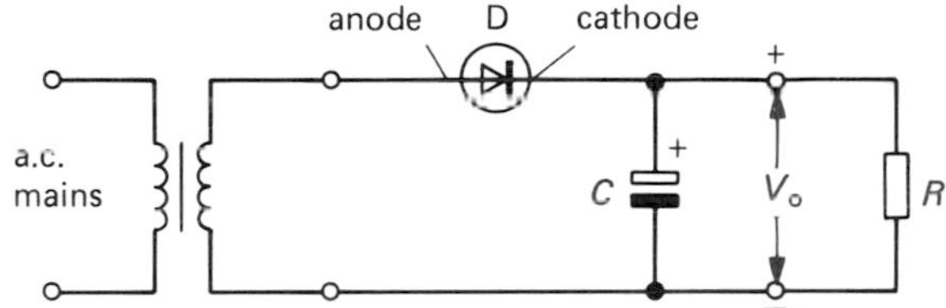

Fig. 41.1

The waveform of the smoothed output voltage V_o for half-wave rectification is shown as a solid line in Fig. 41.2. The corresponding half-cycles of a.c. during which the diode D conducts are drawn dashed. The small variation in the smoothed d.c. is called the *ripple voltage*. It has the same frequency as the a.c. supply and causes 'mains hum'.

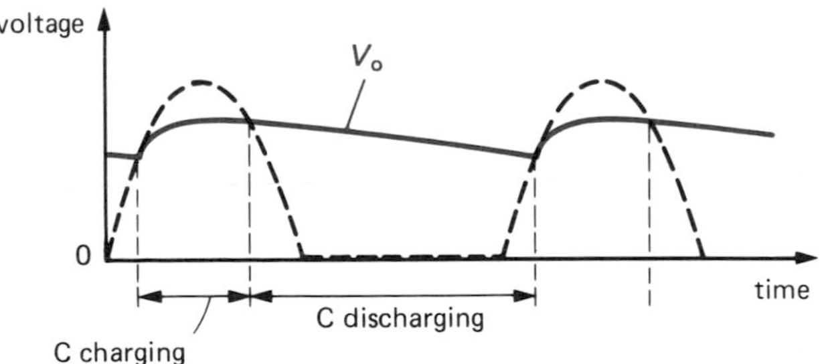

Fig. 41.2

The smoothing action of C is explained as follows. During part of the half-cycle of a.c. when D is forward biased, there is a current pulse which charges up C to *near* the peak value of the a.c. During the rest of the cycle C keeps the load R supplied with current by partly discharging through it. While this occurs, the output p.d. V_o falls until the next pulse of rectified current tops up the charge on C. It does this near the peak of the half-cycle. And so for most of each cycle the load current is supplied by C acting as a reservoir of charge.

The ripple voltage in a full-wave rectifier circuit is smaller, giving better smoothing. Waveforms are shown in Fig. 41.3; in this case the ripple frequency is twice that of the a.c. supply.

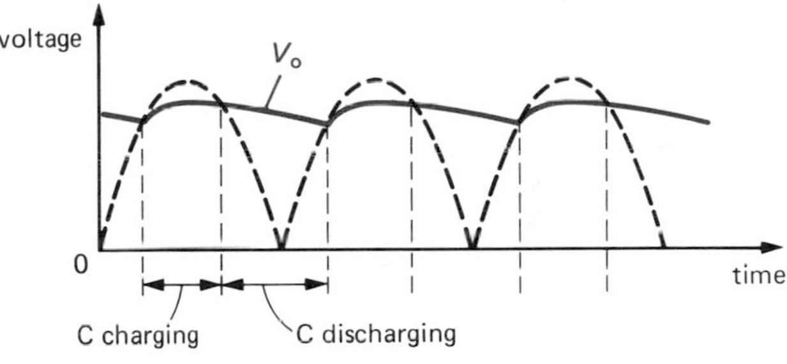

Fig. 41.3

Capacitor and diode ratings

(a) Capacitor. The smoothing action of a reservoir capacitor is due to its large capacitance C making the time constant CR large (p. 33), where R is the resistance of the load, compared with the time for one cycle of the a.c. mains (1/50 s). The larger C is, the better the smoothing but the smaller the fall in V_o and the briefer will be the rectified current pulse to charge up the capacitor again. Consequently, the peak value will have to be greater and damage may occur if the peak current rating of the diode is exceeded. Manufacturers sometimes state the maximum value of reservoir capacitor to be used with a given diode.

The capacitor should have a voltage rating at least equal to the peak value of the transformer's secondary voltage.

(b) Diode. Consideration must be given to the maximum reverse voltage V_{RRM} of the diode D (p. 58). *When it is not conducting*, Fig. 41.1 shows that the total voltage across it is the voltage across C plus that across the transformer secondary. For example if the lower plate of C is at 0 V and the top plate is near the positive peak of the a.c. input, say +12 V, then the cathode of D is at +12 V. Its anode will be at −12 V, i.e. the peak negative value of the voltage across the transformer on a negative half-cycle. The total voltage across D is 24 V in this case. In practice it is wise to use a diode whose V_{RRM} is at least four times the r.m.s. value of the secondary p.d. of the transformer to allow for voltage 'spikes' picked up by the a.c. mains supply.

Capacitor-input filter

The smoothing produced by a reservoir capacitor C_1 can be increased and ripple reduced further by adding a filter circuit. This consists of an inductor or choke L and another large capacitor C_2, arranged as in Fig. 41.4.

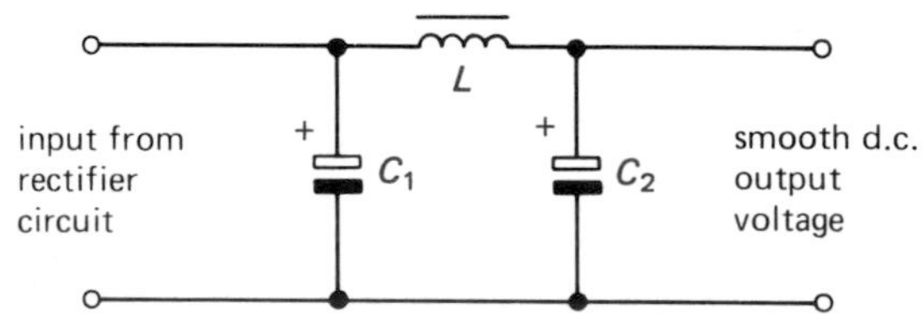

Fig. 41.4

We can regard the varying d.c. voltage produced across C_1 (V_o in Figs. 41.2 and 41.3) as a steady d.c. voltage (the d.c. component) plus a small ripple voltage (the a.c. component). L offers a much greater impedance than C_2 to the a.c. component and so most of the unwanted ripple voltage is developed across L. For the d.c. component the situation is reversed and most of it appears across C_2. The filter thus acts as a voltage divider, separating d.c. from a.c. (i.e. acting as a filter) and producing a d.c. output voltage across C_2 with less ripple. A resistor may replace the inductor when the current to be supplied is small.

Filter circuits have certain disadvantages which make them less suitable for semiconductor circuits which operate at low voltages and high currents. These are due to (a) the risk of large currents in chokes causing magnetic saturation of the iron cores, (b) chokes being large and heavy, and (c) resistors reducing the output voltage.

Questions

1. What is meant by smoothing?
2. If the load current were zero in a rectifier circuit with a reservoir capacitor, to what p.d. would the capacitor charge up? Illustrate your answer with a graph.
3. In the half-wave rectifier circuit of Fig. 41.1 if the transformer secondary p.d. is 7 V r.m.s., what is the maximum p.d. across (i) C, (ii) D?

42 Stabilizing circuits

Regulation

Because of its internal (source) resistance (p. 17), the output voltage from an ordinary power supply (and a battery) decreases as the current it supplies to the load increases, i.e. the 'lost' volts increase and the terminal p.d. decreases. The greater the decrease the worse is the *regulation* of the supply. Good and bad regulation curves are shown in Fig. 42.1.

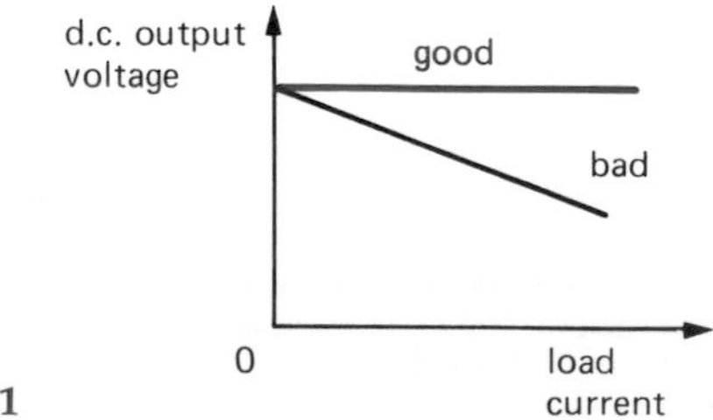

Fig. 42.1

There are many occasions when a d.c. voltage is required which remains constant and is not affected by load current changes. In these cases a stabilizing or regulating circuit is added to the power supply, as shown by the block diagram in Fig. 42.2 of a stabilized power supply.

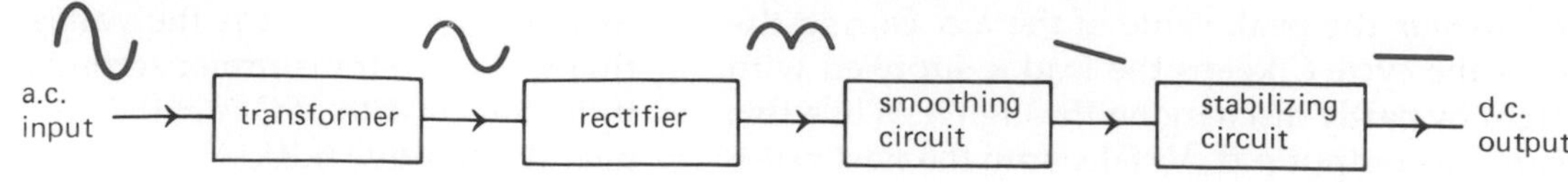

Fig. 42.2

Stabilized voltage circuit

(a) Action. When a Zener diode is reverse biased to its breakdown (Zener) voltage, the voltage across it stays almost constant for a wide range of reverse currents, as its characteristic shows (Figs. 28.2, p. 59). Because of this property, a Zener diode can be used to stabilize the output voltage from a smoothing circuit and reduce fluctuations due to load current changes.

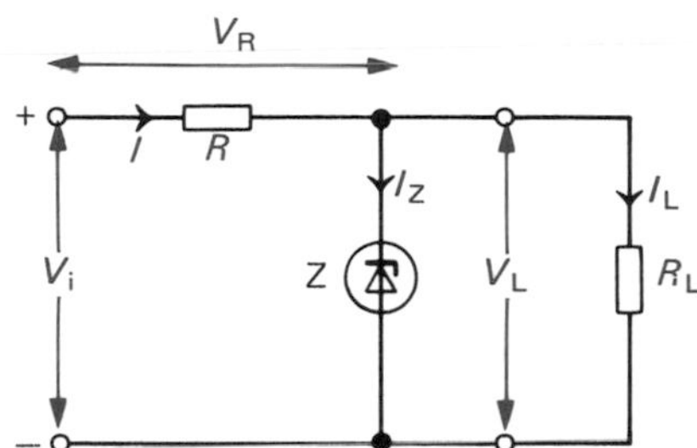

Fig. 42.3

In the circuit of Fig. 42.3, V_i is the smoothed voltage to be stabilized and R_L is the load which changes and draws different currents I_L. The total current I supplied, divides into I_Z through the Zener diode Z and I_L through R_L. Hence

$$I = I_Z + I_L$$

From Ohm's law, since $V_i = V_R + V_L$

$$I = \frac{V_R}{R} = \frac{V_i - V_L}{R}$$

But $V_L \approx V_Z$, i.e. the output voltage ≈ Zener voltage (p. 60),

$$\therefore I = \frac{V_i - V_Z}{R}$$

For a given V_i, I is therefore more or less constant since V_Z and R are constant. This means that if I_L increases, I_Z decreases by the same amount and vice versa. Thus if I_L varies due to R_L varying, V_L is virtually constant (as is V_R).

(b) Further points. If I_L = O (i.e. there is no load drawing current), then $I_Z = I$ and Z has to have the power rating to carry this current safely. On the other hand when I_L is a maximum, I_Z has its minimum value which must be large enough (at least 5 mA for a small diode) to ensure Z works on the breakdown part of its characteristic (not at or above the 'knee' in Fig. 28.2 where the p.d. across the Zener falls to zero). This is achieved by making V_i several volts greater than the required output voltage.

Stabilization also helps to reduce the effect of variations in V_i due either to any 'ripple' on it after smoothing or to fluctuations of the a.c. mains supply. For example, if V_i rises, the increase in I is such that the rise in p.d. occurs across R, leaving V_L about the same. If V_i varies over a wide range, the value of the safety resistor R must be such that when V_i is a minimum there is enough current available to both Z and R_L.

Stabilized current circuit

A constant current supply is sometimes needed, e.g. to recharge Nicad cells (p. 86). In the circuit of Fig. 42.4 the Zener diode Z and its safety resistor R

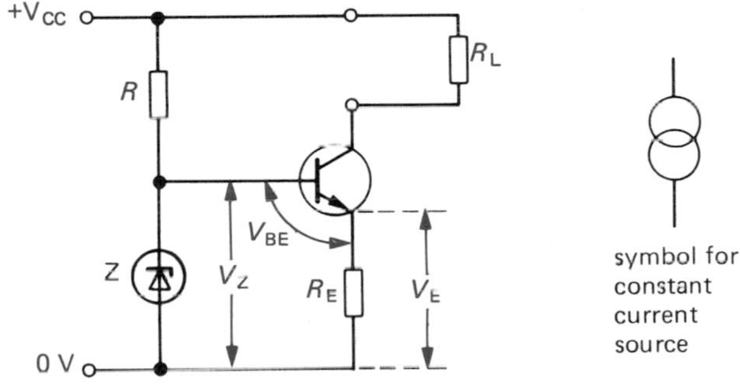

Fig. 42.4

keep the base of the silicon transistor at voltage V_Z. The p.d. V_E across resistor R_E in the emitter circuit is held constant at $V_Z - V_{BE}$ (where $V_{BE} \approx 0.6$ V). So long as $V_{CC} > V_E$, the current through the load R_L (e.g. a Nicad cell) in the collector circuit, whatever its value, is steady and equal to V_E/R_E. The circuit uses the fact the collector current I_C of a transistor is constant for a wide range of collector-emitter voltages V_{CE}, as its $I_C - V_{CE}$ output characteristic shows, Fig. 32.2c, p. 69.

Integrated circuit voltage regulators

A range of integrated circuit stabilizers, called *voltage regulators*, are now available in various packages, Fig. 42.5. They use more complex circuits than the simple Zener diode one and protection is also provided against overloading and overheating. Many are designed for one voltage, e.g. 5 V or 15 V.

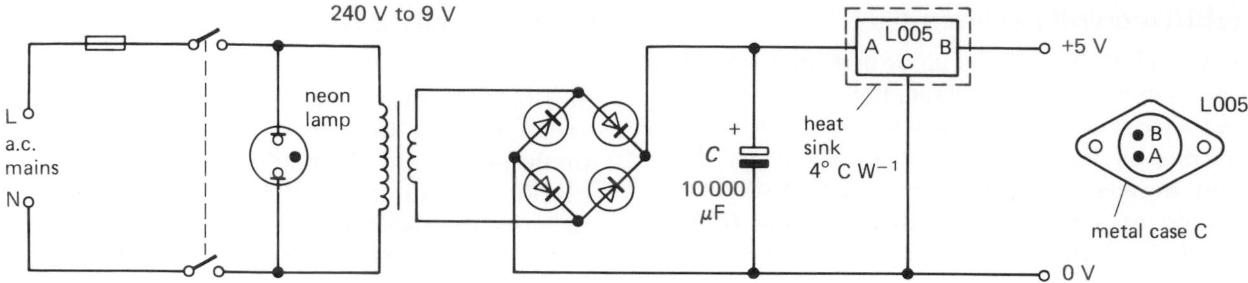

Fig. 42.6

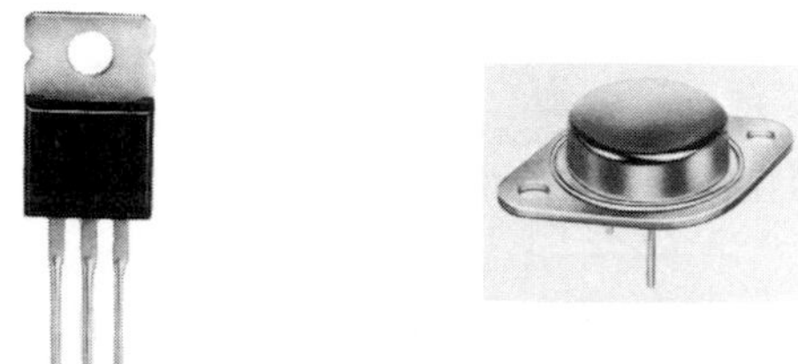

Fig. 42.5

The circuit for a stable 5 V 600 mA supply using the L005 regulator mounted on a 4°C W^{-1} heat sink, is given in Fig. 42.6. *C* is the reservoir capacitor. The neon lamp across the primary of the transformer shows when the mains is on and has a very low power consumption.

Questions

1. What is meant by the *regulation* of a power supply?

2. In the circuit of Fig. 42.7 the Zener diode Z has a breakdown voltage of 3 V.

(a) What is the output p.d. V_o when the input p.d. V_i is (i) 2 V, (ii) 4 V, (iii) 6 V?

(b) What is the current through Z when V_i is (i) 2 V, (ii) 4 V, (iii) 6 V?

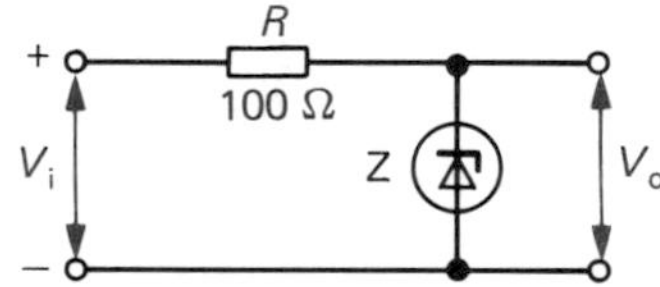

Fig. 42.7

3. In the circuit of Fig. 42.3 the Zener diode Z has a breakdown p.d. of 3 V, the input voltage V_i is 8 V and R = 50 Ω.

(a) If R_L = 300 Ω calculate I, I_L and I_Z

(b) If R_L = 100 Ω, calculate I, I_L and I_Z

(c) If V_L is to be kept steady why must R_L not fall below a certain value? What is this value?

(d) Why is there a maximum value for R?

43 Power control

Thyristor

A thyristor is a four-layer, three-terminal semiconducting device, Fig. 43.1. It used to be called a silicon controlled rectifier (SCR) because it is a rectifier which can control the power supplied to a load with little waste of energy.

When forward biased, a thyristor does not conduct until a positive voltage is applied to the gate. Conduction continues when the gate voltage is removed and stops only if the supply voltage is switched off or is reversed.

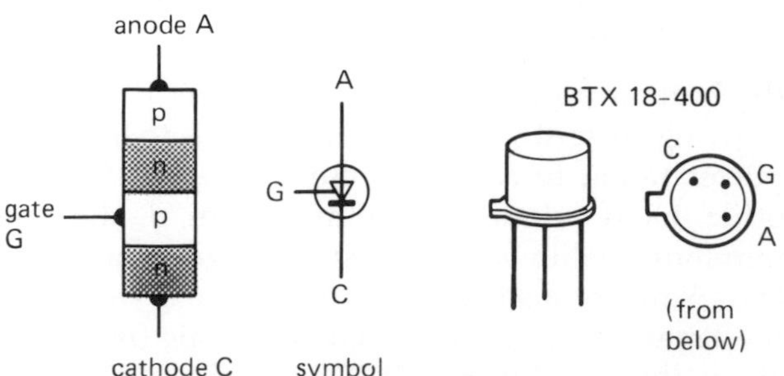

Fig. 43.1

(a) d.c. power control. The circuit of Fig. 43.2 can be used as a simple example. When S_1 is closed, the

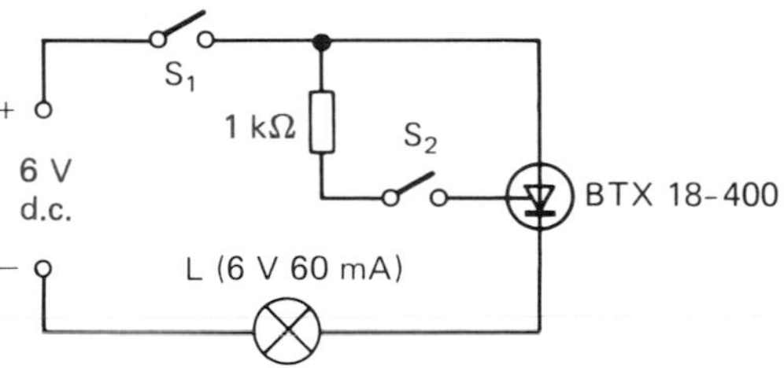

Fig. 43.2

lamp L stays off. When S_2 is closed as well, gate current flows and the thyristor switches on, i.e. 'fires'. The anode current is large enough to light L, which stays alight if S_2 is opened.

(b) a.c. power control. The control of a.c. power can be achieved by allowing current to be supplied to the load during only part of each cycle. A gate pulse is applied automatically at a certain chosen stage during each positive half-cycle of input. This lets the thyristor conduct and the load receives power.

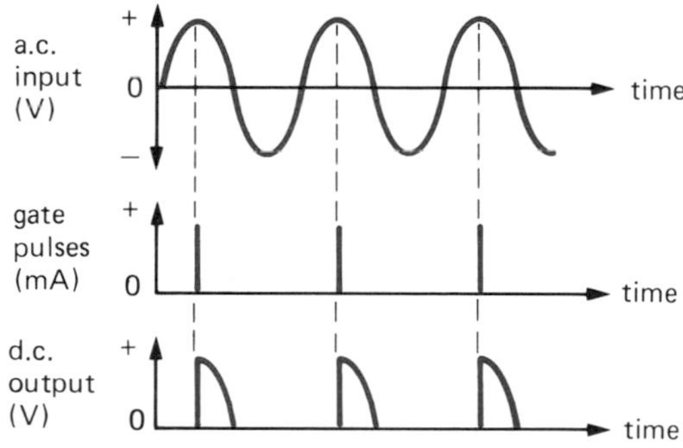

Fig. 43.3

For the case shown in Fig. 43.3 the pulse occurs at the peak of the a.c. input. During negative half-cycles the thyristor is non-conducting and remains so till half-way through the next positive half-cycle. Current flows for only a quarter of a cycle but by changing the timing of the gate pulses, this can be decreased or increased. The power supplied to the load is thus varied from zero to that due to half-wave rectified d.c.

Triac

The thyristor, being in effect a diode, is a half-wave device and only allows half the power to be available. A triac consists of two thyristors connected in parallel but in opposition and controlled by the same gate. It is a two-directional thyristor which is triggered on both halves of each cycle of a.c. input by either positive or negative gate pulses. The power obtained by a load can therefore be varied between zero and full-wave d.c.

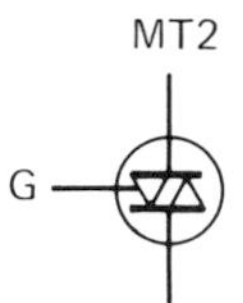

Fig. 43.4

The connections are called *main terminals* 1 and 2 (MT1 and MT2) and *gate* (G), Fig. 43.4. The triggering pulses are applied between G and MT1. The gate current for a triac handling up to 100 A may be no more than about 50 mA. Triacs are used in lamp dimmer circuits and for motor speed control, e.g. in an electric drill.

Questions

1. Explain the action of a thyristor using Fig. 43.5a,b, which is its transistor equivalent.

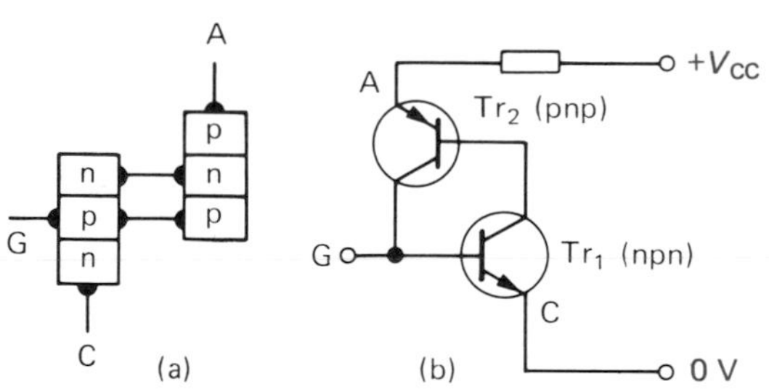

Fig. 43.5

2. What are the advantages of the thyristor in control circuits?

44 Progress questions

1. Describe, with the aid of a labelled diagram, the basic features of an *electric filament lamp*.
Give a circuit diagram of a house lighting system. Explain (i) why the lights in a house do not dim as more of them are switched on, (ii) why the lights are not all extinguished when one lamp fails.
An electric lamp is marked 250 V, 100 W. Interpret these numbers, and discuss how this lamp would behave if the mains voltage was permanently (a) 200 V, (b) 300 V.
The efficiency of a tungsten lamp is 12.5 lumen per watt. Given that 1 watt = 625 lumen, calculate the percentage efficiency of the lamp, and account for its low value.
(*C. part qn.*)

2. Give an account of the production of light within a *fluorescent lamp*. Draw a circuit for the operation of such a lamp, and explain how it works.
A 40 W tubular fluorescent lamp has an efficiency of 50 lumen per watt. Explain what that means.
(*C. part qn.*)

3. (a) In a 13 A mains plug which colour of wire in the supply cable should be connected to the terminal marked (i) L, (ii) N, (iii) E?
(b) What is the difference between the neutral and earth wires in a supply cable?

4. (a) Name *two* fossil fuels and *two* other energy sources used in the large-scale generation of electricity in the U.K. Give *one* advantage and *one* disadvantage of each.
(b) A power station uses 400 kg of coal per minute and its output is 66 MW. Calculate its efficiency if the coal supplies 33 MJ/kg.

5. What is the difference between a primary and a secondary cell? Name an example of each.
Describe the structure and action of a *lead-acid accumulator*. Explain clearly what chemical changes occur as the cell discharges. State the main features in the care and maintenance of a lead-acid accumulator.
Draw a clearly labelled circuit diagram of a system to be used to recharge a 12 V car battery at a suitable rate from the a.c. mains supply. Explain how the circuit works.
(C.)

6. (a) Explain the action of the circuit in Fig. 44.1, and draw sketch graphs of the input p.d. and the p.d. across the load. (You should draw the graphs so that they can be compared.) If the peak voltage induced between X and Z is 10 V what is the peak p.d. across (i) *R*, (ii) D_1?
(b) Explain, using p.d.-time sketch graphs, the effects of adding to the circuit the smoothing capacitor as shown in Fig. 44.2.

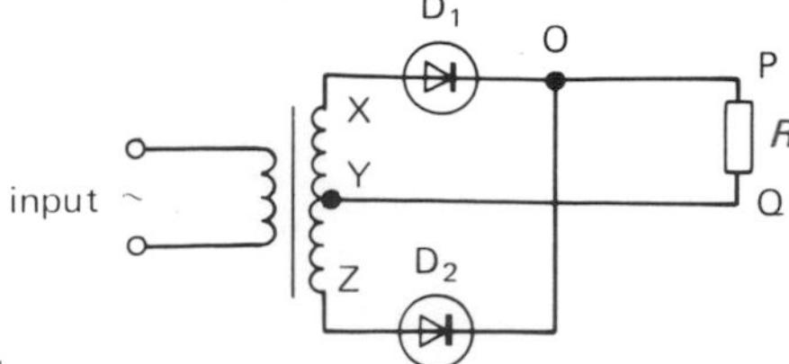

Fig. 44.1

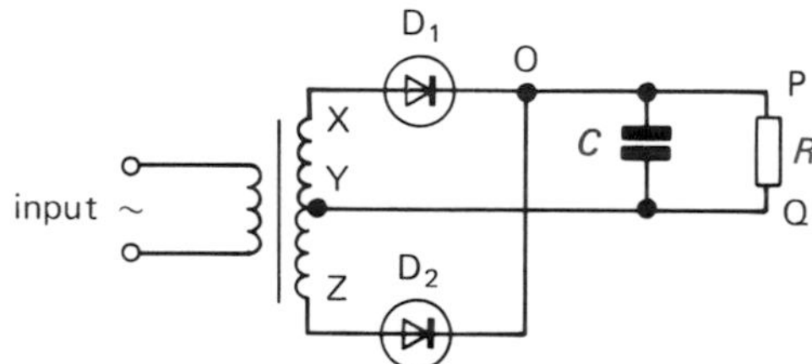

Fig. 44.2

Given that the value of *R* is 100 Ω and the frequency of the supply is 50 Hz, calculate a suitable value for *C*.
What would be the p.d. between P and Q if *R* became open circuit?
(*L.*)

7. A simple voltage stabilizer using a Zener diode is shown in Fig. 44.3. The Zener voltage is 5.6 V.

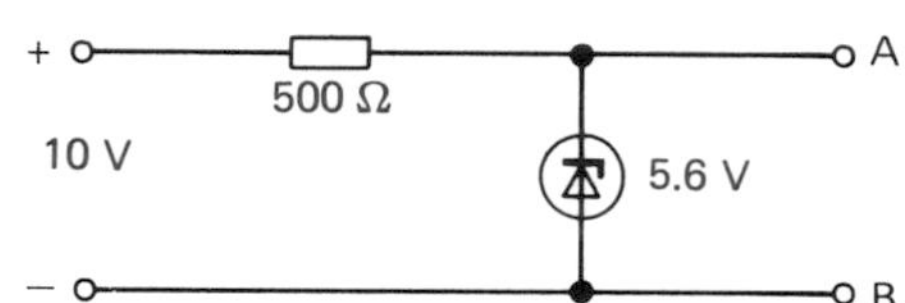

Fig. 44.3

(i) What is the current through the 500 Ω resistor?
(ii) What changes take place in the circuit if a resistive load of 1000 Ω is connected?
(iii) Are there any limitations on the resistive load if the stabilized voltage is to be maintained across AB?
(*O. and C.*)

8. (a) The Zener diode Z in Fig. 44.4 has a breakdown voltage of 4 V. Calculate the current in Z.

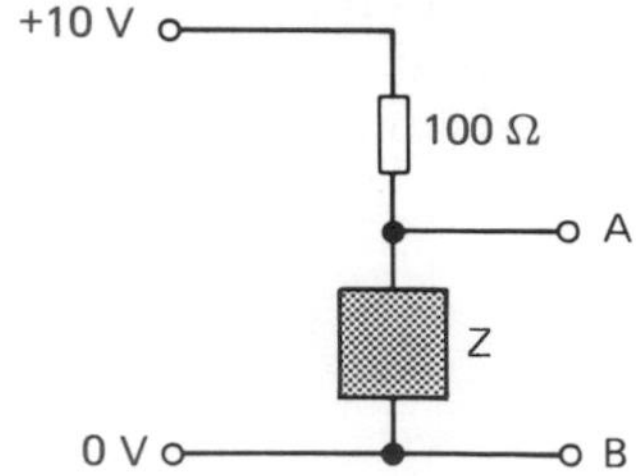

Fig. 44.4

Explain, giving values, the effect on the current in Z of
(i) a rise in supply voltage to 12 V,
(ii) connecting a load of 100 Ω across AB (with a supply voltage of 10 V).
If the circuit of Fig. 44.4 is used to provide a stable voltage supply what is the minimum resistance which may be connected across AB?

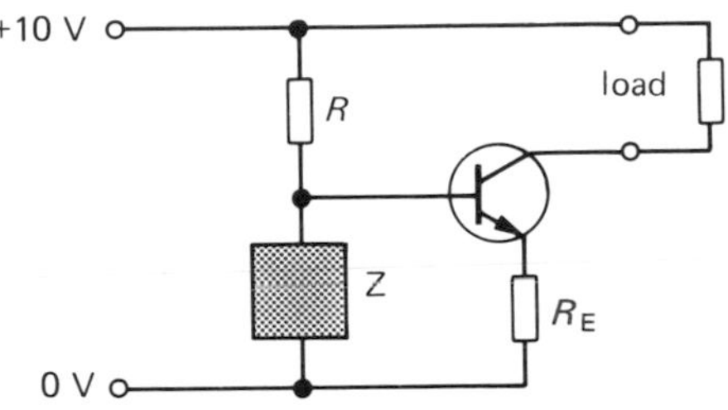

Fig. 44.5

(b) The circuit of Fig. 44.5 provides a stabilized *current* source. Explain
(i) why a rise in supply voltage does not lead to a rise in current through the load, and
(ii) how the circuit responds to a change in the resistance of the load.
Discuss briefly what would happen if
(iii) the load were reduced to zero resistance, and
(iv) the load were removed (i.e. increased to infinite resistance). (*L*)

9. As batteries become increasingly expensive more people are using battery eliminators that they can plug into the mains. The circuit diagram in Fig. 44.6 was suggested to someone wishing to construct such a device.

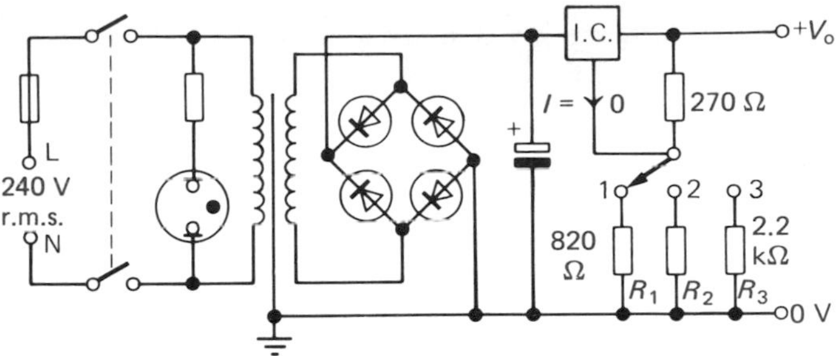

Fig. 44.6

(a) Explain why the fuse and the switches are on the high voltage side of the transformer.
(b) Explain why a lamp is placed in parallel with the primary coil of the transformer. Why is a neon lamp used in preference to a filament lamp?
(c) If the output of the secondary of the transformer is 12 V r.m.s., calculate the ratio of the turns on the primary and secondary coils.
(d) Explain, briefly, how the 12 V a.c. from the transformer is converted into the 16 V d.c. which is fed into the integrated circuit IC.
The integrated circuit is a voltage regulator. It provides a constant voltage of 1.25 V across the 270 Ω resistor. So, with the rotary switch in position 1, the voltage across the 820 Ω resistor will be 3.80 V and the total output, therefore, 5.05 V, approximating to 5 V.
(e) Calculate the value of resistor R_2 if an output of 9 V is required when the switch is in position 2.
(f) Calculate the output voltage when the switch is in position 3.
(g) Give one example of a use of this type of battery eliminator. What current output would be suitable for your example? (*A.E.B. 1982 Electronics*)

Measuring instruments

45 Multimeters

A multimeter measures current, voltage and resistance, each on several ranges.

Analogue multimeter

In this type, Fig. 45.1, the deflection of a pointer over a scale represents the value of the quantity being measured. Basically it consists of a *current-measuring* moving coil meter.

Fig. 45.1

(a) Current and voltage. As an ammeter, different low resistance shunts S (p. 14) are switched in for different current ranges, as shown in Fig. 45.2. As a voltmeter, high resistance multipliers M (p. 14) are connected in series with the coil of the meter.

For a.c. measurements a diode rectifier produces pulses of varying d.c. to which the meter responds.

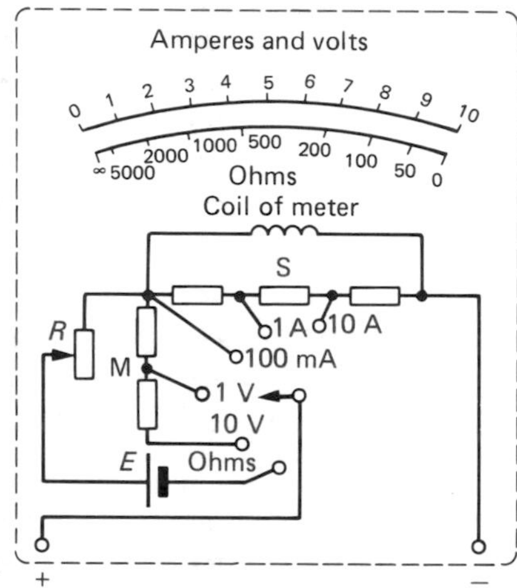

Fig. 45.2

Its a.c. scales are calibrated to read r.m.s. values of currents and p.ds that have sine waveforms.

(b) Resistance. As an ohmmeter, an internal rheostat R (often marked 'ohms adjust' or 'zero Ω') and a battery E are brought into the circuit.

To take a measurement, the meter leads are first held together, making the external resistance between the terminals zero and the current through the meter a maximum. If necessary, R is adjusted (to allow for battery p.d. changes) until the pointer gives a full-scale deflection to the right, i.e. is on the zero of the ohms scale.

The leads are then connected across the unknown resistance R_X. The current I is now less and the pointer indicates R_X in ohms. If the meter is on open circuit, $I = 0$ and R_X is infinite and the pointer sets on the ∞ mark at the left end of the ohms scale.

An ohmmeter has a non-linear scale because $I = E/(R_M + R_X)$ where E is the e.m.f. of the internal battery (1.5 or 15 V) and R_M is the ohmmeter resistance. The expression shows that as R_X increases, I decreases but not uniformly, i.e. doubling R_X does not halve I.

Digital multimeter

The measurement is shown on an LCD or LED digital decimal display (p. 47), often with four figures, Fig. 45.3. It is made by a high resistance *voltage-measuring* analogue-to-digital (A/D) converter (p. 208).

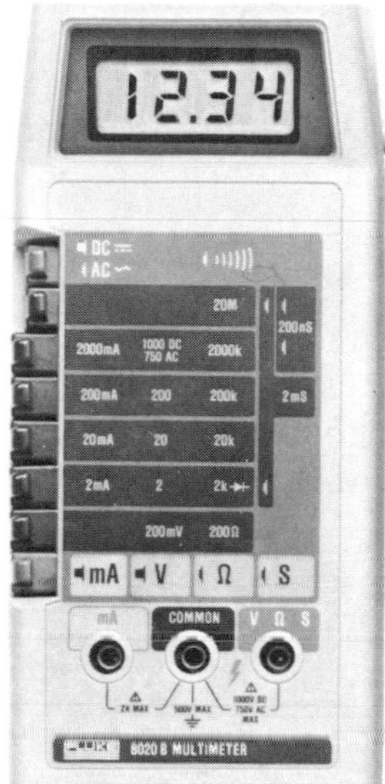

Fig. 45.3

(a) Voltage and current. As a voltmeter, low voltages are measured directly and higher ones via a potential divider which supplies the input to the A/D converter. As an ammeter, a parallel resistor R

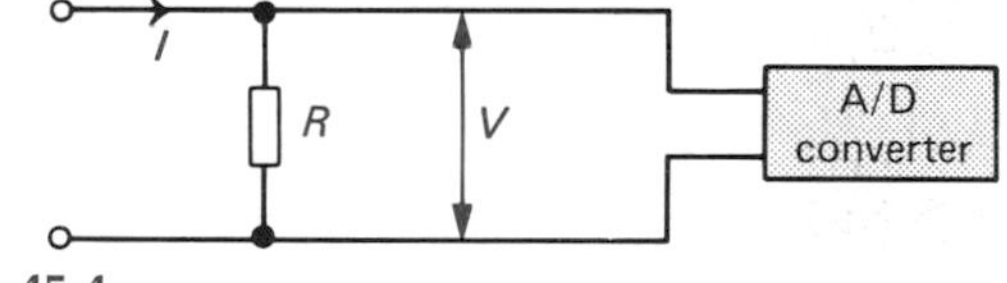

Fig. 45.4

produces a p.d. V across the input to the converter that is proportional to the input current I, Fig. 45.4.

(b) Resistance. As an ohmmeter, a stabilized current supply (p. 91) passes a known constant current I through the unknown resistance R_X, Fig. 45.5. The p.d. across R_X is measured by the A/D converter and is proportional to R_X since $V = IR_X$, i.e. the 'scale' is linear.

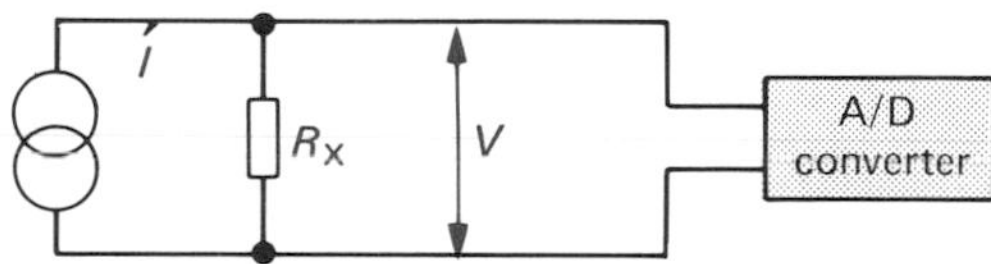

Fig. 45.5

Practical points

1. Before use, check that the reading is zero; if it isn't, correct it by the 'zero adjustment' screw.
2. For d.c. currents and p.ds, the positive terminal must be connected to the circuit so that it is nearer the supply's positive terminal than is the negative terminal of the meter.
3. Switch to the highest range first.
4. Check you are reading the correct scale.
5. When finished leave the multimeter on the highest d.c. voltage range.

Fault-finding

Multimeters are useful for locating faults in electronic circuits and components. Sometimes in resistance measurements, use is made of the fact that in analogue meters, the negative (black) terminal has positive (+) polarity (due to the inter-

Comparison of multimeters

Property	**Analogue** (moving coil)	**Digital** (electronic)
Reading errors	Can occur especially when pointer between marks	Less likely
Input resistance as a voltmeter	Moderate, e.g. $20\,V_R$ kΩ/V where V_R is the f.s.d. voltage. Varies with range	High, e.g. 10 MΩ on d.c.; 0.5 MΩ on a.c. Same on all ranges.
Scale/display	Scale continuous	Display changes by 1 digit
Response to input	Continuous	Samples taken regularly
Power used	None except as an ohmmeter	Small if LCD

nal battery) and the positive (red) terminal has negative (−) polarity. With digital meters the polarity is given in the instruction booklet (but often the terminal marked 'mA' is positive).

(a) Diodes. A diode should have a low resistance when the cathode (the end with the band round it) is connected to the terminal with negative polarity and the anode to the terminal of positive polarity. The connections using an analogue multimeter are shown in Fig. 45.6. Reversing the connections gives a high resistance reading in a 'healthy' diode.

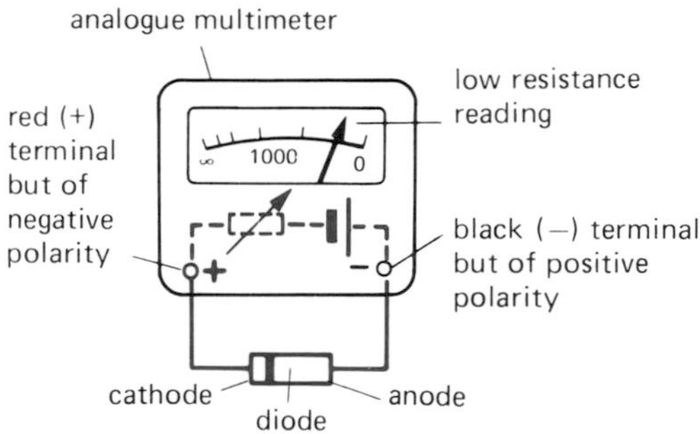

Fig. 45.6

(b) Transistors. In a junction transistor the resistance should be high between the collector and emitter for *both* methods of connection to the multimeter. This enables the base to be identified since it gives a low resistance one way with either the collector or emitter and a high reading the other way.

For an n-p-n transistor the resistance is lower when the multimeter terminal of positive polarity (i.e. the

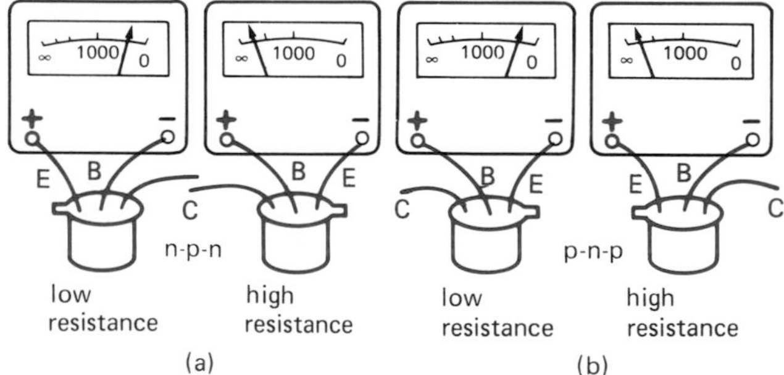

Fig. 45.7

black one marked −) is connected to the base and the other terminal to the emitter or collector than it is with the leads reversed, Fig. 45.7a. For a p-n-p type the opposite is true, Fig. 45.7b. In each case the resistance is low when the p-n junction concerned is forward biased and use is made of the fact that a transistor behaves as two back-to-back diodes.

(c) Non-polarized capacitors (p. 26). If the resistance of the capacitor is less than about 1 MΩ, it is allowing d.c. to pass (from the battery in the multimeter) i.e. it is 'leaking' and is faulty. With large value capacitors there may be a short initial burst of current as the capacitor charges up.

(d) Polarized capacitors (p. 26). For the dielectric to form in these, a positive voltage must be applied to the positive lead of the capacitor (marked by a +) from the multimeter terminal of positive polarity. When first connected, the resistance is low but rises as the dielectric forms.

(e) 'Dry' joints. These are badly soldered joints (p. 215) which have a high resistance. In the circuit of Fig. 45.8 if there is a 'dry' joint at X, the voltmeter reading to the right of X will be +6 V and to the left 0 V because there is infinite resistance at X.

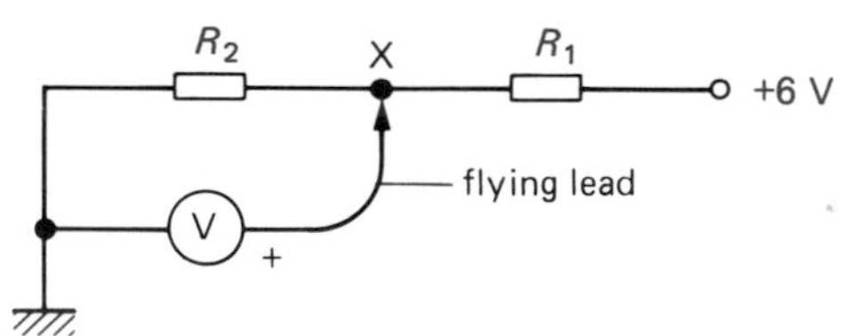

Fig. 45.8

(f) Bulbs, fuses, connecting wire. Their resistance should be low, if not they are faulty.

Questions

1. Draw *three* basic circuits to show how a moving-coil meter is adapted for use as (a) an ammeter, (b) a voltmeter and (c) an ohmmeter.

2. What would be the p.d. across the 10 kΩ resistor in the circuit of Fig. 45.9 if the voltmeter were not connected into the circuit?

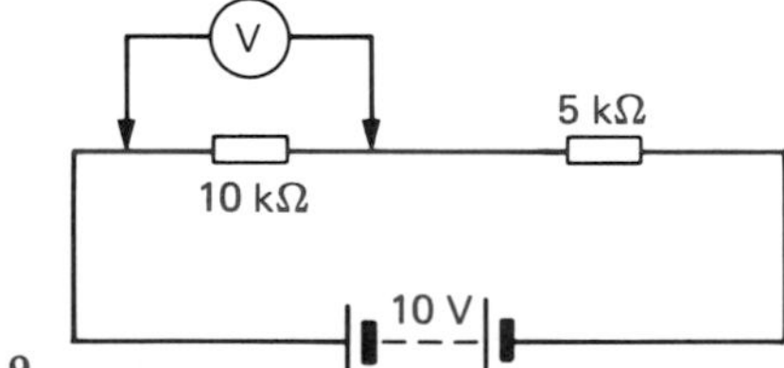

Fig. 45.9

If the voltmeter has a resistance of 1 kΩ per volt and is set on its 10 V scale what would it read when connected across the 10 kΩ resistor?
What conclusion do you draw from these values?
(L. part qn.)

3. Why does a diode have a low resistance when connected one way to an ohmmeter but a high resistance the other way round?

46 Oscilloscopes

The cathode ray oscilloscope (CRO) is one of the most important instruments ever to be developed. It is used mainly as a 'graph-plotter' to display the waveform of the voltage applied to its input, i.e. to show how it varies with time. Non-electrical effects can also be studied since almost any measurement can be changed into a voltage by an appropriate transducer. A simple model is shown in Fig. 46.1.

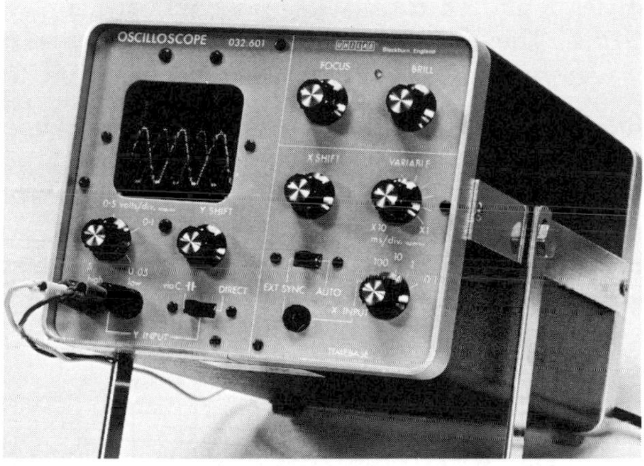

Fig. 46.1

Description

The block diagram in Fig. 46.2 shows that a CRO consists of

(i) a *cathode ray tube* (CRT), which employs electrostatic deflection and incorporates brilliance and focus controls (p. 48),

(ii) a *power supply* and

(iii) *circuits to the X- and Y-plates* of the CRT.

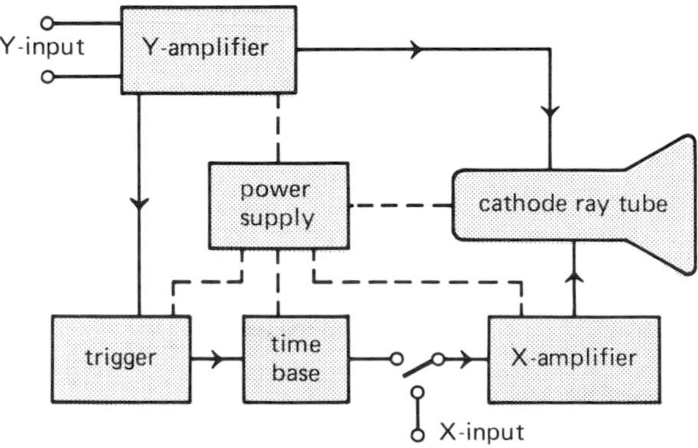

Fig. 46.2

(a) Y-input circuit. The input p.d. is connected to the *Y-input* terminals, which may be marked 'high' and 'low'. If the *a.c./d.c. selector* switch is in the 'a.c.' (or 'via C') position, a capacitor blocks any d.c. in the input but allows a.c. to pass; in the 'd.c.' (or 'direct') position, both d.c. and a.c. pass. The input is amplified, via the *Y-amp gain* control (often marked 'volts/cm or div'), by the *Y-amplifier* before it is applied to the Y-plates. It is then large enough to deflect the electron beam vertically.

The *Y-shift* control allows the spot (or waveform) on the screen to be moved 'manually' in the Y-direction by applying a positive or negative voltage to one of the Y-plates.

(b) X-input circuit. The p.d. applied to the X-plates via the *X-amplifier* can either be from an external source connected to the *X-input* terminal or, what is commoner, from the *time base circuit* in the CRO.

The *time base* deflects the beam horizontally in the X-direction and makes the spot sweep across the screen from left to right at a steady speed determined by the setting of the time base controls. The coarse control (usually marked 'time or ms/cm or div') gives several preset (fixed) sweep speeds each variable within limits by the fine control (sometimes marked 'variable'). It must then make the spot 'fly' back very rapidly to its starting position, ready for the next sweep. The time base voltage should therefore have a sawtooth waveform like that in Fig. 46.3.

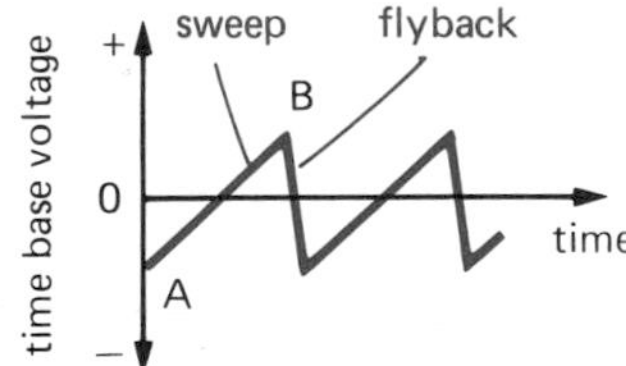

Fig. 46.3

Since AB is linear, the distance moved by the spot is directly proportional to time and the horizontal deflection becomes a measure of time, i.e. a time axis or base. If the Y-input p.d. represents a quantity which varies with time, its waveform is displayed.

To keep the waveform (trace) steady on the screen, each horizontal sweep must start at the same point on the Y-input waveform. The time base frequency should therefore be that of the input or a sub-multiple of it. Successive traces then coincide and, owing to the persistence of vision and the phosphor on the screen, the trace appears at rest.

One way to obtain synchronization of the time base and input frequencies is to feed part of the Y-amplifier output to a *trigger* circuit. This gives a pulse at a certain point on the input (selected in some CROs by a *trig level* control) which starts the sweep of the time base sawtooth, i.e. it 'triggers' the time base. In many CROs automatic triggering by the Y-input occurs if they are set on 'auto' (instead of 'ext sync' in the CRO shown) but in some cases, no trace is obtained until there is an input.

The *X-shift* control allows 'manual' shift in the X-direction.

Uses of the CRO

(a) Practical points. The *brilliance* (brightness or intensity) control, which is usually the *on/off* switch as well, should be as low as possible when there is just a spot on the screen. Otherwise screen burn occurs and the phosphor is damaged. If possible it is best to defocus the spot or draw it into a line by running the time base.

When preparing the CRO for use, set the *brilliance, focus, X-* and *Y-shift* controls to their mid positions. The *time base* and *Y-amp gain* controls can then be adjusted to suit the input.

(b) Voltage measurements. A CRO has a large impedance and can be used as a d.c./a.c. voltmeter if the p.d. to be measured is connected across the Y-input terminals. With d.c., the spot (time base off) or line (time base on) is deflected vertically, Fig. 46.4a,b. With a.c. (time base off) the spot moves up and down producing a vertical line if the motion is fast enough, Fig. 46.4c.

When the *Y-amp gain* control is on, say, 1 V/div, a deflection of 1 division on the screen graticule

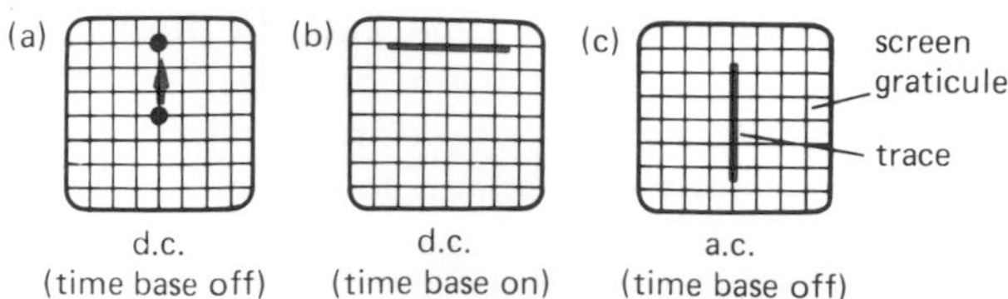

Fig. 46.4

would be given by a 1 V d.c. Y-input; a line 1 div long would be produced by an a.c. input of 1 V peak-to-peak, i.e. peak voltage = 0.5 V and r.m.s. voltage = 0.7 × peak voltage = 0.7 × 0.5 = 0.35 V. On a setting of 0.05 V/div, a Y-input of 0.05 V causes a deflection of 1 div.

While there are more accurate ways of measuring p.ds, the CRO can measure alternating p.ds at frequencies of several megahertz.

(c) Time measurements. The CRO must have a calibrated time base. When set on, for example, 1 ms/div the spot takes 1 millisecond to move 1 division and so travels 10 horizontal divisions in 10 ms. The period and so the frequency of a waveform can be found in this way.

(d) Waveform display. The a.c. voltage whose waveform is required is connected to the Y-input with the time base on. When the time base frequency equals that of the input, one complete wave is displayed; if it is half that of the input, two waves are formed. These and other effects are shown in Fig. 46.5.

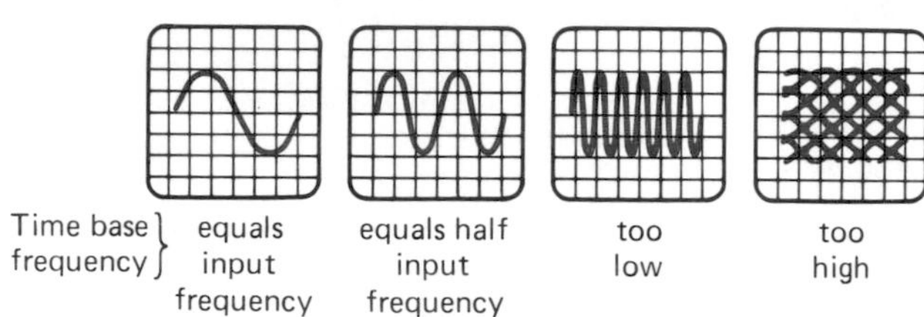

Fig. 46.5

(e) Phase relationships. If two sine wave p.ds of the same frequency and amplitude are applied simultaneously to the X- and Y-inputs (time base off), traces like those in Fig. 46.6 are obtained which depend on their phase difference. The method can be used to study the phase relationships between p.ds and currents in a.c. circuits.

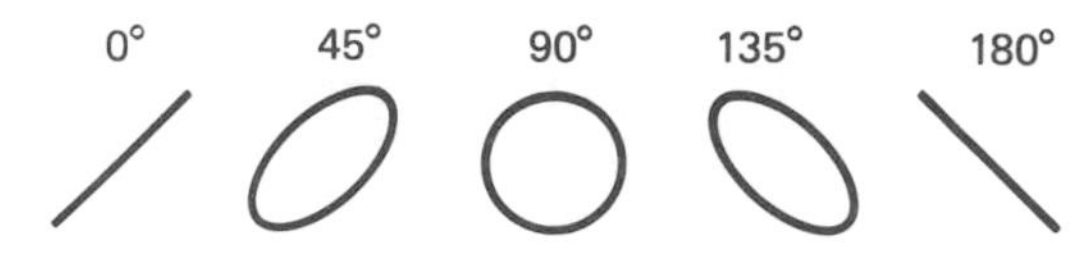

Fig. 46.6

(f) Frequency comparison. When two p.ds of different frequencies f_x and f_y are applied to the X- and Y-inputs (time base off), traces called *Lissajous'*

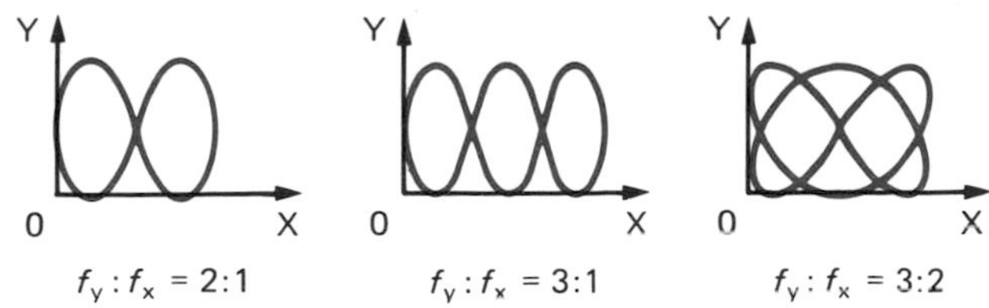

Fig. 46.7

figures are obtained. Some are shown in Fig. 46.7. The frequency ratio is given by

$$\frac{f_y}{f_x} = \frac{\text{no. of loops touching OX}}{\text{no. of loops touching OY}}$$

Dual beam and dual trace CROs

These are useful for comparing two traces simultaneously. In the dual beam type the CRT has two electron guns, each with its own Y-plates but having the same time base. In the commoner dual trace type, electronic switching produces two traces from a single beam. The switching is done by a square wave having a frequency much higher than that of either input. The top of the wave connects one input (Y_1) to the Y-plates and the bottom connects the other (Y_2). This occurs so rapidly that there appears to be two separate traces, one above the other on the screen.

Questions

1. When the Y-amp gain control on a CRO is set on 2 V/cm, an a.c. input produces a 10 cm long vertical trace. What is (a) the peak voltage, (b) the r.m.s. voltage, of the input?

2. What is the frequency of an alternating p.d. which is applied to the Y-plates of a CRO and produces five complete waves on a 10 cm length of the screen when the time base setting is 10 ms/cm?

47 Signal generators

A signal generator is an oscillator (p. 116) which produces a signal in the form of a.c. of known but variable frequency. It is used for testing, fault-finding and experimental work. There are two types.

Audio frequency generator

A typical instrument, Fig. 47.1, covers frequencies from 0.1 Hz to 100 kHz in several ranges. The output may have a sine, square or triangular waveform.

To allow matching to different loads (p. 113), there are usually 'low' and 'high' impedance outputs, e.g. 1 Ω and 100 Ω respectively. The former would be used for direct connection to a 4 Ω loudspeaker and the latter when providing a signal to an oscilloscope (p. 99).

If the maximum output voltage (e.g. 6 V r.m.s.) is not required it can be fed into an attenuator which controls the output by (i) reducing it in steps by a factor of 10 or 100 and (ii) allowing it to be varied continuously from 0 V up to the maximum.

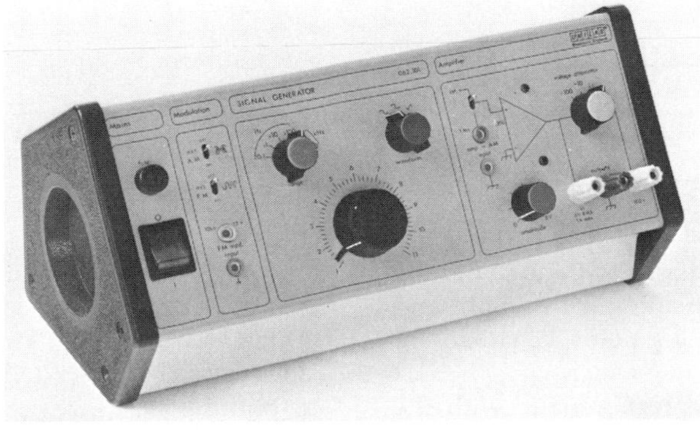

Fig. 47.1

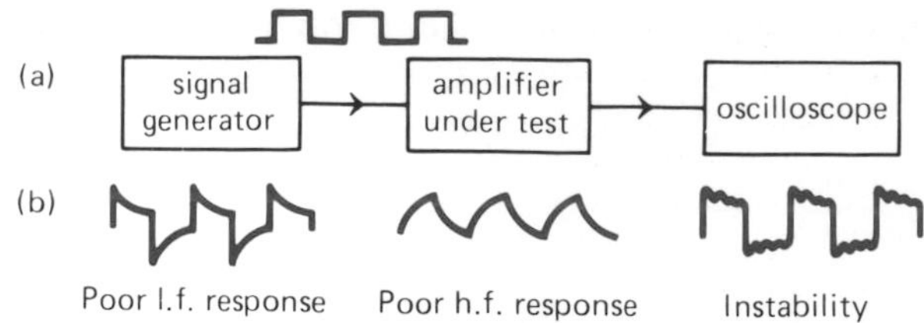

Fig. 47.2

One use for an a.f. signal generator is to test the response of an audio amplifier (p. 172) to different frequencies by supplying it with a square wave, Fig. 47.2a. If all the many (sine-wave) frequencies making up the square wave (p. 20) are amplified equally, the output from the amplifier is seen on the oscilloscope (p. 99) to be a square wave. The waveforms obtained for other responses are given in Fig. 47.2b.

Radio frequency generator

The frequency coverage of this type is from 100 kHz to 300 MHz or higher. The output is a pure sine wave which can usually be amplitude or frequency modulated (p. 168) for fault-finding in radio receivers. An attenuator allows the output, normally 200 to 300 mV, to be reduced when smaller signals are required.

The oscillator producing the r.f. usually contains some type of *LC* circuit (p. 37), coarse adjustment of the frequency being obtained by switching in different coils and fine adjustment by altering a variable capacitor.

48 Progress questions

1. A multimeter measures resistance by incorporating its meter in series with an internal battery of e.m.f. 4.5 V and a variable resistor. When the arrangement is 'zeroed' for a resistance measurement by short-circuiting the leads, 1.0 mA passes through the meter. When a resistor *R* is connected between the leads, the meter current is 0.5 mA. What is the resistance of *R*? *(O. and C.)*

2. The circuit diagram in Fig. 48.1 represents a simple ohmmeter. The device to be tested is connected to terminals X and Y.

(a) Describe the procedure that should be undertaken before any resistance measurements are attempted.

(b) A device is fitted across X and Y and no movement of the pointer occurs. What value of resistance for the device has been indicated?

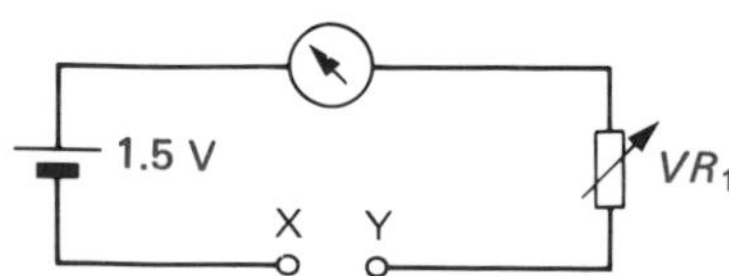

Fig. 48.1

(c) A different device is fitted across X and Y and the meter pointer gives a small reading to the right and then returns slowly to its starting point. What kind of device would give this result?

(d) Describe how you would test a diode with the simple ohmmeter in Fig. 48.1. Give an indication of the kind of readings you would expect for a good diode. *(A.E.B. 1982 Control Tech.)*

3. Explain how you would use an ohmmeter (a) to determine which terminal of a transistor is the base, and (b) to distinguish between n-p-n and p-n-p types of transistor.
Why is an ohmmeter scale non-linear? *(L)*

4. An ohmmeter was used to test for a fault in the section of the circuit in Fig. 48.2. Connection of the meter to A and B indicated a resistance of 10 kΩ whichever way round the meter was connected. There is known to be only one simple fault. Which of the components *could* be faulty and which component or components (if any) can be assumed not to be faulty? Justify your answers. *(A.E.B.)*

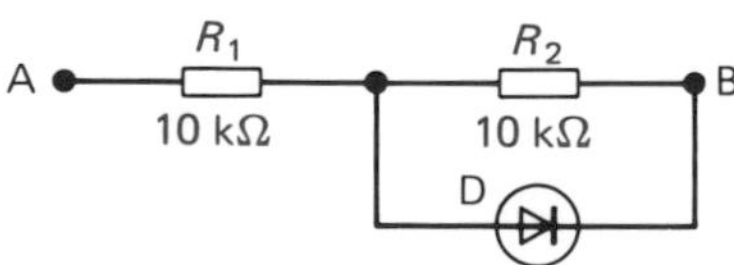

Fig. 48.2

5. An oscilloscope has a sensitivity of 5 V/cm and a time base adjusted to give a sweep time of 1 ms/cm. Show, on graticules like that in Fig. 48.3, the trace for (a) a 250 Hz sine wave of amplitude 10 V, (b) a 500 Hz square wave of amplitude 5 V. (C.)

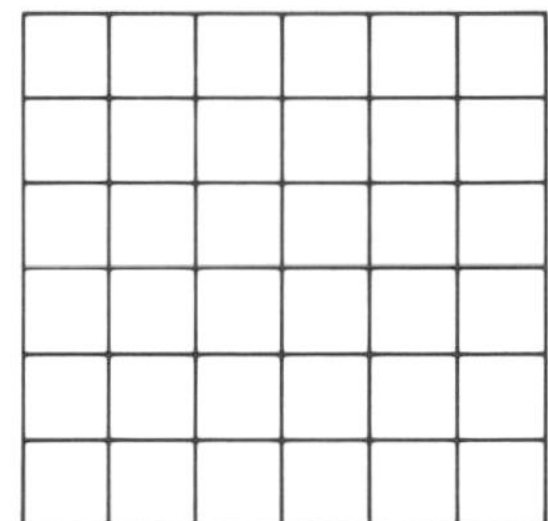

Fig. 48.3

6. Fig. 48.4 shows a waveform observed on the face of a cathode ray oscilloscope. The centimetre markings of the graticule are also shown. The y-deflection sensitivity was set at 0.1 V cm^{-1} and the time base was set at 100 μs cm^{-1}.

Give as much information as you can about the waveform. *(O. and C.)*

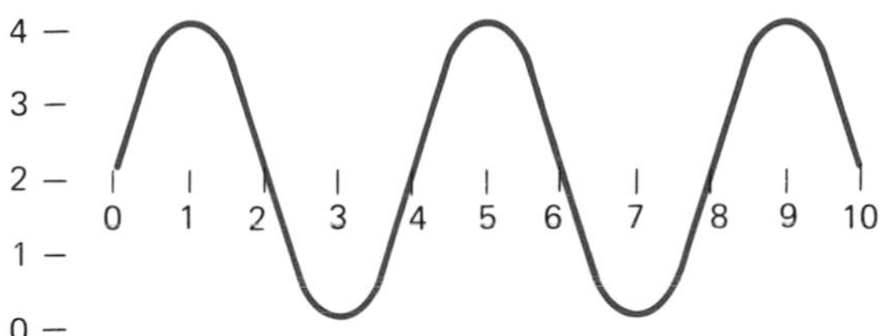

Fig. 48.4

7. The waveform of a p.d. applied internally to the X-plates of an oscilloscope is shown in Fig. 48.5. What is the function of this waveform?

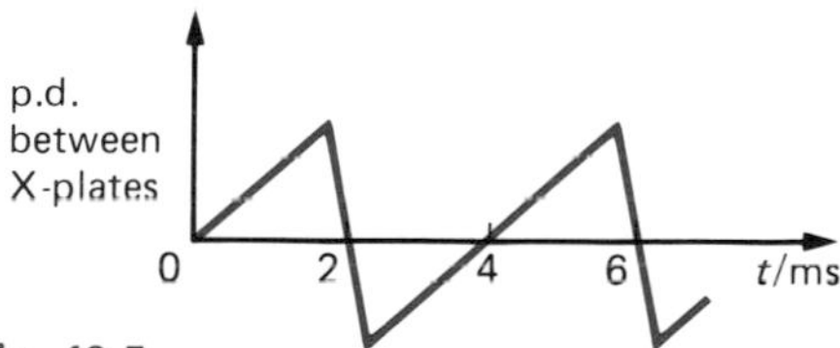

Fig. 48.5

Sketch the appearance of the trace on the screen of the oscilloscope of a sinusoidal alternating p.d. of frequency 1 kHz applied to the Y-input (with the p.d. of Fig. 48.5 on the X-plates).

8. (a) Explain the terms (i) frequency, (ii) phase difference. State *three* uses of an oscilloscope. Explain briefly how, in one of these applications, measurements are made.

(b) A sinusoidal voltage of frequency 300 Hz is applied to the X-plates of an oscilloscope, and a sinusoidal voltage of frequency f is applied to the Y-plates. What is f when the trace on the screen is that shown in (i) Fig. 48.6a, (ii) 48.6b? (C.)

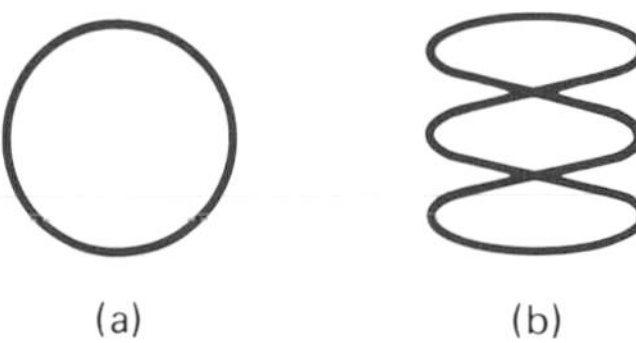

Fig. 48.6

9. (a) Explain the action of (i) the time base, (ii) the synchronization control, of a cathode ray oscilloscope.

(b) How would you use an oscilloscope to measure (i) a d.c. voltage, (ii) the r.m.s. value of a sinusoidal waveform and (iii) the period of a square wave?

10. A sine wave p.d. of peak value 6 V is applied to the circuit of Fig. 48.7. An oscilloscope, set for d.c. input, is connected across YE so that when Y is positive relative to E, the spot is deflected upwards.

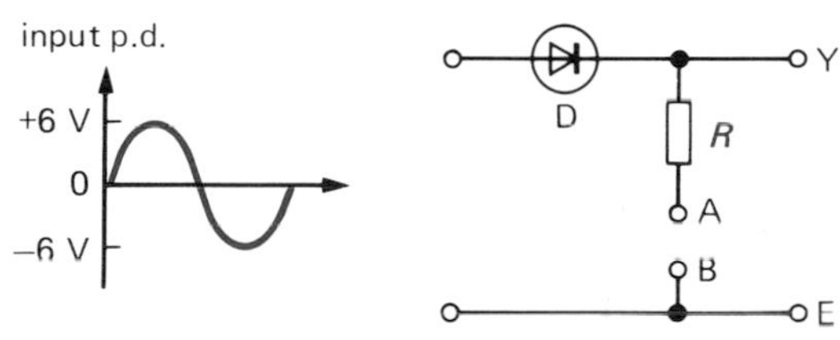

Fig. 48.7

Sketch the waveform you would see when (a) A and B are connected by a wire and (b) A is made 3 V positive relative to B. Mark voltage values on the waveforms. (Ignore the voltage drop across D when it is conducting.)

Analogue electronics

49 Transistor voltage amplifiers I

Introduction

(a) Analogue electronics. This is one of the two main branches of electronics, the other is *digital electronics* (p. 132). In analogue electronics, the signals being processed are electrical representations (analogues) of physical quantities which vary continuously over a range of values. The information they carry, e.g. the loudness and pitch of a sound, is in the amplitude and shape of their waveforms, Fig. 49.1. Most audio and radio signals are in this category.

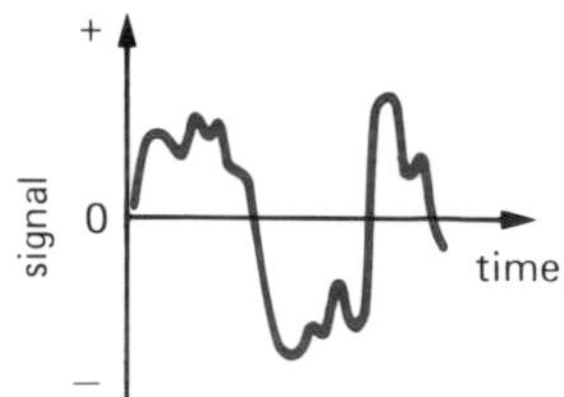

Fig. 49.1

Analogue circuits are *amplifier-type* circuits, designed so that a linear relationship exists between the input and output signals, i.e. input and output are directly proportional. They are also called *linear* circuits.

(b) Amplifiers. In general, the job of an amplifier is to produce an output which is an enlarged copy of the input. Amplification of one or more of the input voltage, current and power occurs. In the last case, the extra power at the output is provided by an external energy source, e.g. a battery. An amplifier differs from a transformer in that it can have a power gain greater than 1.

An *audio frequency* (a.f.) amplifier amplifies a.c. signals in the audio frequency range (20 Hz to 20 kHz), and is the type considered in this and the next two chapters. *Radio frequency* (r.f.) amplifiers (p. 181) operate above 20 kHz, at radio and television signal frequencies.

The symbol for an amplifier is shown in Fig. 49.2; in effect it has three terminals, viz. input, output and common. 'Common' or 'ground' is connected to

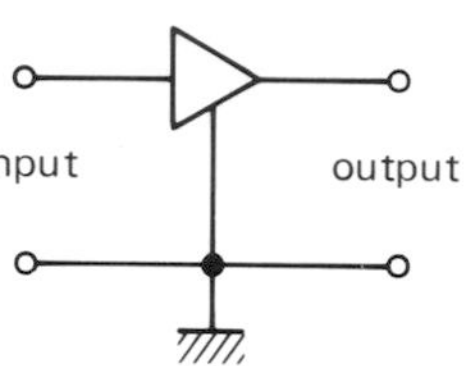

Fig. 49.2

one side of input, output and the power supply (not shown). It is usually taken as the reference point i.e. 0 V, for all circuit voltages and if connected to earth (symbol ⏚), it is called 'earth'.

Voltage amplification by a transistor

Many transducers (p. 42) produce a.c. voltages which must be amplified before use. The junction transistor, despite being a current-amplifying device, is used mainly as a voltage amplifier.

(a) Action. The basic circuit is the same as the one in which a transistor with a collector load R_L acts as a switch (Fig. 31.1, p. 65) and is given again with its voltage characteristic in Fig. 49.3. The latter shows that if the input voltage V_i changes from 0.8 to 1.2 V, the corresponding variation of the output voltage V_o is from 5 to 1 V. That is, a change of 0.4 V in V_i produces a change of 4 V in V_o.

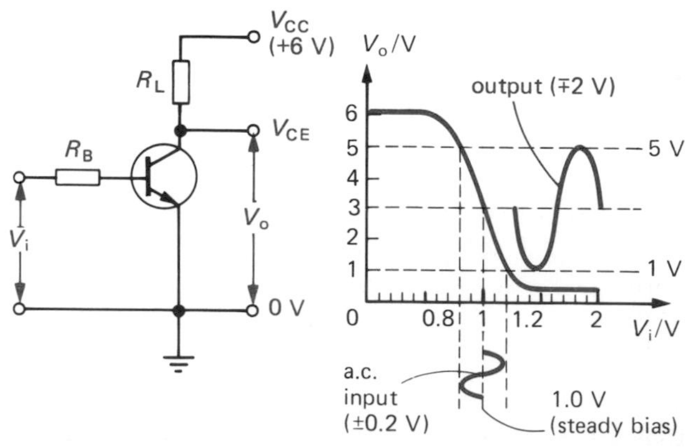

Fig. 49.3

The *voltage gain A* is given by

$$A = \frac{\text{change in } V_o}{\text{change in } V_i} = \frac{4}{0.4} = 10$$

(b) Bias. If small a.c. input voltages are to be amplified, a steady bias voltage must also be applied to the base to forward bias the base-emitter junction. It should produce a base current which creates a collector current that sets the collector voltage (V_{CE}) at about *half the supply voltage* (V_{CC} = 6.0 V) in the absence of an a.c. input voltage (called the 'quiescent' state). This allows V_{CE} to have its maximum 'swing' capability from (in theory) 0 V to V_{CC} when the a.c. input is applied.

In the case we are considering, a base bias of +1 V makes V_{CE} = +3 V. An a.c. input of ±0.2 V peak superimposed in this bias, would cause V_{CE} to 'swing' down to 1 V and up to 5 V. The output variation would be ∓2 V and if the characteristic is linear over this range, the output is a ten times amplified undistorted copy of the a.c. input.

The distortion occurring in the output when the bias voltage is too low (or too high) or the a.c. input too large, is shown by the graphs in Fig. 49.4a,b.

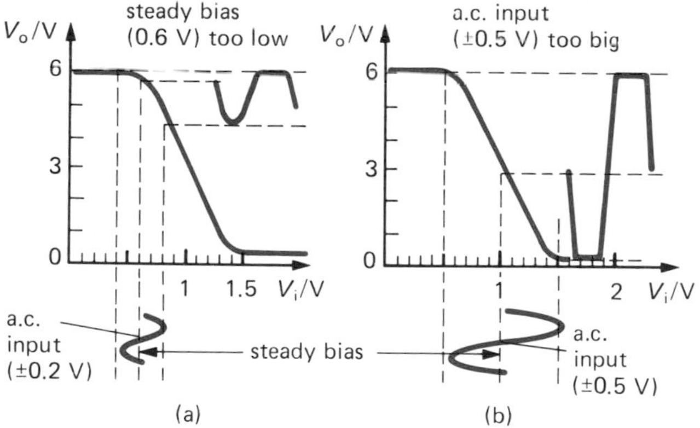

Fig. 49.4

(c) Further points. The transistor and load *together* bring about voltage amplification. It is the small variations of base current which cause the larger variations of collector current to produce varying voltages across the load.

The output is 180° out of phase with the input so that if the input increases, the output decreases, as the sine wave graphs in Fig. 49.3 show.

Common emitter amplifier

In this circuit, Fig. 49.5, the bias for the base is conveniently obtained via a resistor R_B connected to the collector power supply V_{CC}. In the quiescent state a steady d.c. base current I_B flows from V_{CC}, through R_B into the base and back to 0 V via the emitter.

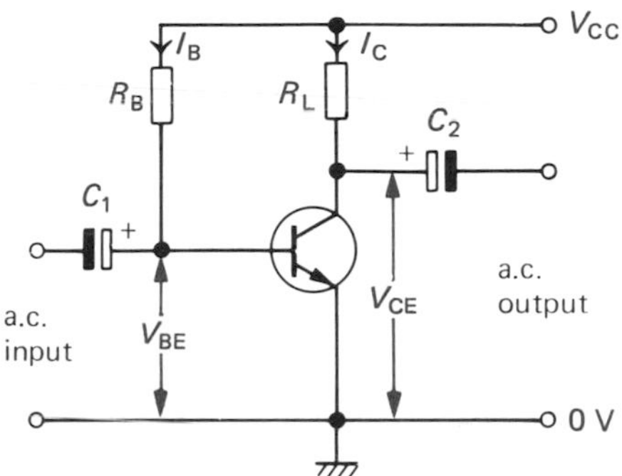

Fig. 49.5

(a) Calculation of R_L and R_B. Suppose (i) the n-p-n transistor is a silicon type which works satisfactorily on a quiescent (no a.c. input) collector current I_C of 3 mA, (ii) h_{FE} = 100 and (iii) V_{CC} = 6 V.

The collector-emitter circuit equation is (p. 66)

$$V_{CC} = I_C R_L + V_{CE}$$

That is,

$$R_L = (V_{CC} - V_{CE})/I_C \qquad (1)$$

For an undistorted, amplified output V_{CE} must be $\frac{1}{2}V_{CC}$, i.e. 3 V.

Substituting values in (1) we get

$$R_L = (6-3)\text{V}/3\text{ mA} = 1\text{ k}\Omega$$

Since $h_{FE} = I_C/I_B$ we have

$$I_B = I_C/h_{FE} = 3\text{ mA}/100 = 0.03\text{ mA}$$

The base-emitter circuit equation is

$$V_{CC} = I_B R_B + V_{BE}$$

where V_{BE} = 0.6 V.

Rearranging gives

$$R_B = (V_{CC} - V_{BE})/I_B \qquad (2)$$

Substituting values in (2) and we get

$$R_B = (6-0.6)\text{ V}/0.03\text{ mA} = 5.4\text{ V}/0.03\text{ mA}$$

$$= 540/3 = 180\text{ k}\Omega$$

(b) Coupling capacitors. When an a.c. input voltage is applied, the collector-emitter voltage becomes a varying direct voltage and may be regarded as an alternating voltage superimposed on a steady direct voltage, i.e. on the quiescent value of V_{CE}, Fig. 49.6. Only the a.c. part is wanted and capacitor C_2 blocks the d.c. part but allows the a.c. part to pass, i.e. it couples the a.c. to the next stage of the circuit. Capacitor C_1 stops any external d.c. voltage upsetting the quiescent base voltage but couples the a.c. input voltage to the transistor (p. 28).

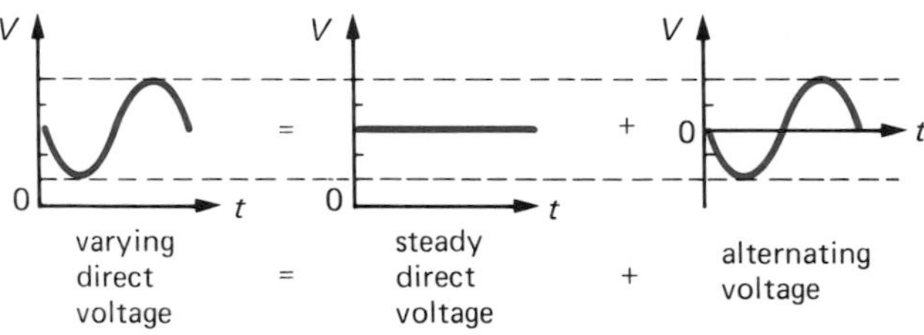

Fig. 49.6

C_1 and C_2 have a low reactance ($X_C = 1/(2\pi fC)$) at the audio frequencies involved. For frequencies down to 10 Hz, 10 μF electrolytics are suitable, connected with the polarities shown.

Summing up, a transistor acts as a voltage amplifier if (i) it has a suitable collector load and (ii) it is biased so that the quiescent value of $V_{CE} \approx \frac{1}{2} V_{CC}$.

(c) Frequency response. The *voltage gain A* of a capacitor-coupled amplifier is fairly constant over most of the a.f. range but it falls off at the lower and upper limits. At low frequencies, the reactances of the coupling capacitors increase and less of the low frequency part is passed on. At high frequencies various stray capacitances in the circuit can cause the fall.

A typical *voltage gain-frequency* curve is shown in Fig. 49.7. (To fit in the large frequency range, frequencies are not plotted on the usual linear scale but

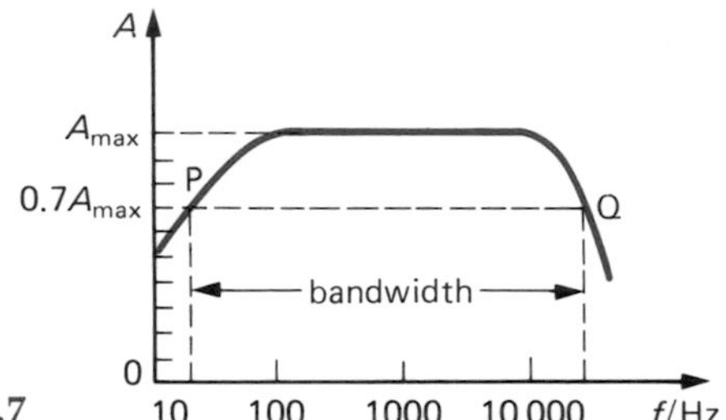

Fig. 49.7

on a logarithmic scale in which equal divisions represent equal changes in the 'log of the frequency f'.) The bandwidth is the range of frequencies within which A does not fall below $1/\sqrt{2}$ (i.e. 0.7 of its maximum value A_{max}). The two points P and Q at which this happens are called the '3 dB' points. (The decibel (dB) scale compares signal levels.)

Questions

1. In the circuit of Fig. 49.5 what is the purpose of (i) R_L, (ii) R_B, (iii) C_1 and (iv) C_2?

2. In the circuit of Fig. 49.5 if $V_{CC} = 9$ V, $R_L = 2$ kΩ, $I_C = 2$ mA and $h_{FE} = 100$ for the transistor, calculate (a) V_{CE}, (b) I_B and (c) R_B if $V_{BE} = 0.6$ V.

3. Explain the following terms: (a) analogue electronics, (b) voltage amplifier, (c) a.f. amplifier, (d) 'common' or 'ground', (e) voltage gain and (f) quiescent state.

50 Transistor voltage amplifiers II

Load lines

When designing a voltage amplifier it is important to ensure that as well as obtaining the desired voltage gain, there is minimum distortion of the output. The choice of the quiescent or *d.c. operating point* (i.e. the values of I_C and V_{CE}) determines whether these requirements will be met and is made by constructing a *load line*.

(a) Drawing a load line. The output characteristics of a transistor (Fig. 32.2c, p. 69) show the relation between V_{CE} and I_C with *no load* in the collector circuit. With a load R_L, the equation connecting them is (p. 66):

$$V_{CC} = I_C R_L + V_{CE}$$

Rearranging we get:

$$V_{CE} = V_{CC} - I_C R_L \quad (1)$$

Knowing V_{CC} and R_L this equation enables us to calculate V_{CE} for different values of I_C. If a graph of I_C (on the y-axis) is plotted against V_{CE} (on the x-axis), the straight line so obtained is the *load line*. It can be drawn knowing just two points, the easiest being the end points A and B where the line cuts the V_{CE}- and I_C-axes.

For A we put $I_C = 0$ in (1) and get $V_{CE} = V_{CC} = 6\,V$ (say).

For B we put $V_{CE} = 0$ in (1) and get $I_C = V_{CC}/R_L$. If $R_L = 1\,k\Omega$ say, then $I_C = 6\,V/1\,k\Omega = 6\,mA$.

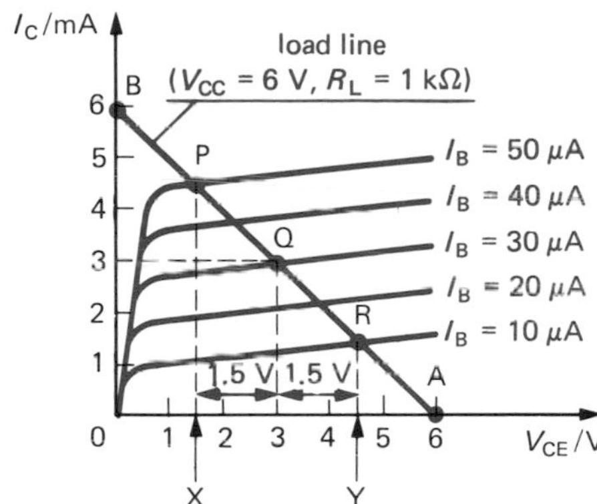

Fig. 50.1

In Fig. 50.1, AB is the load line for $V_{CC} = 6\,V$ and $R_L = 1\,k\Omega$. It is shown superimposed on the output characteristics of the transistor. We can regard a load line as the output characteristic of the *transistor and load* for particular values of V_{CC} and R_L. Different values of either give a different load line.

(b) Choosing the d.c. operating point. The best 'position' for the d.c. operating point is near the *middle* of a load line which cuts the output characteristics where they are straight and equally spaced. Otherwise the output has a distorted waveform and its 'swing' capability (from near V_{CC} to near 0 V) is reduced.

If Q is chosen in Fig. 50.1, the quiescent value of V_{CE} is 3 V (i.e. half V_{CC}) and I_C is 3 mA. The value of I_B which gives these values is seen from the output characteristic passing through Q to be 30 μA. R_B can then be calculated as before (p. 105).

(c) Voltage gain A. This is given by

$$A = \frac{\text{output voltage}}{\text{input voltage}} = \frac{\text{change in collector voltage}}{\text{change in base voltage}}$$

It can be obtained from the load line by noting that when the input causes I_B to vary from 10 to 50 μA (from R to P), V_{CE} varies from 4.5 to 1.5 V (from Y to X). If the input characteristic of the transistor (Fig. 32.2b, p. 69) shows that, for example, V_{BE} has to change by 40 mV (0.04 V) to cause a 40 μA change in I_B, then

$$A = \frac{4.5-1.5}{0.04} = \frac{3.0}{0.04} = 75$$

Stability

The simple common emitter voltage amplifier circuit of Fig. 49.5 (p. 105) has two serious defects.

(a) Effect of h_{FE}. The first arises from the fact that it does not work satisfactorily if h_{FE} varies widely (as it does even for transistors of the same type, p. 70) from the value used when designing it.

For example, if h_{FE} is greater, then I_C is greater and causes a larger voltage drop across the collector load R_L. Consequently the quiescent value of V_{CE} is much less than half V_{CC}, thus upsetting the operating point and in this case reducing the swing capability in the negative-going direction. If h_{FE} is smaller, the swing capability in the positive-going direction becomes smaller. In both events the output is distorted.

(b) Thermal runaway. The second problem occurs when the temperature of the transistor (junction) rises, due to the heating effect of I_C or to an increase in the surrounding (ambient) temperature. There is greater vibration of the semiconductor atoms and more free electrons and holes are produced. This increases I_C and causes further heating and so on until the transistor is destroyed by 'thermal runaway'. Transistors carrying large currents are therefore used with heat sinks (p. 88) to minimize this effect.

In an effort to compensate automatically for increases of I_C (for whatever reason), and so stabilize the d.c. operating point, special bias circuits have been developed.

Collector-to-base bias

This is the simplest circuit but it is a useful general purpose voltage amplifier which gives adequate stability. The improvement is achieved by halving the value of R_B in the circuit of Fig. 49.5 and connecting it between the collector and base as in Fig. 50.2 rather than between V_{CC} and base. The quiescent value of I_B then depends on the quiescent value of V_{CE}.

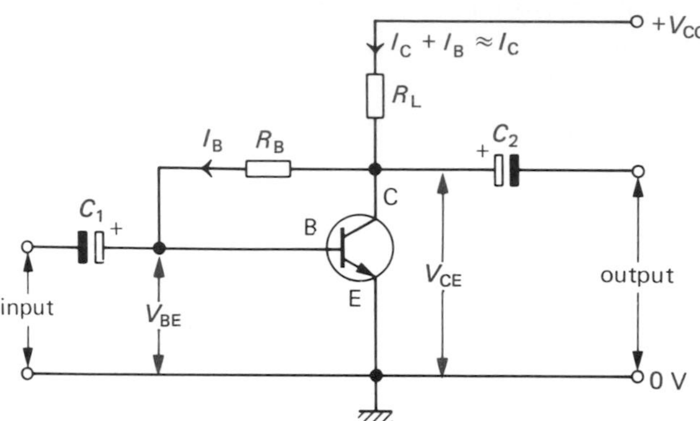

Fig. 50.2

For the circuit in Fig. 50.2 we can write (since I_C is much greater than I_B):

$$V_{CC} = I_C R_L + V_{CE} \quad (2)$$

where

$$V_{CE} = I_B R_B + V_{BE} \quad (3)$$

From (2) you can see that if I_C increases, V_{CE} decreases since V_{CC} is fixed. From (3) it therefore follows that since V_{BE} is constant (0.6 V or so for silicon), I_B must also decrease and in so doing tends to bring back I_C to its original value.

Taking the quiescent conditions as $V_{CE} = \frac{1}{2} V_{CC} = 3\,V$, $I_C = 3\,mA$ and $I_B = 0.03\,mA$ (as in the 'Calculation for R_L and R_B' on p. 105), the value of R_B in Fig. 50.2 is found by rearranging equation (3) to give:

$$R_B = \frac{V_{CE} - V_{BE}}{I_B} = \frac{(3-0.6)\,V}{0.03\,mA}$$

$$= 2.4\,V/0.03\,mA = 80\,k\Omega$$

This is about half the value of 180 kΩ for R_B in the unstabilized circuit of Fig. 49.5.

Questions

1. The load line for the amplifier in Fig. 50.2 is shown in Fig. 50.3.

(a) Use it to find the values of V_{CC} and R_L.

(b) If Q is the d.c. operating point, what are the quiescent values of V_{CE}, I_C and I_B?

(c) Calculate R_B if the transistor is a silicon type.

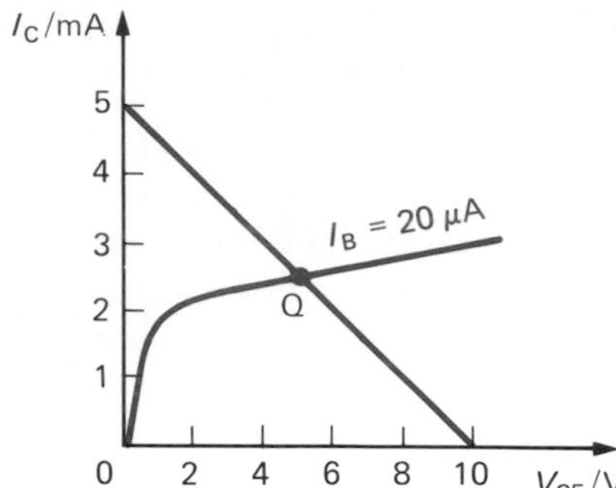

Fig. 50.3

2. The output characteristics of a junction transistor in common-emitter connection are shown in Fig. 50.4. The transistor is used in an amplifier with a 9 V supply and a load resistor of 1.8 kΩ.

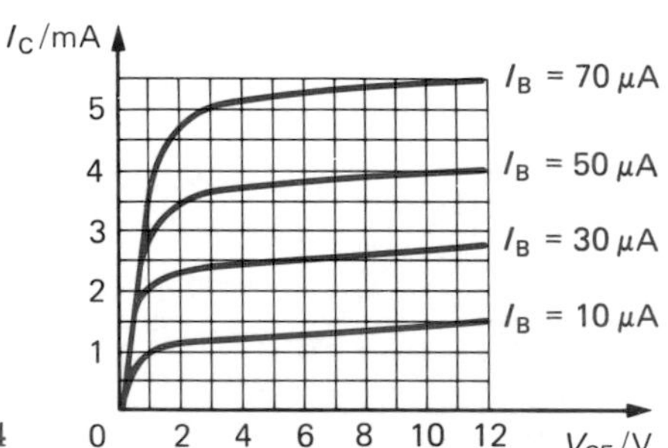

Fig. 50.4

(a) What is the value of I_C when $V_{CE} = 0$?

(b) Lay a ruler along the load line and choose a suitable d.c. operating point. Read off the quiescent values of I_C, I_B and V_{CE}.

(c) What is the quiescent power consumption of the amplifier?

(d) If an alternating input voltage varies the base current by $\pm 20\,\mu A$ about its quiescent value, what is (i) the variation in the collector-emitter voltage and (ii) the peak output voltage?

(e) An input characteristic of the transistor is given in Fig. 50.5. Use it to find the base-emitter voltage variation which causes a change of $\pm 20\,\mu A$ in the quiescent base current.

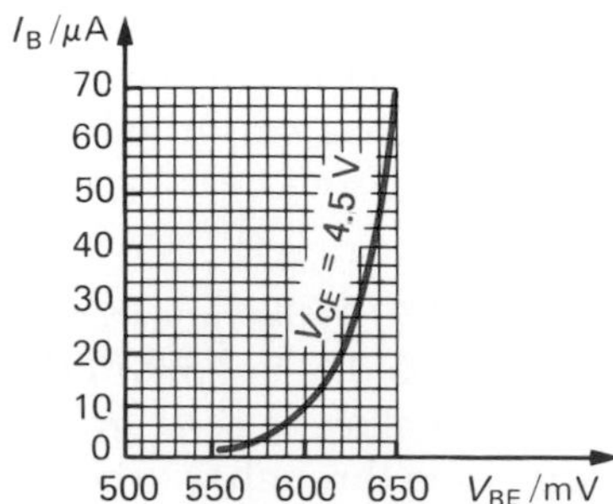

Fig. 50.5

(f) Using your answers from (d) and (e) find the voltage gain A of the amplifier.

(g) If the amplifier uses the collector-to-base bias circuit of Fig. 50.2, calculate the value of R_B to give the quiescent value of I_B. (Assume $V_{BE} = 0.6\,V$.)

51 Transistor voltage amplifiers III

Simple two-stage amplifier

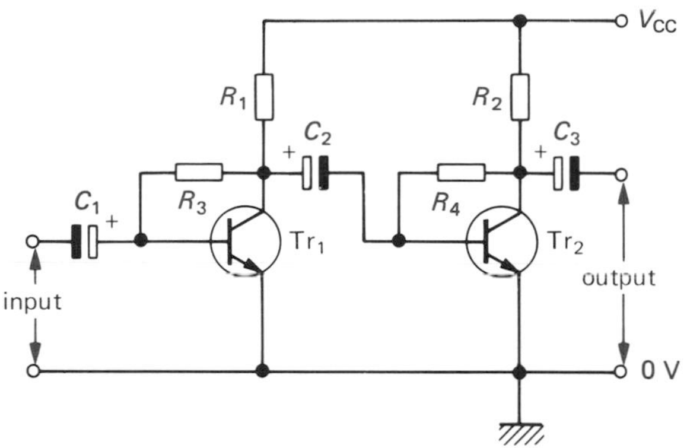

Fig. 51.1

To obtain greater gain, two or more amplifier stages are coupled. The circuit in Fig. 51.1 is for a two-stage capacitor-coupled a.f. voltage amplifier using collector-to-base bias. The output from Tr_1 is applied to the input of Tr_2 via C_2 which connects the collector of Tr_1 (it needs a quiescent d.c. voltage of $\frac{1}{2}V_{CC}$ for correct operation) and the base of Tr_2 (it is +0.6 V above the emitter at 0 V, being a forward biased junction). A direct connection, without C_2, would fix the collector of Tr_1 at only +0.6 V and would also allow a base current of several mA through Tr_2 (via R_1), saturating it permanently.

Fully-stabilized voltage amplifier

When complete stabilization of the d.c. operating point is required, *voltage divider and emitter resistor bias* is used. The circuit for a junction transistor is shown in Fig. 51.2. It has *three* main features.

(a) Voltage divider (R_1, R_2). The junction of R_1 and R_2 fixes the base *voltage* at a value sufficient to forward bias the base-emitter junction. (In collector-to-base bias the base *current* was fixed.) R_1 and R_2 are chosen so that the current through them is about ten times greater than the quiescent base current so

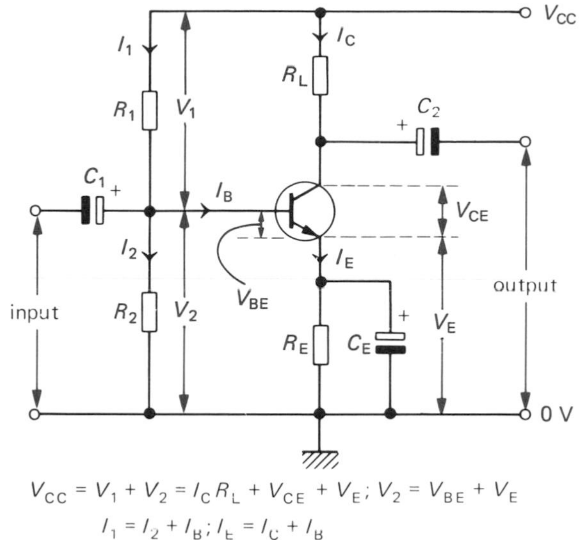

$V_{CC} = V_1 + V_2 = I_C R_L + V_{CE} + V_E;\ V_2 = V_{BE} + V_E$

$I_1 = I_2 + I_B;\ I_E = I_C + I_B$

Fig. 51.2

that if the latter changes the base voltage is hardly affected (see p. 16).

(b) Emitter resistor (R_E). If I_C increases (due to h_{FE} increasing or a temperature rise), then I_E also increases, as does V_E (= $I_E R_E$). The voltage V_2 across R_2 equals $V_{BE} + V_E$. Therefore if V_2 is fixed (by the voltage divider), V_{BE} must *decrease* when V_E increases. As a result I_B decreases, the fall being enough to restore I_C to its original value.

(c) Decoupling capacitor (C_E). When an a.c. input is applied, I_C and I_E become varying d.c., i.e. they consist of steady d.c. + a.c. The large capacitor C_E across R_E offers an easy path for the a.c. The combined impedance of C_E and R_E in parallel is negligible to the a.c. Without C_E an a.c. voltage would be developed across R_E and reduce the gain (see p. 112).

(d) Further point. The output voltage can only swing between V_{CC} and V_E (not between V_{CC} and 0 V). The value of R_E is therefore chosen to make V_E about one-fifth V_{CC}, e.g. 1 V on a 6 V supply.

52 FET voltage amplifier

The transfer characteristic of a FET (p. 71) shows that changes in the gate voltage V_{GS} cause changes in the drain current I_D. In a FET a.f. voltage amplifier, Fig. 52.1, these changes are converted into larger voltage changes by a load resistor R_L in the drain (output) circuit. The values of R_L and the supply voltage V_{DD} (note the symbol: V_{CC} is used for the supply voltage to the collector of a bipolar transistor) are both higher than for a bipolar transistor to obtain reasonable gain. (Typical values are 22 kΩ and 18 V.)

Fig. 52.1

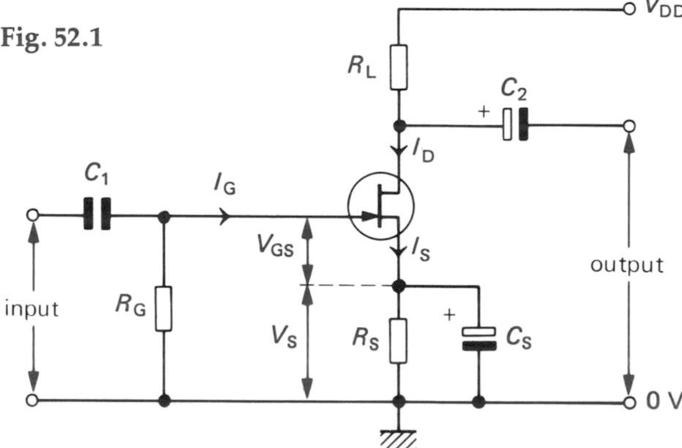

FET voltage amplifiers give lower gains than junction transistor types but they draw negligible currents from the a.c. input source, i.e. they have a high input resistance. They are therefore suited to operation with devices such as crystal microphones (p. 43) which can only supply small currents (because of their high internal resistance).

Bias and stability

(a) Bias. The d.c. operating point for distortionless amplification is chosen from the load line (p. 106). It is realized in practice by applying the correct quiescent bias voltage to the gate which, as we saw earlier (p. 71), should be negative relative to the source.

In the quiescent state the source current I_S $(=I_D + I_G \approx I_D$ since $I_G \approx 0)$ is steady and causes a voltage drop across the source resistor R_S (typically 8.2 kΩ). The source end of R_S is therefore positive with respect to the other end connected to ground. R_G (e.g. 2.2 MΩ) ensures that the gate has the same potential as the lower end of R_S in the circuit. This is so because there is negligible current through R_G (i.e. $I_G \approx 0$) and hence practically no voltage across it. Both ends of R_G have the same potential, namely that of the lower end of R_S, i.e. 'ground' (0 V). The source is thus at a higher potential than the gate, i.e. the gate is negative relative to the source.

(b) Stability. The circuit automatically compensates for any change of I_S and helps to stabilize the d.c. operating point because any increase in I_S increases the voltage V_S $(=I_S R_S)$ across R_S. The potential of the source end of R_S (and so of the source) rises and since the lower end of R_S and the gate are tied to 0 V, the gate-source voltage V_{GS} goes more negative, tending to reduce I_S to its previous value.

Decoupling and coupling

(a) Decoupling. The large capacitor C_S (e.g. 100 μF) provides a bypass round R_S for the a.c. part of the source current (which becomes a varying d.c.) when the a.c. input is applied. Otherwise a varying voltage would be developed across R_S and reduce the gain (see p. 112).

(b) Coupling. C_1 blocks any d.c. voltage at the input (and stops it upsetting the quiescent bias) but couples the a.c. signal. The voltage this develops across R_G is applied to the gate for amplification and for it to be large, R_G should be large compared with the reactance of C_1. Since it is, C_1 can be small, e.g. 0.1 μF.

53 Amplifiers and feedback

The performance of an amplifier can be changed by feeding part or all of the output back to the input. The feedback is *positive* if it is in phase with the input, i.e. adds to it so that it reinforces changes at the output, as in Fig. 53.1a. As a result the overall gain of the amplifier increases.

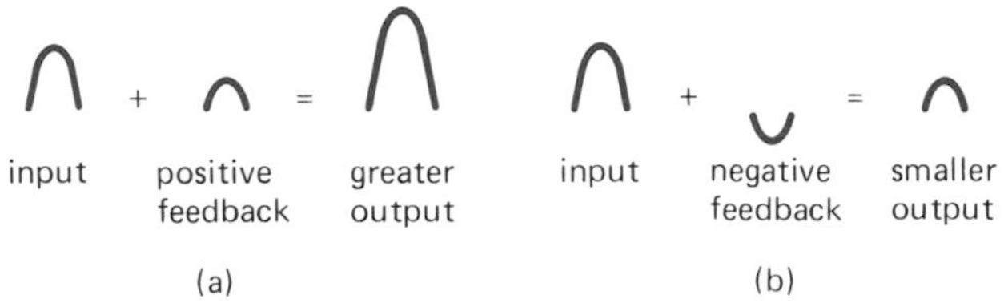

input + positive feedback = greater output

(a)

input + negative feedback = smaller output

(b)

Fig. 53.1

The feedback is *negative* if it is 180° out of phase (in antiphase) with the input, i.e. subtracts from it, as in Fig. 53.1b. In this case the gain is reduced but as we see presently, other advantages arise.

The feedback equation

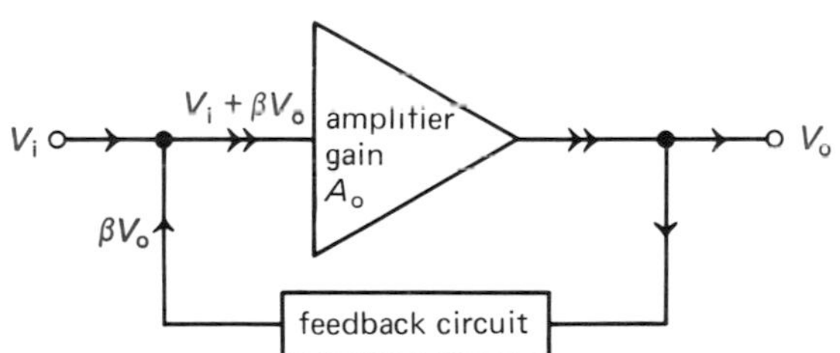

Fig. 53.2

Suppose the amplifier in Fig. 53.2 has a voltage gain of A_o, called the *open-loop gain*, with no feedback. If a fraction β of the output V_o is fed back so as to add to the *actual* input signal to be amplified V_i (i.e. positive feedback), then the *effective* input to the amplifier becomes $V_i + \beta V_o$. β is the *feedback factor*.

The gain of the amplifier itself is still A_o and so

$$V_o = A_o(V_i + \beta V_o)$$

Rearranging

$$V_o - \beta A_o V_o = A_o V_i$$

That is

$$V_o(1 - \beta A_o) = A_o V_i$$

$$\therefore \frac{V_o}{V_i} = \frac{A_o}{1 - \beta A_o}$$

The voltage gain A of the whole amplifier with feedback, called the *closed-loop gain*, is given by

$$A = \frac{V_o}{V_i}$$

Hence

$$A = \frac{A_o}{1 - \beta A_o} \qquad (1)$$

This is the general equation for an amplifier with feedback, and as it is, applies to positive feedback, i.e. β is positive. It shows that (i) $A > A_o$ and (ii) if $\beta A_o = 1$, $A = A_o/0$, i.e. A is infinitely large. The amplifier then gives an output with no input, which is what oscillators (p. 116) do using positive feedback.

For negative feedback, β is negative and the equation becomes

$$A = \frac{A_o}{1 + \beta A_o} \qquad (2)$$

In this case $A < A_o$ and if $\beta A_o \gg 1$, then

$$A \approx \frac{A_o}{\beta A_o} \approx \frac{1}{\beta} \qquad (3)$$

This result has important consequences.

Advantages of negative feedback (n.f.b.)

Although n.f.b. reduces the gain of an amplifier and makes more stages of amplification necessary, it has several very desirable results, provided $A_o \gg A$.

(i) *Voltage gain is accurately predictable and constant* because, as equation (3) shows, A does not depend on A_o (which can vary widely from one amplifying device, e.g. a transistor, to another, even of the same type). A depends only on β and this is usually determined by the values of two resistors, often in a potential divider. Resistors of high stability and known accuracy are available (p. 23) and by using n.f.b., these qualities can also become those of the gain of the whole amplifier.

(ii) *Distortion of the output is reduced*, i.e. the output waveform is a truer copy of the input waveform.

(iii) *Frequency response is better*, i.e. a wider range of frequencies is amplified by the same amount, so increasing the bandwidth (p. 106). (It is roughly true that *gain* × *bandwidth* is constant.)

Negative feedback circuits

(a) Collector-to-base resistor. The collector-to-base bias circuit of Fig. 50.2 is shown again in Fig. 53.3. R_B not only provides the correct quiescent d.c. bias to the base-emitter junction (as well as stabilizing the d.c. operating point by what is in effect d.c. negative feedback), but it also supplies negative feedback for the a.c. input as follows.

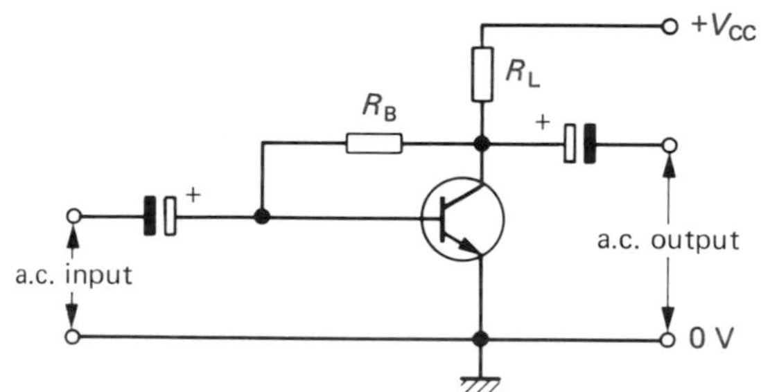

Fig. 53.3

When an alternating input is applied, an alternating current is fed back through R_B to the base, being superimposed on the d.c. base current. It is in antiphase with the alternating input because the output voltage in a common-emitter amplifier is 180° out of phase with the input (see p. 105), i.e. the feedback is negative.

It can be shown that the feedback factor β is given by

$$\beta = \frac{R_L}{R_B}$$

If the value of R_B to provide the correct d.c. bias is not the same as that to give the required amount of a.c. negative feedback, the bias is obtained by other means (e.g. by the 'voltage divider and emitter resistor' method) and a capacitor is joined in series with R_B to stop d.c. reaching the base.

(b) Source resistor. The circuit is shown in Fig. 53.4 for an FET voltage amplifier. It is the same as Fig. 52.1 but with the source decoupling capacitor C_S omitted.

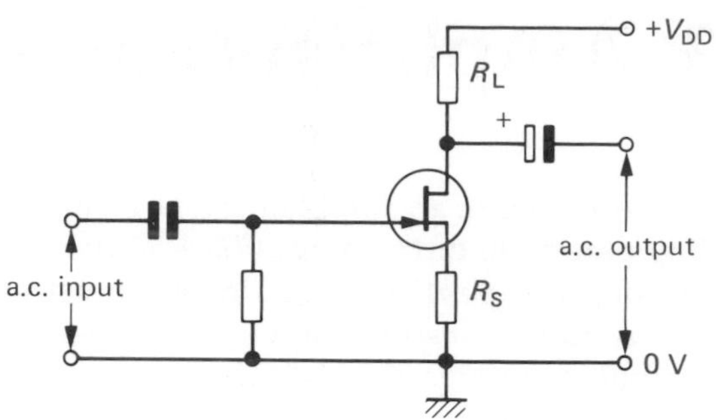

Fig. 53.4

When an alternating input voltage is applied, alternating components are superimposed on the d.c. drain and source currents. The one on the drain current creates the output voltage across R_L; the one on the source current produces an alternating voltage across the source resistor R_S which is fed back to the gate in antiphase with the input voltage (since there is no bypass capacitor to decouple R_S as in Fig. 52.1). The action is briefly as follows.

A positive half-cycle of the input makes the gate less negative, the source current increases, as does the p.d. across R_S. The source therefore becomes more positive with respect to the gate (as explained on p. 110), i.e. the gate tends to go more negative and thereby reduce the effective input to the amplifier.

It can be shown that β is given by

$$\beta = \frac{R_S}{R_L}$$

With a junction transistor the n.f.b. resistor is in the emitter circuit.

Questions

1. Explain (a) positive feedback and (b) negative feedback.

2. (a) State *one* disadvantage and *three* advantages of n.f.b.
(b) Write down the n.f.b. equation, stating the meaning of each symbol.

3. The open-loop gain of an amplifier is 200. What is the closed-loop gain when the negative feedback factor is 1/50?

4. (a) Calculate β for Fig. 53.3 if $R_L = 1\,\text{k}\Omega$ and $R_B = 100\,\text{k}\Omega$.
(b)What is β for Fig. 53.4 if $R_L = 10\,\text{k}\Omega$ and $R_S = 1\,\text{k}\Omega$?

54 Amplifiers and matching

Input and output impedance

An amplifier has to be 'matched' to the transducer or circuit supplying its input or receiving its output. Usually this means ensuring that either the maximum voltage or the maximum power is transferred to or from the amplifier. In all cases, the input or output impedance (since we are concerned with a.c.) of the amplifier is an important factor.

The *input impedance* Z_i equals V_i/I_i where I_i is the a.c. flowing into the amplifier when voltage V_i is applied to the input. It depends not only on the a.c. input resistance r_i (p. 69) of the transistor but also on the presence of capacitors, resistors, etc. in the amplifier circuit. In effect, the amplifier behaves as if it had an impedance Z_i connected across its input terminals.

The *output impedance* Z_o is the a.c. equivalent of the internal or source resistance of a battery (p. 17). It causes a 'loss' of voltage at the output terminals when the amplifier is supplying current. We can think of an amplifier as an a.c. generator of voltage V which, on open circuit equals the voltage V_o at the output terminals. On closed circuit, when an output current flows, V_o is less than V by the voltage dropped across Z_o, as in the d.c. case.

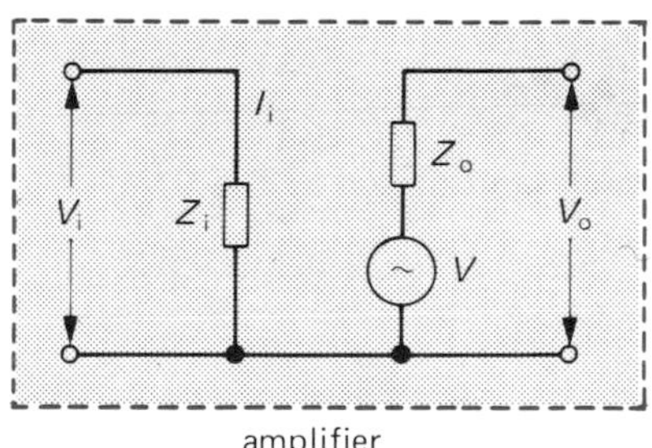

Fig. 54.1

The diagram in Fig. 54.1, called the *equivalent circuit* of an amplifier, is useful when considering matching problems. Z_i and Z_o can be measured (in ohms) by experiment.

Matching to signal and load

(a) Signal. The source supplying the a.c. signal to the *input* of an amplifier can also be regarded as an a.c. generator, of voltage V_S, having an output impedance Z_S, as shown by its equivalent circuit in Fig. 54.2. If the input current is I_i, we can write

$$V_S = I_i(Z_S + Z_i) \text{ where } V_i = I_iZ_i$$

It follows that

$$V_i = V_S\left(\frac{Z_i}{Z_S + Z_i}\right) \quad (1)$$

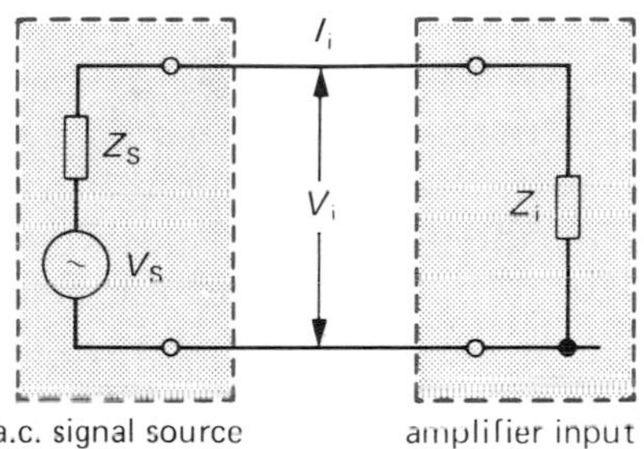

Fig. 54.2

The most common requirement is for *maximum voltage transfer* from the signal source to the input of the amplifier. That is, V_i should be as large as possible. Equation (1) shows that if Z_i is large compared with Z_S (say, ten times larger), then $V_i \approx V_S$ and very little of V_S is 'lost' across Z_S.

In practice, things may be very different if for example, a common emitter amplifier with an input impedance Z_i of 1 kΩ is supplied with an a.c. input signal from a crystal microphone having an output impedance Z_o of 1 MΩ. Only about 1/1000 of the voltage V_S generated by the microphone would be available at the amplifier input (since $Z_i/(Z_S + Z_i) \approx$ 1/1000). Obviously the impedance matching needs to be improved. Ways of doing so will be considered in the next chapter.

(b) Load. If the *output* of the amplifier supplies a load of impedance Z_L with current I_o, then from Fig. 54.3 we have

$$V = I_o(Z_o + Z_L) \text{ where } V_o = I_oZ_L$$

Hence

$$V_o = V\left(\frac{Z_L}{Z_o + Z_L}\right) \quad (2)$$

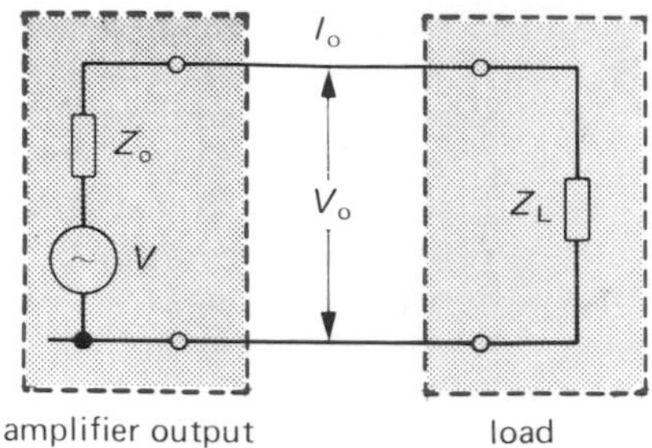

Fig. 54.3

For *maximum voltage transfer* equation (2) shows that (as in (a)), we need $Z_L \gg Z_o$ if the voltage loss across Z_o is to be small.

For *maximum power transfer* the maximum power theorem (p. 18) states that this occurs when $Z_L = Z_o$. But here too there are problems in practice. For example, a common emitter amplifier with its output connected directly to a loudspeaker with a typical impedance Z_L of 8Ω would not deliver maximum power to it because the output impedance Z_o of the amplifier is very much greater. Again steps need to be taken to improve matters.

Worked example

Calculate (a) the voltage across and (b) the power developed in the load Z_L in the circuit of Fig. 54.4 if V_S = 15 mV, Z_S = 500Ω and Z_L = 8Ω. The amplifier characteristics are Z_i = 1000Ω, A_o = 100 and Z_o = 8Ω.

(a) From equation (1), the voltage V_i applied to the input is

$$V_i = 15\left(\frac{1000}{500 + 1000}\right) = \frac{15 \times 2}{3} = 10\,\text{mV}$$

The amplifier output voltage V is given by

$$V = A_o \times V_i = 100 \times 10 = 1000\,\text{mV} = 1\,\text{V}$$

From equation (2), the voltage V_o across the load Z_L of 8Ω, is

$$V_o = 1\left(\frac{8}{8 + 8}\right) = 0.5\,\text{V}$$

(b) Power in load $= \dfrac{V^2_o}{Z_L} = \dfrac{(0.5)^2}{8} \approx 0.03\,\text{W}$

Questions

1. (a) Why does a signal generator designed to supply an alternating voltage *to* a circuit under test have a low output impedance (e.g. less than 100Ω)?
(b) Why does a CRO intended for studying voltages *from* a circuit under test have a high input impedance (e.g. 1 MΩ)?

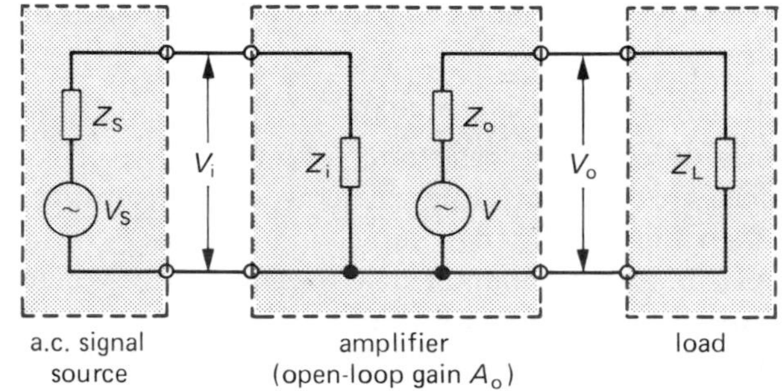

Fig. 54.4

2. In the circuit of Fig. 54.4 calculate
(a) V_i if V_S = 100 mV, Z_S = 1 kΩ and Z_i = 4 kΩ,
(b) V if A_o = 50,
(c) V_o if $Z_o = Z_L$ = 4Ω, and
(d) the power developed in Z_L.

55 Impedance matching circuits

Emitter-follower

An emitter-follower has a *high* input impedance (e.g. 0.5 MΩ) and a *low* output impedance (e.g. 30Ω) It can therefore provide 'step-down' impedance matching when connected between a high impedance signal source and a low impedance load. One of its main roles is as the output stage i.e. the power amplifier, of a multistage amplifier feeding a loudspeaker (p. 174).

The basic circuit is shown in Fig. 55.1; it uses 100% n.f.b.

(a) Action. The input V_i is applied between base and ground. The emitter resistor R_L, across which

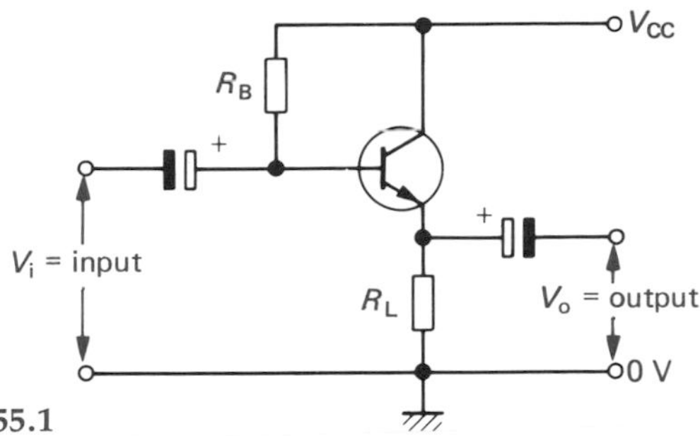

Fig. 55.1

the output voltage V_o is developed, is in the input circuit as well as the output circuit. As a result, V_o is in series with V_i but has opposite polarity; *all* V_o thus opposes V_i and acts as the n.f.b. voltage (since R_L is not decoupled) with a feedback factor $\beta = 1$. The input current is reduced (for the same V_i), thereby in effect increasing the input impedance of the circuit. R_B fixes the d.c. operating point.

The circuit gets its name from the fact that the emitter voltage 'follows' the input voltage closely since any increase in V_i increases V_o (due to the emitter current increasing).

The low output impedance arises from the fact that the input and output are connected by the forward biased, low resistance base-emitter junction. (In a common-emitter amplifier, the input and output circuits are separated by the high resistance of the reversed biased collector-base junction, giving it a much higher output impedance.)

(b) Voltage and current gain. The *voltage gain* is always just less than one, because V_i is applied across the voltage divider formed by the small resistance r_e of the base-emitter junction in series with the much larger resistance of R_L. V_o on the other hand is the p.d. across R_L alone and so is slightly smaller than V_i.

The *current gain* is high, as in the common emitter amplifier, since the emitter (output) current is greater than the base (input) current.

(c) Common collector amplifier. This is the other name given to the circuit. From the point of view of a.c., the collector is common to both input and output circuits, being connected to ground by the negligible resistance of the power supply.

Source-follower

This is the FET version of the emitter follower; it is also called the *common drain amplifier*. Due to the FET it has an even higher input impedance, e.g. 100 MΩ or more; the output impedance is also higher, e.g. 500 Ω. It makes a useful *buffer amplifier*, giving current amplification, to match a very high impedance input transducer such as a crystal microphone to a conventional amplifier.

The three transistor amplifier circuits

A transistor can be connected as an amplifier in three ways. The common emitter (CE) and common collector (CC) modes have been considered.

The third method is the *common base* (CB) circuit, which has a *low* input impedance (e.g. 50 Ω) and a *high* output impedance (e.g. 1 MΩ). The voltage gain is high but the current gain is just less than one. Fig. 55.2 shows the basic connections. Because it performs better at high frequencies than the common emitter amplifier it is used to amplify weak signals from a television aerial.

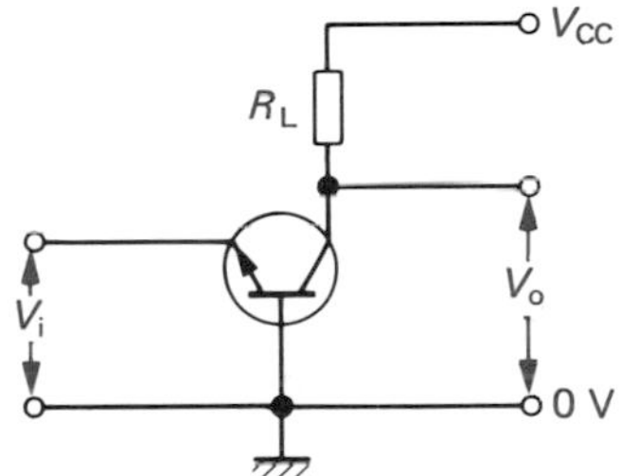

Fig. 55.2

The choice of amplifier for a particular job often depends on the impedance matching requirements. The main properties of the three one-stage transistor amplifiers are summarized in Table 55.1.

Table 55.1

Property	CE	CC	CB
Input impedance	medium	high	low
Output impedance	medium	low	high
Voltage gain	high	≈ 1	high
Current gain	high	high	≈ 1
Power gain	high	low	medium
Phase inversion (output w.r.t. input)	180°	0°	0°

Transformer matching

Transformers can be used either to step-up or step-down impedances.

Suppose a signal, produced by a 'generator' of out-

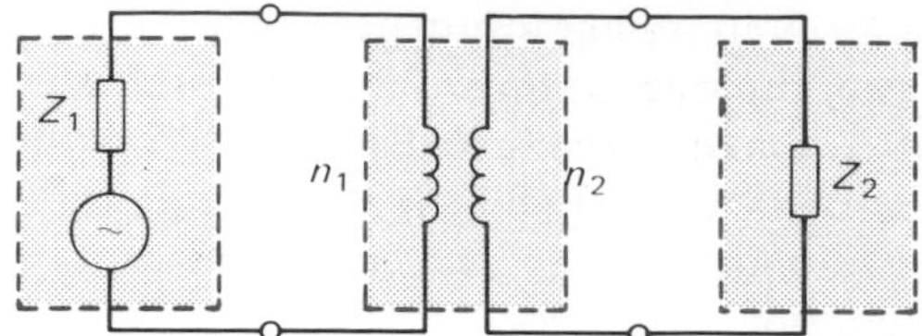

Fig. 55.3

put impedance Z_1, has to be matched to a load of input impedance Z_2. It can be shown that by connecting a transformer as in Fig. 55.3, Z_2 can be made to 'look' equal to Z_1, and the maximum power transferred, if the turns ratio n_1/n_2 is chosen so that

$$\frac{n_1}{n_2} = \sqrt{\frac{Z_1}{Z_2}}$$

Hence if an amplifier with an output impedance Z_1 is to be matched to a loudspeaker of lower impedance Z_2 by this method, a suitable *step-down* transformer is required since $Z_1 > Z_2$. On the other hand, if a transformer is used to match a microphone of low impedance Z_1 (e.g. a ribbon type) to the input of an amplifier of high impedance Z_2, a *step-up* ratio is necessary.

Transformer matching at radio frequencies is much more common than at audio frequencies because a.f. transformers tend to be large, heavy and expensive and also introduce distortion.

Questions

1. In the circuit of Fig. 55.4, why must the output voltage V_o be less than the input voltage V_i?

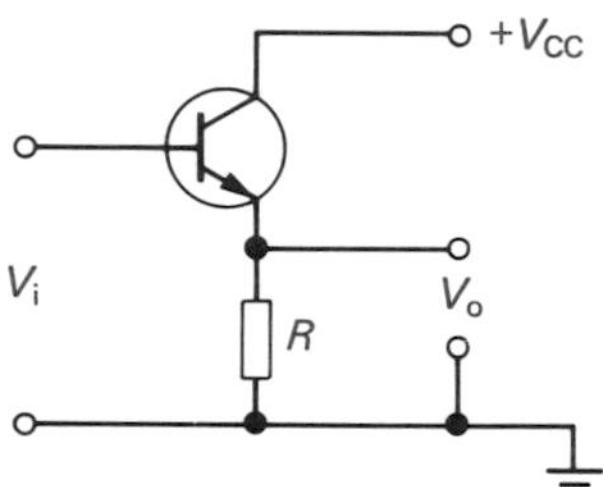

Fig. 55.4

Why does the circuit present to the input a resistance far greater than R? *(L.)*

2. An audio-frequency generator with an output impedance of 2 kΩ is used to test a loudspeaker of impedance 5Ω. What fraction of the open-circuit signal from the generator appears at the terminals of the loudspeaker if the connection is direct? If connection is made using a transformer as intermediary, what properties should it have to give optimum power transfer to the loudspeaker? *(O. and C. part qn.)*

56 Transistor oscillators

Introduction

An electronic oscillator generates a.c. having a sine, square or some other shape of waveform, depending on the particular circuit employed. Audio frequency (a.f.) oscillators are used in signal generators (p. 101), as are radio frequency (r.f.) types. The latter are also important in radio and television transmitters and receivers (pp. 180 and 184).

Most oscillators are *amplifiers with a feedback loop* from output to input which ensures that the feedback is (i) in phase with the input, i.e. is *positive*, and (ii) sufficient to make good the inevitable energy losses in the circuit, otherwise the oscillations of the electrons forming the a.c. are damped (p. 37). In effect, oscillators supply their own input and convert d.c. from the power supply into a.c.

Tuned (LC) oscillator

The circuit in Fig. 56.1 is for a tuned, sine wave oscillator. Basically it is that of an amplifier with d.c. bias and stabilization being provided by R_1, R_2 and R_3 and decoupling by C_2 and C_3. The tuned circuit,

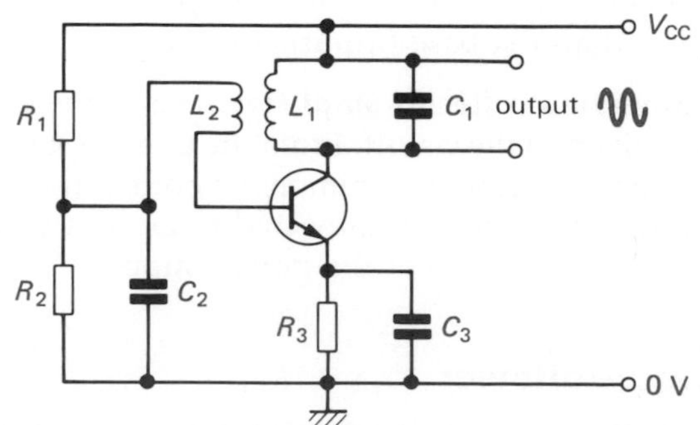

Fig. 56.1

consisting of inductor L_1 and capacitor C_1 is the collector load (C_1 may be variable) and it determines the frequency f of the oscillations, given by (p. 37).

$$f = \frac{1}{2\pi\sqrt{L_1C_1}}\,\text{Hz} \qquad (1)$$

where L_1 is in henrys (H) and C_1 is in farads (F).

In practice, to start oscillations, we do not have to apply a sine wave input. Switching on the power supply produces a current pulse which charges C_1 and starts oscillations automatically. Left to themselves these would decay but feedback occurs due to the oscillatory current in L_1 inducing, by transformer action, a voltage of the same frequency in L_2, arranged close to L_1. This voltage is applied to the base of the transistor as the input and is amplified to cause a larger oscillatory current in L_1 and hence a larger voltage in L_2 and so on. The coupling between L_1 and L_2 should be just enough to maintain oscillations.

A single-stage common emitter amplifier produces a 180° phase shift between its output and input (p. 105). The feedback circuit L_1L_2 must therefore introduce another 180° shift (if it is to be positive) by being connected the 'right way round' to the transistor base.

Tuned oscillators are most suitable for generating radio frequencies. At audio frequencies coils and capacitors to give the large values of L_1 and C_1 required by equation (1) are too bulky. An a.f. sine wave oscillator will be considered later (p. 128).

When a sine wave r.f. oscillator with a fixed, very stable frequency is required, use is made of a piezoelectric quartz crystal (p. 43). The crystal behaves like a tuned circuit and a.c. is produced with a frequency which depends on the shape and size of the crystal.

Unijunction oscillator

(a) Relaxation oscillators. These produce non-sine wave output voltages which, in two important cases, have sawtooth and square waveforms. The former are used as time bases in oscilloscopes (p. 99) and television receivers. The latter are used to produce electronic music since they are rich in harmonics (p. 20) and are useful for testing a.f. amplifiers (p. 102). Both sawtooth and square waves are available (along with sine waves) from a.f. signal generators (p. 101).

The action of a relaxation oscillator depends on the charging of a capacitor followed by a period of 'relaxation' when the capacitor discharges through a resistor.

(b) Unijunction transistor (UJT). The construction of a UJT and its symbol are shown in Fig. 56.2. It consists of a bar of n-type semiconductor with connections at opposite ends, called *base 1* and *base 2*,

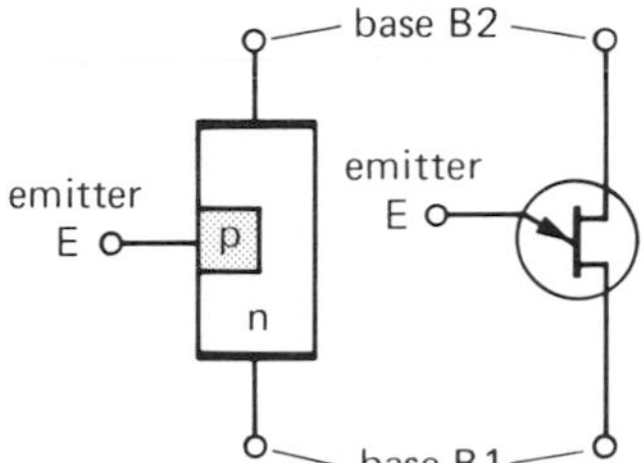

Fig. 56.2

and a p-type region, the *emitter* E about half-way along. Therefore, although it is a three-terminal device, it has just one p-n junction (hence unijunction) and is really a diode (not a transistor) with two base connections.

In use, B_2 is made positive with respect to B_1 via a load resistor R_L, Fig. 56.3a. If the voltage at the emitter is less than that on the n-type bar where the emitter is located, the p-n junction is reversed biased and no emitter current flows. In this condition the bar is acting as a potential divider consisting of two resistors R_{B2} and R_{B1} and $V_E < V_1$, Fig. 56.3b.

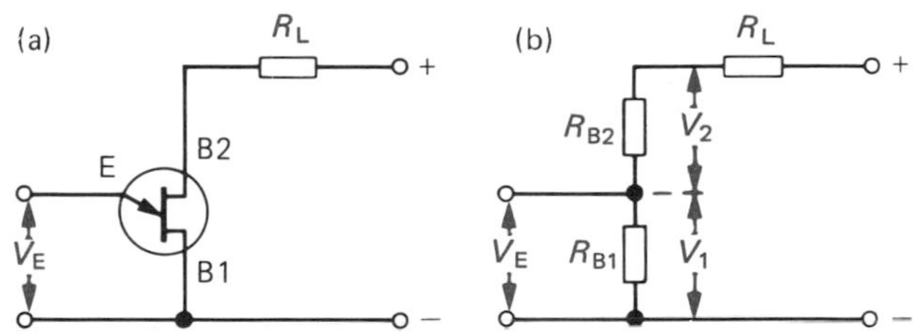

Fig. 56.3

When $V_E > V_1$, the emitter junction becomes forward biased, causing a fall in the value of R_{B1} and a sharp rise in the emitter current from E to B_1.

(c) Sawtooth oscillator. The circuit for a simple, sawtooth relaxation oscillator using a unijunction transistor is shown in Fig. 56.4. Capacitor C charges up from the supply through the large value resistor R_C until it gets to the potential which forward biases the emitter junction and makes it conduct. C then

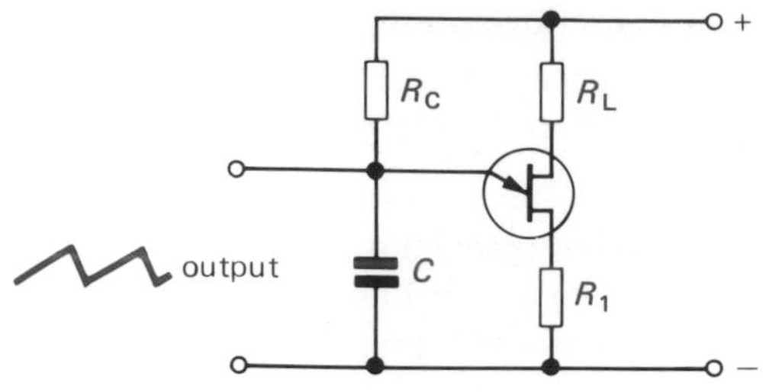

Fig. 56.4

discharges rapidly through the small value resistor R_1 and the p.d. across C falls to zero.

The charge–discharge action is repeated again and again at a rate determined by the values of R_C and C, to give a sawtooth voltage as the output across C.

Questions

1. Calculate the maximum and minimum frequencies of the oscillations which could be generated by a variable capacitor of maximum capacitance 500 pF and minimum capacitance one-tenth of this value connected to a coil of inductance $10\,\mu$H. ($1\,\text{pF} = 10^{-12}\,\text{F}$).

2. An LC oscillator has a fixed inductance of $50\,\mu$H and is required to produce oscillations over the band 1 MHz to 2 MHz. Calculate the maximum and minimum values of the variable capacitor required. (Take $\pi^2 = 10$.)

57 Progress questions

1. The circuit for a very simple audio amplifier is shown in Fig. 57.1. State

(i) the type of transistor used,
(ii) where you would connect a power supply (give polarities), and
(iii) where you would connect a microphone.

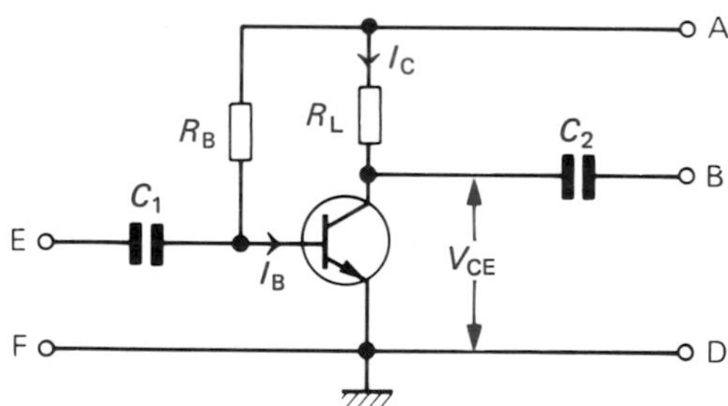

Fig. 57.1

2. Fig. 57.2 shows a load line on the $I_C - V_{CE}$ characteristics of a transistor used in a simple amplifier circuit. Point P represents the steady-state condition of the transistor.
Determine (a) the value of the load resistance, (b) the power dissipated in the load resistor in the steady state, and (c) the power dissipated in the transistor in the steady state.
Is the total power dissipation increased or decreased if the bias is shifted to point Q on Fig. 57.2. Give reasons, or calculations, to support your answer. (*O. and C.*)

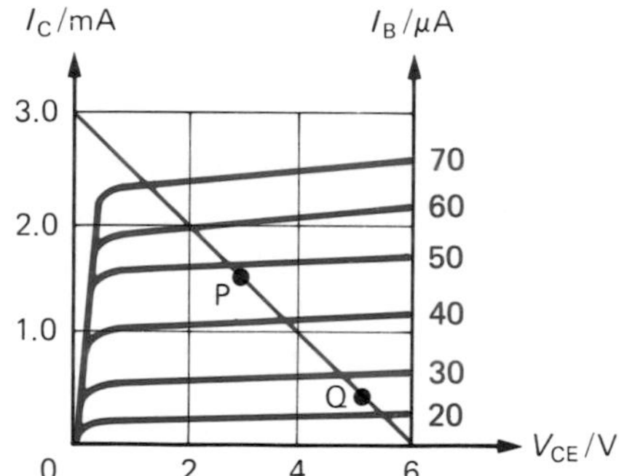

Fig. 57.2

3. The quiescent (no input signal) value of the current through R_S in Fig. 57.3 is 2 mA. What is the voltage with respect to ground (0 V) at (a) S, (b) D and (c) G?

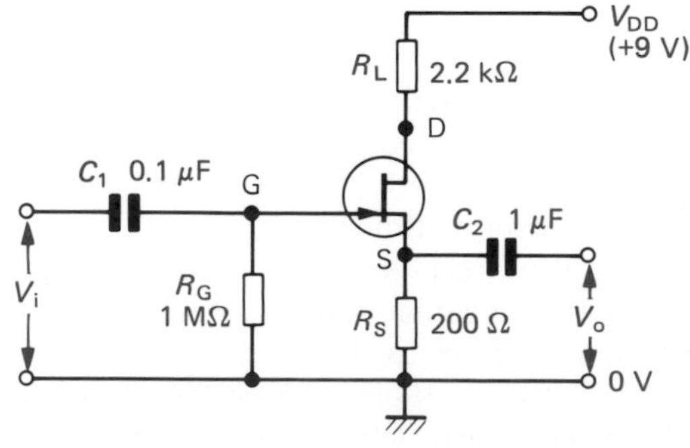

Fig. 57.3

4. Explain what is meant by *negative feedback*. When negative feedback is used in an amplifier what effect does it have on (a) the gain, (b) the distortion produced? (*C.*)

5. (a) Draw a circuit with negative feedback. Explain how the feedback is produced and state its effect.
(b) Draw a circuit with positive feedback. Explain how the feedback is produced and state its effect. (*L.*)

6. How does the 500 Ω resistor introduce negative feedback into the circuit shown in Fig. 57.4? What is the gain of the amplifier? (*L.*)

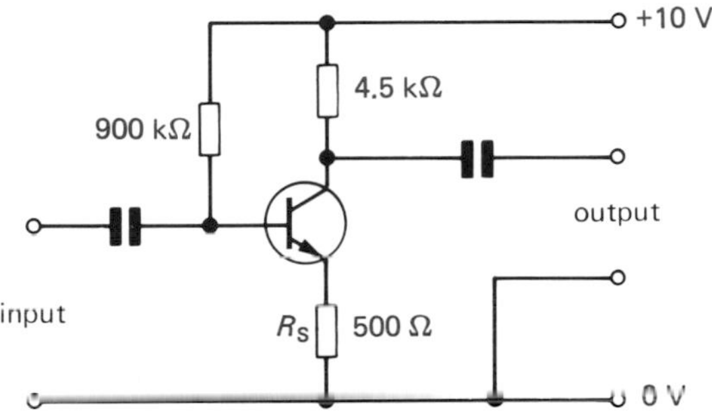

Fig. 57.4

7. (a) Transistors are subject to thermal runaway—that is, the collector current causes a slight rise in the temperature which causes greater current to flow and so on. This can lead ultimately to destruction of the transistor. Explain how a low value resistor placed in the emitter lead helps to overcome the problem.
(b) In an audio amplifier this resistor is often by-passed with a large value capacitor. Explain the reason for this. (*L.*)

8. An amplifier has the following properties:

open loop voltage gain	= 100
input impedance	= 1.5 kΩ
output impedance	= 4 Ω

Calculate the voltage which would be developed across the load in the circuit shown in Fig. 57.5.

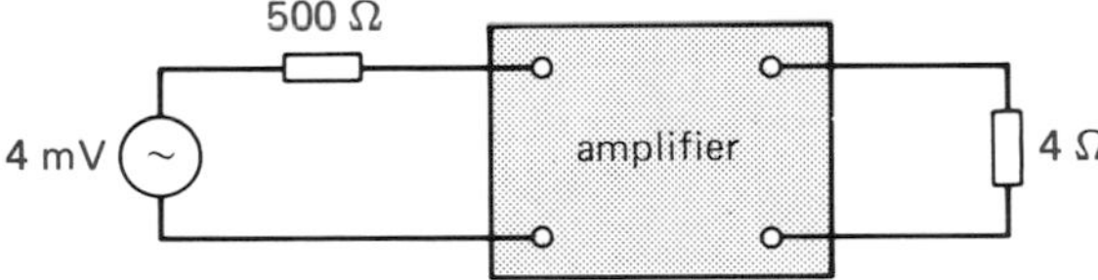

Fig. 57.5

Calculate the power developed in the load and explain the significance of the value of the load shown for this particular amplifier.
By considering the above system, explain the advantage of (a) a high input impedance, and (b) a low output impedance. (*L.*)

9. The input resistance of an electronic system is 10 kΩ. A transducer with a source voltage of 3 V and a source resistance of 2 kΩ is connected to the system.
(a) Calculate the voltage across the input.
(b) Calculate the power transferred to the system.
(c) The system is now to be modified. State the new input resistance if the maximum possible power is to be transferred from the transducer to the system. (*A.E.B.*)

10. (a) Copy and correctly complete the characteristics of an emitter follower, using: 50; low; high; 0.95.

Input resistance	Current gain ......
Output resistance	Voltage gain ....

(b) Explain one use of an emitter follower. (*O.L.E.*)

11. If there is no input signal, state the values in Fig. 57.6 of
(i) the voltage relative to ground (0 V) at B,
(ii) the voltage relative to ground (0 V) at E, and
(iii) the collector current I_C.

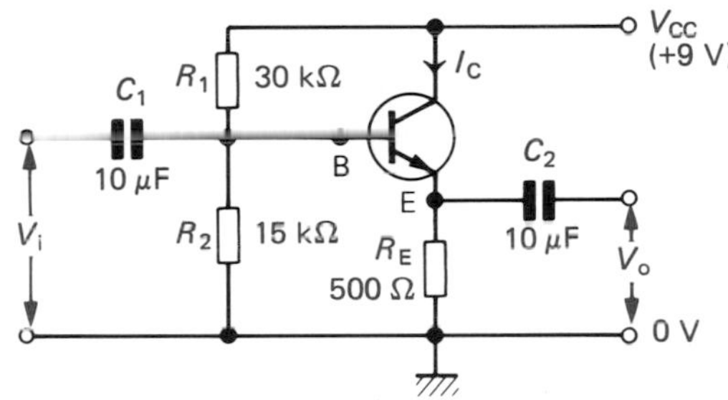

Fig. 57.6

12. The circuit diagram in Fig. 57.7 shows a simple transistor oscillator. Give a full description of how it works, with particular reference to the components labelled A, B, C, D and E.
What extra features are normally added to convert such an oscillator to a *signal generator*? Discuss the uses of the signal generator.
An oscillator has a tuned circuit consisting of a 100 mH inductor and a capacitor. The output frequency is 5 kHz. What order of magnitude of capacitance must be used? (*C. part qn.*)

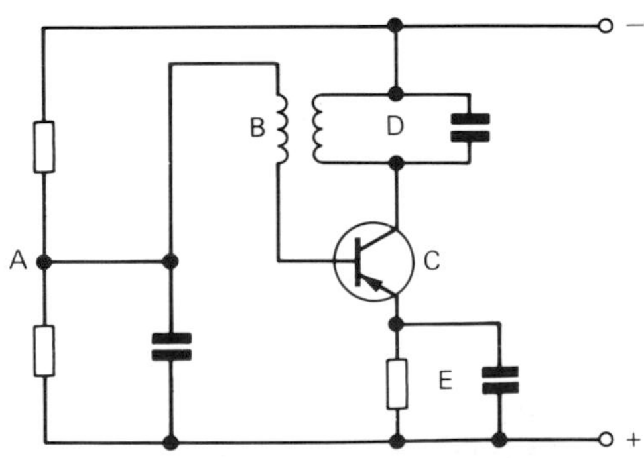

Fig. 57.7

58 Operational amplifier

Introduction

Operational amplifiers (op amps) were originally made from discrete components. They were designed to solve mathematical equations electronically, by performing operations such as addition, etc. in analogue computers (p. 127). Nowadays in IC form they have many uses, one of the most important being as high-gain d.c. and a.c. voltage amplifiers. A typical op amp contains twenty transistors as well as resistors and small capacitors.

(a) Properties. The chief properties are:

(i) a very *high open-loop voltage gain* A_o of about 10^5 for d.c. and low frequency a.c., which decreases as the frequency increases;
(ii) a very *high input impedance*, typically 10^6 to $10^{12}\,\Omega$, so that the current drawn from the device or circuit supplying it is minute and the input voltage is passed on to the op amp with little loss (p. 113);
(iii) a very *low output impedance*, commonly $100\,\Omega$, which means its output voltage is transferred efficiently to any load greater than a few kilohms.

(b) Description. An op amp has one output and two inputs. The *non-inverting* input is marked + and the *inverting input* is marked −, as shown on the amplifier symbol in Fig. 58.1. Operation is most

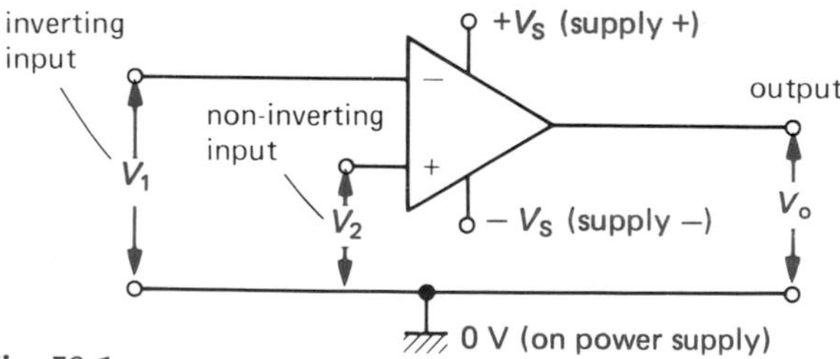

Fig. 58.1

convenient from a dual balanced d.c. power supply giving equal positive and negative voltages $\pm V_s$, (i.e. $+V_s$, 0 and $-V_s$) in the range ± 5 V to ± 15 V. The centre point of the power supply, i.e. 0 V, is common to the input and output circuits and is taken as their voltage reference level.

Do not confuse the input signs with those for the power supply polarities, which for clarity, are often left off circuit diagrams.

Action and characteristic

(a) Action. If the voltage V_2 applied to the non-inverting (+) input is positive relative to the other input, the output voltage V_o is positive; similarly if V_2 is negative, V_o is negative, i.e. V_2 and V_o are in phase. At the inverting (−) input, a positive voltage V_1 relative to the other input causes a negative output voltage V_o and vice versa, i.e. V_1 and V_o are in antiphase.

Basically an op amp is a *differential* voltage amplifier, i.e. it amplifies the difference between the voltages V_1 and V_2 at its inputs. There are three cases:

(i) if $V_2 > V_1$, V_o is positive,
(ii) if $V_2 < V_1$, V_o is negative,
(iii) if $V_2 = V_1$, V_o is zero.

In general, the output V_o is given by

$$V_o = A_o(V_2 - V_1)$$

where A_o is the open-loop voltage gain.

In most applications, but not all (see p. 124), single-input working is used.

(b) Transfer characteristic. A typical voltage characteristic showing how the output V_o in volts (V) varies with the input $(V_2 - V_1)$ in microvolts (μV) is given in Fig. 58.2. It reveals that it is only within

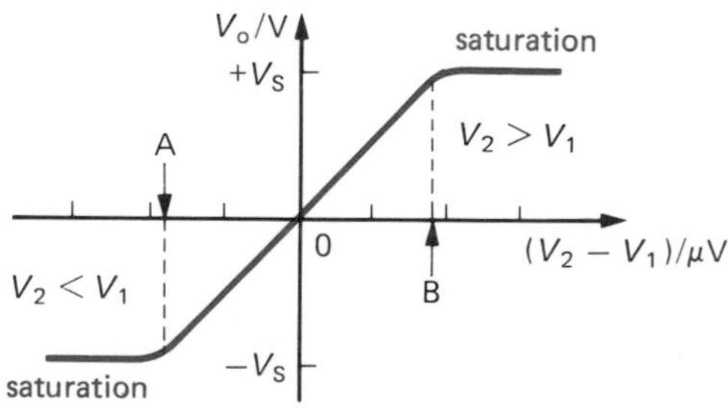

Fig. 58.2

the very small input range A0B that the output is directly proportional to the input, i.e. when the op amp behaves more or less *linearly* and there is minimum distortion of the amplifier output. Inputs outside the linear range cause *saturation* and the output is then close to the maximum value it can have, i.e. $+V_s$ or $-V_s$.

The limited linear behaviour is due to the very high open-loop gain A_o and the higher it is, the greater is the limitation. For example, on a ± 9 V supply, since the output voltages can never exceed these values, and have maximum values near either $+9$ V or -9 V, the maximum input voltage swing (for linear amplification) is, if $A_o = 10^5$, $\pm 9\,\text{V}/10^5 = \pm 90\,\mu\text{V}$. A smaller value of A_o would allow a greater input.

Negative feedback

Op amps almost always use negative feedback (n.f.b.), obtained by feeding back some (or all) of the output to the inverting (−) input. The feedback produces a voltage at the output that opposes the one from which it is taken, thereby reducing the new output of the amplifier, i.e. the one with n.f.b. The resulting closed-loop gain A is then less than the open-loop gain A_o but a wider range of voltages can be applied to the input for amplification. (Feedback applied to the non-inverting (+) input would be positive and would increase the overall output.)

In addition, as stated for discrete component amplifiers (p. 111), n.f.b. gives (provided $A_o \gg A$),

(i) predictable and constant voltage gain A,
(ii) reduced distortion of the output, and
(iii) better frequency response, i.e. increased bandwidth which depends on the amount of feedback, as shown by the frequency response curve in Fig. 58.3 for a 741 op amp using various values of feedback factor β. For example, when $\beta = 0.1$, the gain is 10 and constant for frequencies up to 10^5 Hz (100 kHz). (Note that *gain* × *bandwidth* $\approx 10^6$.)

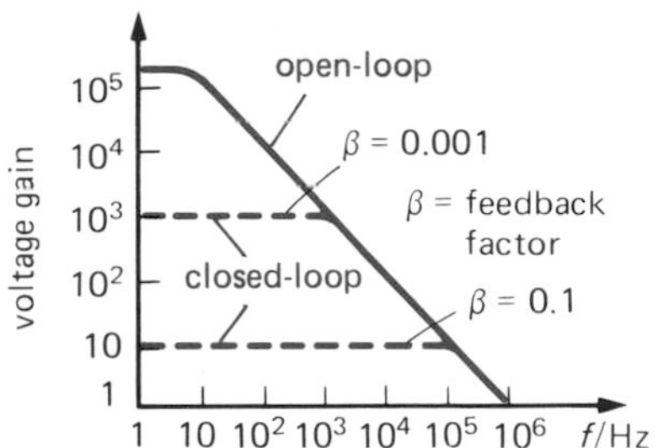

Fig. 58.3

The advantages of using n.f.b. outweigh the accompanying loss of gain which is easily increased by using two or more op amp stages.

Direct coupled amplifiers

In the transistor amplifiers considered before (pp. 104–10), capacitors were used at the input and output of each stage to couple a.c. signals but also to block any d.c. that might upset the operation. However 'level shifting' circuits have now been developed which omit capacitors and allow direct coupling. Both a.c. and d.c. signals can then be amplified. Op amps are direct coupled amplifiers.

Questions

1. Explain the following terms used in connection with op amps: (a) open-loop gain, (b) differential input and (c) saturation output voltage.

2. If A_o for an op amp is 10^5, calculate the maximum input voltage swing that can be applied for linear operation on a ± 15 V supply.

59 Op amp voltage amplifiers

Inverting amplifier

(a) Basic circuit, Fig. 59.1. The input voltage V_i (a.c. or d.c.) to be amplified is applied via resistor R_i to the inverting (−) terminal. The output voltage V_o is therefore in antiphase with the input. The non-inverting (+) terminal is held at 0 V. Negative feedback is provided by R_f, the *feedback resistor*, feeding back a certain fraction of the output voltage to the inverting (−) terminal.

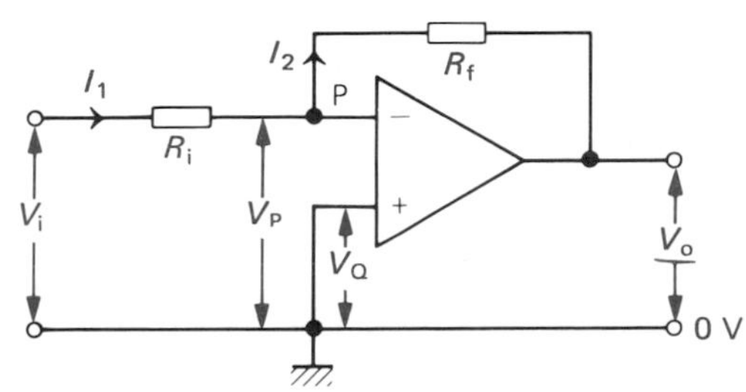

Fig. 59.1

(b) Gain. Two assumptions are made to simplify the theoretical derivation of an expression for the gain of the amplifier, i.e. we consider an *ideal op amp*. They are:

(i) each input draws zero current from the signal source (it is less than 0.1 μA for the 741 op amp), i.e. their input impedances are infinite, and

(ii) the inputs are both at the same potential if the op amp is not saturated, i.e. $V_P = V_Q$. (The greater the open-loop gain A_o is, the truer will this be, because as we saw earlier (p. 121), if the supply voltage $V_S = \pm 9\,V$ and $A_o = 10^5$, then maximum $V_o \approx \pm 9\,V$ and maximum $V_i \approx \pm 9\,V/10^5 \approx \pm 90\,\mu V$.)

If A_o is made infinite, we are making the 'infinite gain approximation'.

In the above circuit $V_Q = 0$, therefore $V_P = 0$ and P is called a *virtual earth* (or ground) point, though of course it is not connected to ground. Hence

$$I_1 = (V_i - 0)/R_i \text{ and } I_2 = (0 - V_o)/R_f$$

But $I_1 = I_2$ since by assumption **(i)** the input takes no current, therefore

$$\frac{V_i}{R_i} = \frac{-V_o}{R_f}$$

The negative sign shows that V_o is negative when V_i is positive and vice versa. The closed-loop gain A is given by

$$A = \frac{V_o}{V_i} = \frac{-R_f}{R_i} \quad (1)$$

For example, if $R_f = 100\,k\Omega$ and $R_i = 10\,k\Omega$, $A = -10$ exactly and an input of 0.1 V will cause an output change of 1.0 V.

Equation (1) shows that the gain of the amplifier depends only on the two resistors (which can be made with precise values) and not on the characteristics of the op amp (which vary from sample to sample).

(c) Input impedance. The other versatile feature of this circuit is the way its input impedance can be controlled. Since point P is a virtual earth (i.e. at 0 V), R_i may be considered to be connected between the inverting (−) input terminal and 0 V. The input impedance of the *circuit* is therefore R_i in parallel with the much greater input impedance of the op amp, i.e. effectively R_i, whose value can be changed.

Non-inverting amplifier

(a) Basic circuit, Fig. 59.2. The input voltage V_i (a.c. or d.c.) is applied to the non-inverting (+) terminal of the op amp. This produces an output V_o that is in phase with the input. Negative feedback is obtained by feeding back to the inverting (−) terminal, the fraction of V_o developed across R_i in the voltage divider formed by R_f and R_i across V_o.

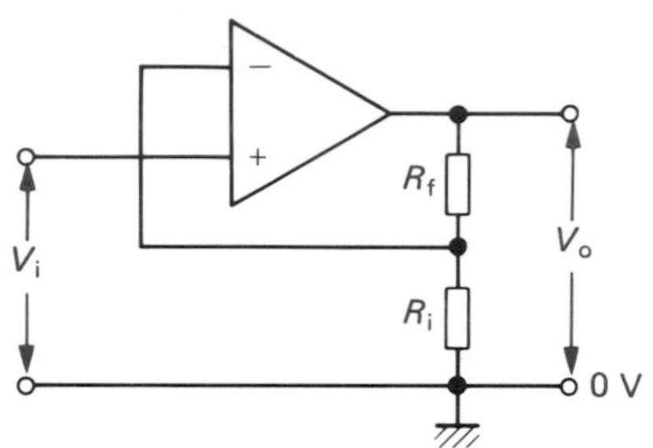

Fig. 59.2

(b) Gain. For the feedback factor β we can write

$$\beta = \frac{R_i}{R_i + R_f}$$

We saw before (p. 111) that for an amplifier with open-loop gain A_o, the closed-loop voltage gain A is given by

$$A = \frac{A_o}{1 + \beta A_o}$$

For a typical op amp $A_o = 10^5$, so βA_o is large compared with 1 and we can say $A = A_o/\beta A_o = 1/\beta$.

Hence

$$A = \frac{V_o}{V_i} = \frac{R_i + R_f}{R_i} = 1 + \frac{R_f}{R_i} \quad (2)$$

For example, if $R_f = 100\,k\Omega$ and $R_i = 10\,k\Omega$, then $A = 110/10 = 11$. As with the inverting amplifier, the gain depends only on the values of R_f and R_i and is independent of the open-loop gain A_o of the op amp.

(c) Input impedance. Since there is no virtual earth at the non-inverting (+) terminal, the input impedance is much higher (typically 50 MΩ) than that of the inverting amplifier. Also it is unaffected if the gain is altered by changing R_f and/or R_i. This circuit gives good matching when the input is supplied by a high impedance source such as a crystal microphone.

Voltage-follower

This is a special case of the non-inverting amplifier in which 100% n.f.b. is obtained by connecting the output directly to the inverting (−) terminal, as shown in Fig. 59.3. Thus $R_f = 0$ and R_i is infinite.

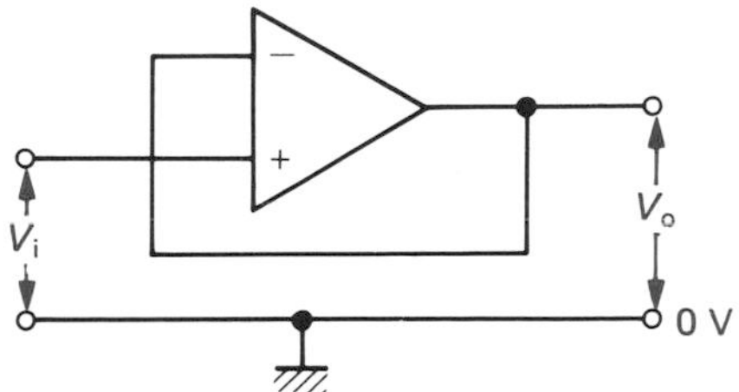

Fig. 59.3

Because all of the output is fed back $\beta = 1$ and since $A = 1/\beta$ (when A_o is very large), then $A \approx 1$. The voltage gain is nearly 1 and V_o is the same as V_i to within a few millivolts.

The circuit is called a *voltage-follower* because, as with its transistor equivalent (the emitter-follower) V_o follows V_i. It has an extremely high input impedance and a low output impedance. Its main use is as a *buffer amplifier*, giving current amplification, to match a high impedance source to a low impedence load. For example, it is used as the input stage of an analogue voltmeter where the highest possible input impedance is required (so as not to disturb the circuit under test) and the output voltage is measured by a relatively low impedance moving coil meter. In such op amps, FETs replace bipolar transistors in the early stages of the IC.

Questions

1. In the circuit of Fig. 59.1 the power supply to the op amp is ±9 V and the input voltage $V_i = +1$ V. What is the value of the output voltage V_o when

(a) $R_f = 20\,\text{k}\Omega$ and $R_i = 10\,\text{k}\Omega$, and

(b) $R_f = 200\,\text{k}\Omega$ and $R_i = 10\,\text{k}\Omega$?

2. Repeat Question 1 for the circuit of Fig. 59.2.

60 Op amp summing amplifier

When connected as a multi-input inverting amplifier, an op amp can be used to add a number of voltages (d.c. or a.c.) because of the existence of the virtual earth point. This in turn is a consequence of the high value of the open-loop voltage gain A_o.

Such circuits are employed as 'mixers' in audio applications to combine the outputs of microphones, electric guitars, pick-ups, special effects, etc. They are also used to perform the mathematical process of addition in analogue computing (p. 127).

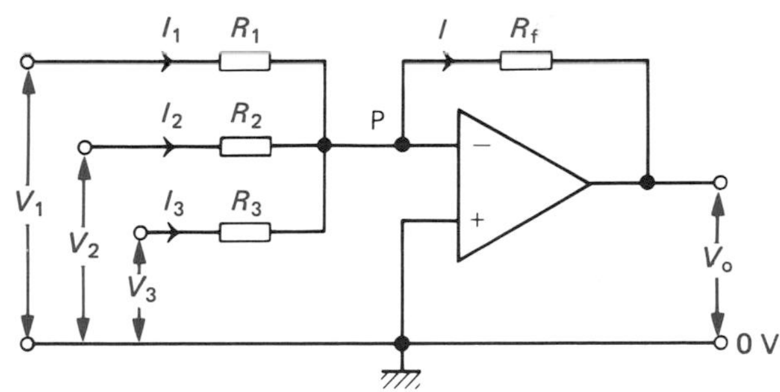

Fig. 60.1

Action

In the circuit of Fig. 60.1 three input voltages V_1, V_2 and V_3 are applied via input resistors R_1, R_2 and R_3 respectively. Assuming that the inverting terminal of the op amp draws no input current, all of it passing through R_f, then

$$I = I_1 + I_2 + I_3$$

Since P is a virtual earth (i.e. at 0 V), it follows that

$$\frac{-V_o}{R_f} = \frac{V_1}{R_1} + \frac{V_2}{R_2} + \frac{V_3}{R_3}$$

Therefore

$$V_o = -\left(\frac{R_f}{R_1} \cdot V_1 + \frac{R_f}{R_2} \cdot V_2 + \frac{R_f}{R_3} \cdot V_3\right)$$

The three input voltages are thus added and amplified if R_f is greater than each of the input resistors, i.e. 'weighted' summation occurs. Alternatively the input voltages are added and attenuated if R_f is less than every input resistor.

For example, if $R_f/R_1 = 3$, $R_f/R_2 = 1$ and $R_f/R_3 = 2$, and $V_1 = V_2 = V_3 = +1\,V$, then addition of 3 V, 1 V and 2 V occurs. That is $V_o = -(3\,V + 1\,V + 2\,V) = -6\,V$.

If $R_1 = R_2 = R_3 = R_i$ (say), the input voltages are amplified or attenuated equally, and

$$V_o = -\frac{R_f}{R_i}(V_1 + V_2 + V_3)$$

Further, if $R_i = R_f$, then

$$V_o = -(V_1 + V_2 + V_3)$$

In this case the output voltage is the sum of the input voltages (but is of opposite polarity).

Summing point

The virtual earth is also called the *summing point* of the amplifier. It isolates the inputs from one another so that each behaves as if none of the others existed and none feeds any of the other inputs even though all the resistors are connected at the inverting input.

Also, the resistors can be selected to produce the best impedance matching with the transducer supplying its input (see p. 113), although compromise may be necessary to obtain both the gain required and the correct input impedance.

Question

1. In the circuit of Fig. 60.2 the power supply is $\pm 15\,V$, $R_f = 30\,k\Omega$ and $R_i = 15\,k\Omega$. Calculate V_o when
(a) $V_1 = +1.0\,V$ and $V_2 = +4\,V$, and
(b) $V_1 = +1.0\,V$ and $V_2 = -4\,V$.

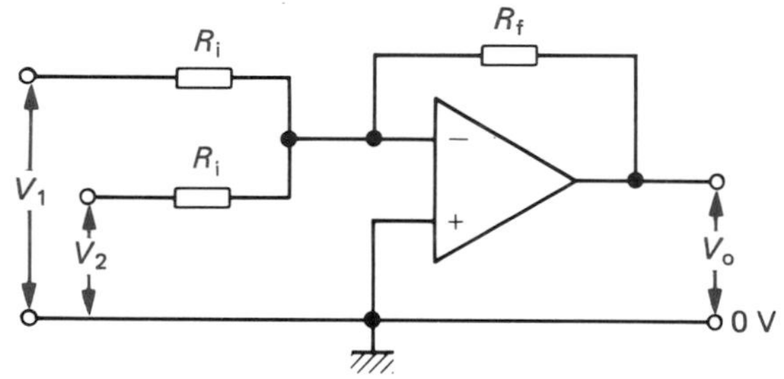

Fig. 60.2

61 Op amp voltage comparator

Action

If both inputs of an op amp are used simultaneously, then, as we saw earlier (p. 120), the output voltage V_o is given by

$$V_o = A_o\,(V_2 - V_1)$$

where V_1 is the inverting (−) input, V_2 the non-inverting (+) input and A_o the open-loop gain, Fig. 61.1. The voltage difference between the inputs, i.e. $(V_2 - V_1)$ is amplified and appears at the output.

When $V_2 > V_1$, V_o is *positive*, its maximum value being the positive supply voltage $+V_s$, which it has

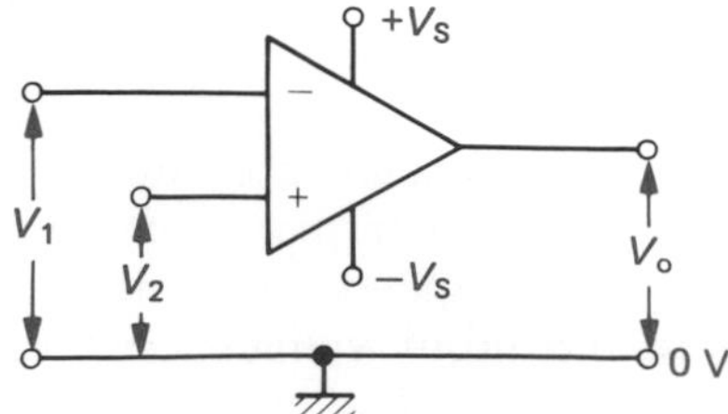

Fig. 61.1

when $(V_2 - V_1) \geqslant V_s/A_o$. The op amp is then saturated. If $V_s = +15\,V$ and $A_o = 10^5$, saturation occurs when $(V_2 - V_1) \geqslant 15\,V/10^5$, i.e. when V_2 exceeds V_1 by $150\,\mu V$ and $V_o \approx 15\,V$.

When $V_1 > V_2$, V_o is *negative* and saturation occurs if V_1 exceeds V_2 by V_s/A_o, typically by $150\,\mu V$. But in this case $V_o \approx -V_s \approx -15\,V$.

A small change in $(V_2 - V_1)$ thus causes V_o to switch between near $+V_s$ and near $-V_s$ and enable the op amp to indicate when V_2 is greater or less than V_1, i.e. to act as a *differential amplifier* and compare two voltages. It does this in an electronic digital voltmeter (p. 208).

Some examples

The waveforms in Fig. 61.2 show what happens if V_2 is an alternating voltage (sufficient to saturate the op amp).

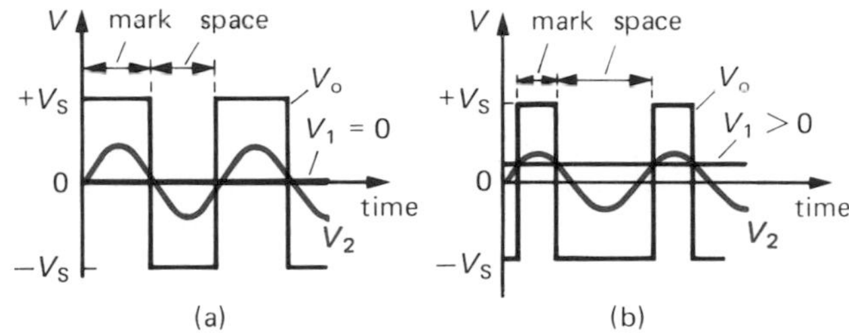

Fig. 61.2

In (a), $V_1 = 0$ and $V_o = +V_s$ when $V_2 > V_1$ (positive half cycle) and $V_o = -V_s$ when $V_2 < V_1$ (negative half cycle). V_o is a 'square' wave with a mark-to-space ratio of 1 (p. 19).

In (b), $V_1 > 0$ and switching of V_o occurs when $V_2 \approx V_1$, the mark-to space ratio now being < 1 and the output a series of 'pulses'.

In effect, the op amp in its saturated condition converts a continuously varying analogue signal (V_2) to a two-state (i.e. 'high'–'low') digital output (V_o).

Alarm circuit

In the light-operated alarm circuit of Fig. 61.3, R and the LDR form a voltage divider across the +15/0/−15 V supply. The op amp compares the voltage V_1 at the voltage divider junction, i.e. at the inverting (−) input, with that at the non-inverting (+) input, i.e. with V_2 which is 0 V. In the dark the

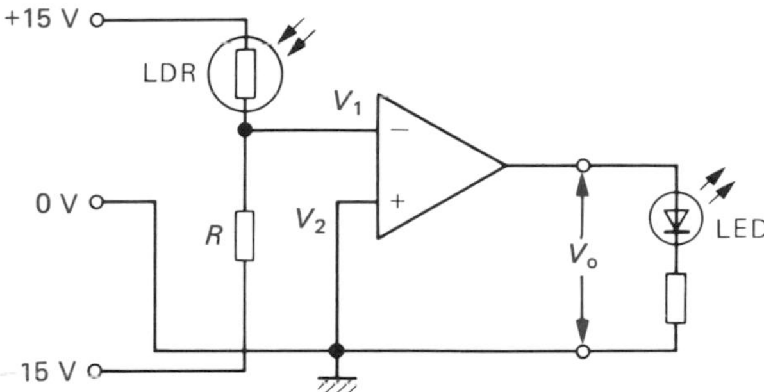

Fig. 61.3

resistance of the LDR is much greater than that of R, so more of the 30 V across the voltage divider is dropped down the LDR making V_1 fall below 0 V. Now $V_2 > V_1$ and the output voltage V_o switches from near −15 V to near +15 V and the LED lights.

Question

1. In Fig. 61.4 V_1 and V_2 are the p.ds at the − and + inputs respectively of an op amp on a ±6 V supply. Copy the graphs and draw on the same axes the output p.d. V_o.

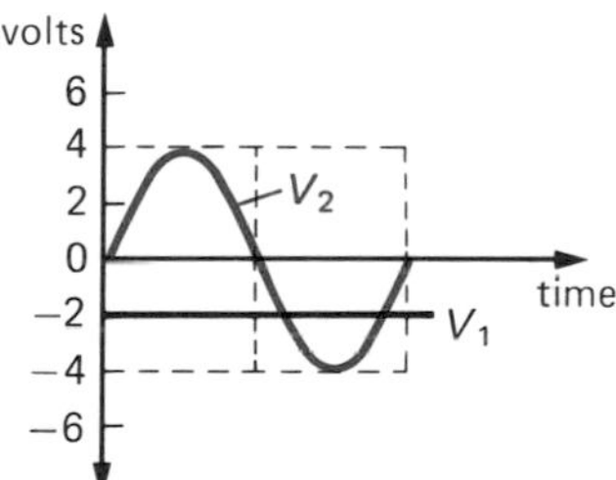

Fig. 61.4

62 Op amp integrator

Action

The circuit is the same as for the op amp inverting amplifier (Fig. 59.1) but feedback occurs via a capacitor C, as in Fig. 62.1, rather than via a resistor. If the input voltage V_i, applied through the input resistor R, is *constant*, and $CR = 1$ s (e.g. $C = 1\,\mu$F, $R = 1\,\text{M}\Omega$), the output voltage V_o after time t (in seconds) is given by

$$V_o = -V_i t$$

The negative sign is inserted because, when the inverting input is used, V_o is negative if V_i is positive and vice versa.

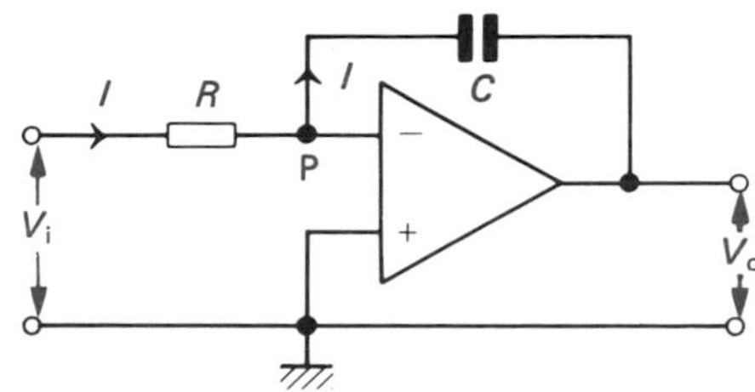

Fig. 62.1

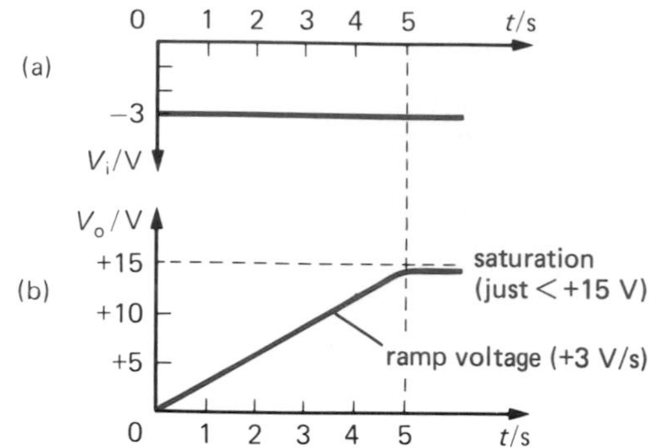

Fig. 62.2

For example, if $V_i = -3\,V$, Fig. 62.2a, V_o *rises steadily* by +3 V/s and, if the power supply is ±15 V, V_o reaches about +15 V after 5 s, when the op amp saturates, Fig. 62.2b. V_i is thus 'added up' or integrated over a time t to give V_o a *ramp voltage* waveform whose slope will be shown to be proportional to V_i.

The integrating action of the circuit on V_i is similar to that of a petrol filling station pump which 'operates' on the rate of flow (in litres per second) and the delivery time (in seconds) to produce as its output, the volume of petrol supplied (in litres).

Theory

Since P is a virtual earth in Fig. 62.1, i.e. at 0 V, the voltage across R is V_i and that across C is V_o. Assuming (as before) that none of the input current I enters the op amp inverting input, then all of I flows 'through' C and charges it up. If V_i is constant, I will be constant, with a value

$$I = \frac{V_i}{R} \qquad (1)$$

C therefore charges at a *constant rate* and the potential of the output side of C (which is V_o since its input side is at 0 V) changes so that the feedback path absorbs I. If Q is the charge on C at time t and the p.d. across it (i.e. the output voltage) changes from 0 to V_o in that time then, since I is constant, we have

$$Q = -V_oC = It \qquad \text{(pp. 7 and 26)}$$

Substituting for I from (1), we get

$$-V_oC = \frac{V_i}{R}\cdot t$$

That is

$$V_o = -\frac{1}{CR}\cdot V_it$$

If $CR = 1\,s$, then

$$V_o = -V_it$$

A more general mathematical treatment shows that if V_i varies then, using calculus notation

$$V_o = -\frac{1}{CR}\int V_i\,dt$$

where $\int V_i\,dt$ is the 'integral of V_i with respect to time t'.

Question

1. In the circuit of Fig. 62.1 if $CR = 1\,s$ and $V_i = +6\,V$
   (a) does V_o rise or fall and at what rate, and
   (b) if the op amp power supply is ±18 V, what is the maximum value of V_o and when is it reached?

63 Analogue computing

Electronic analogues

Analogue computers are used as electronic models or analogues of mechanical and other systems in cases where conducting experiments on the system itself would be costly or time-consuming or dangerous, or where the system has not yet been built. For example, when designing a bridge or an aircraft wing, information is required before construction starts about how it will react to variable factors like wind speed and temperature.

Such predictions involve solving *differential equations* which are often too complex for ordinary mathematical methods. An equation containing first order differential coefficients, e.g. dy/dt (pronounced 'dee-y-by-dee-t') has to be integrated once to obtain a solution, i.e. an expression for y in terms of t. One with second order differential coefficients, e.g. d^2y/dt^2 (pronounced 'dee-two-y-by-dee-t-squared'), requires two integrations.

An analogue computer, consisting of the appropriate number of op amp integrators, will provide the solution if it is set up with voltages representing differential coefficients of the quantities under investigation. (In recent years analogue computers have been less popular because digital computers can now be programmed to simulate moving physical systems.)

Solving a differential equation

Consider the very simple second order differential equation (normally solved mathematically), of

$$\frac{d^2y}{dt^2} = 10$$

We wish to obtain an expression for y in terms of t. Integration can be regarded as the reverse process of differentiation (for both of which there are mathematical rules). For example, the integral of dy/dt is y (i.e. $\int \frac{dy}{dt}\,dt = y$) and the integral of d^2y/dt^2 is dy/dt.

Therefore, if a constant d.c. voltage representing d^2y/dt^2 is supplied as the input to two op amp integrators in series (having $CR = 1$ s for each stage), as in Fig. 63.1, integration (and inversion) occurs twice and the output voltage will represent y in terms of t.

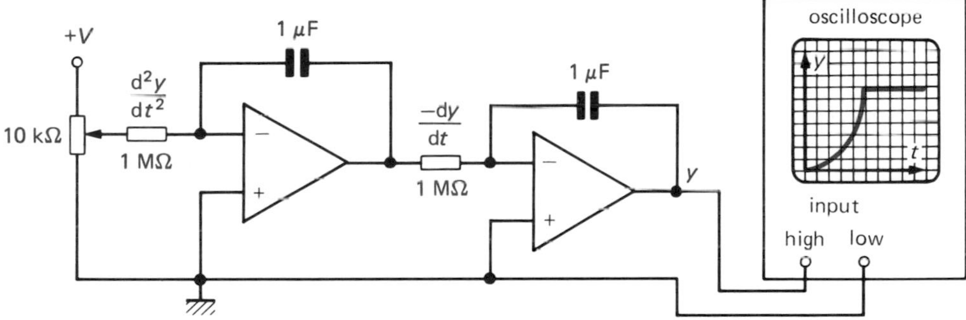

Fig. 63.1

If the output is displayed on an oscilloscope, the graph produced on the screen by the spot shows how y varies with t. In fact, a solution is $y = 5t^2$. Knowing the scales of the voltage and time axes, numerical answers can be obtained.

The equation $d^2y/dt^2 = 10$ represents the motion of an object falling from rest, neglecting air resistance, y being the distance fallen (metres) in time t (seconds). Equations for systems such as car suspensions, spaceship travel and automatic pilots for aircraft and ships are much more complex but they are solved in basically the same way.

Question

1. When and why are analogue computers useful as electronic models of physical systems?

64 Op amp oscillators

Astable multivibrator

An op amp voltage comparator (p. 124) in its *saturated* condition can operate as a relaxation oscillator (p. 117) if suitable external components are connected and *positive* feedback used.

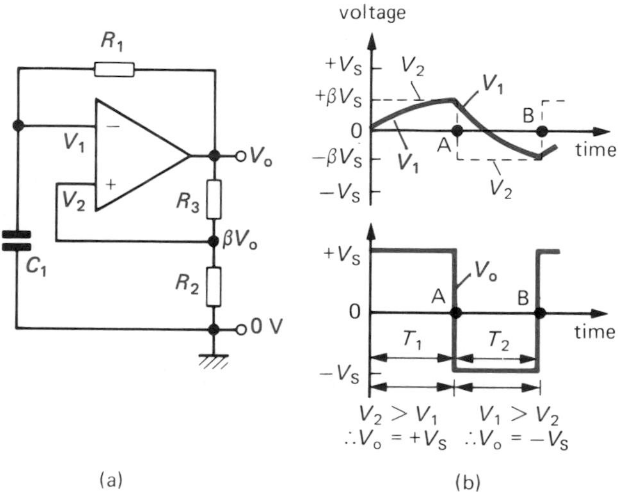

Fig. 64.1

The circuit and various voltage waveforms are shown in Fig. 64.1a, b. Suppose the output voltage V_o is positive at a particular time. A certain fraction β of V_o, is fed back as the non-inverting input voltage V_2 which equals βV_o where $\beta = R_2/(R_2 + R_3)$. V_o is also fed back via R_1 to the inverting terminal and V_1 rises (exponentially) as C_1 is charged.

After a time which depends on the time constant C_1R_1, V_1 exceeds V_2 and the op amp switches into negative saturation, i.e. $V_o = -V_s$ (point A on graphs). Also, the positive feedback makes V_2 go negative (where $V_2 = -\beta V_o = -\beta V_s$).

C_1 now starts to charge up in the opposite direction, making V_1 fall rapidly (during time AB on the graphs) and eventually become more negative than V_2. The op amp therefore switches again to its positive saturated state with $V_o = +V_s$ (point B on graphs). This action continues indefinitely at frequency f given by $f = 1/(T_1 + T_2)$.

The output voltage V_o provides *square* waves and the voltage across C_1 is a source of *triangular* waves. The latter have 'exponential' sides, but they are 'straightened' by extra circuitry which ensures C_1 is charged by a constant current, rather than by the exponential one supplied through R_1.

The term 'astable multivibrator' arises because (i) the circuit has no stable states and switches automatically from a 'high' to a 'low' output, (hence 'astable'), and (ii) the output is a square wave which consists of many sine wave frequencies (p. 20), (hence 'multivibrator').

Wien oscillator

Sine waves in the audio frequency range can be generated by an op amp using a Wien network circuit containing resistors and capacitors.

We saw earlier (Fig. 10.11a, p. 28), that if a.c. is applied to a resistor and capacitor in series, the voltages developed across them are 90° out of phase. In the Wien circuit, a network of two resistors R_1, R_2 (usually equal) and two capacitors C_1, C_2 (also usually equal), arranged as in Fig. 64.2, acts as

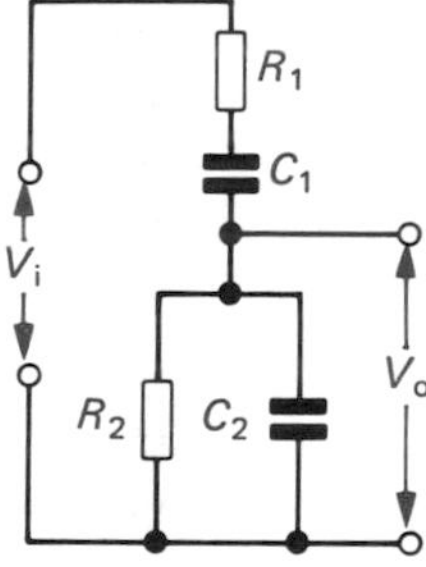

Fig. 64.2

the positive feedback circuit. The network is an a.c. voltage divider and theory shows that the output voltage V_o is *in phase* with the input voltage V_i, i.e. the phase shift is zero, at one frequency f, given by

$$f = \frac{1}{2\pi RC}\ \text{Hz}$$

where R is in ohms and C in farads. At all other frequencies, there is a phase shift.

To obtain oscillations the network must therefore be used with a non-inverting amplifier which gives an output to the Wien network that is in phase with its input.

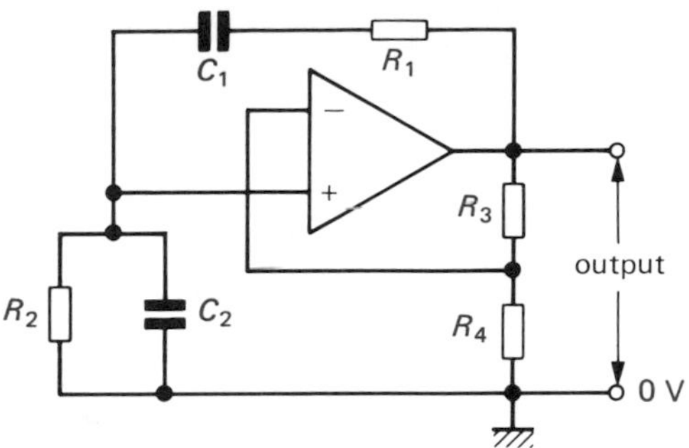

Fig. 64.3

In the circuit of Fig. 64.3, the frequency-selective Wien network, R_1C_1, R_2C_2, applies positive feedback to the non-inverting (+) input. Negative feedback is supplied via R_3 and R_4 to the inverting (−) input. It can be proved that so long as the voltage gain of the amplifier exceeds three, oscillations will be maintained at the desired frequency f.

A variable-frequency output, as in an a.f. signal generator (p. 101), can be obtained if R_1 and R_2 are variable, 'ganged' resistors (i.e. mounted on the same spindle so that they can be altered simultaneously by one control). Also C_1 and C_2 would be pairs of capacitors that are switched in for different frequency ranges.

Waveform generators

Integrated circuit a.f. oscillators, based on the principle just outlined, and called *waveform generators* (e.g. 8038) are now available with outputs having sine, square, triangular, sawtooth and pulse waveforms. The frequency, which is stable over a wide range of temperatures and supply voltage, can be set externally in the range 0.001 Hz to 1 MHz.

Questions

1. In the astable multivibrator circuit of Fig. 64.1, state *two* ways in which the frequency of the output could be (a) increased, (b) decreased.

2. In the Wien oscillator circuit of Fig. 64.3, the op amp could be replaced by a two-stage transistor (common emitter) amplifier but not by a single-stage one. Why?

65 Progress questions

1. An operational amplifier is connected as in Fig. 65.1. Write down the equation which links the output voltage (V_o) and the input voltage (V_i) with the two resistances. The power supplies for the operational amplifier, which are not shown on the circuit diagram, are ±9 V.

Fig. 65.1

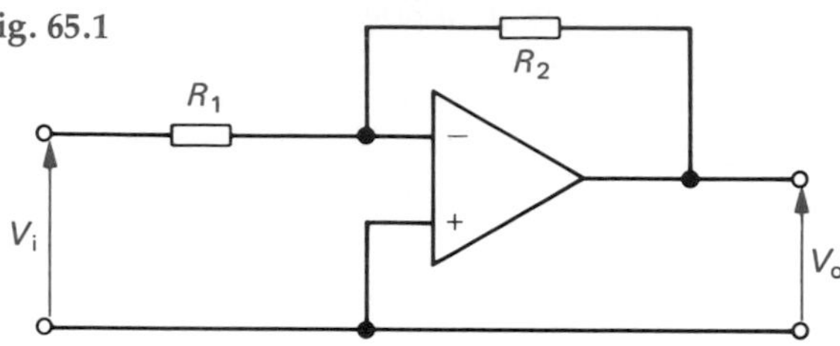

(a) If the values of the two resistors in the above circuit are R_1 = 2.2 kΩ and R_2 = 22 kΩ, calculate the output voltage when (i) the input voltage is +0.5 V, (ii) the input voltage is +1.0 V.

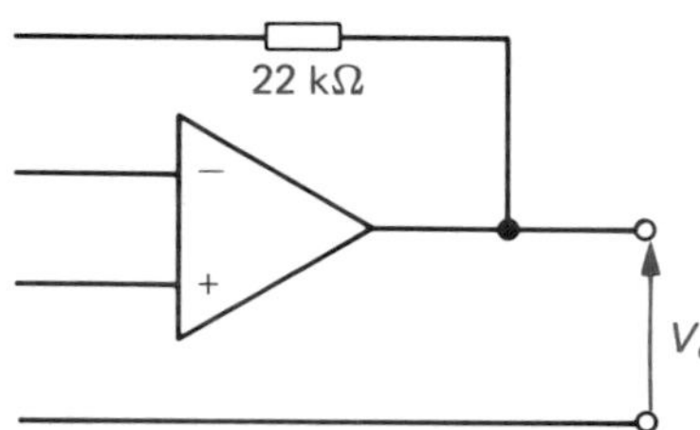

Fig. 65.2

(b) Copy and complete the circuit diagram in Fig. 65.2 to show how you would use the operational amplifier to amplify an input voltage of +0.1 V to an output voltage of +2.3 V.

Calculate the value of any additional components and indicate the values on your circuit diagram.

(A.E.B. 1982 Electronics)

2. The input voltage to the circuit shown in Fig. 65.3 is a sine wave with a peak-to-peak value of 1 V. Using the same set of axes, sketch the voltage waveforms at J, K and L. (You may assume that the open-loop gain of the amplifier is very large.)

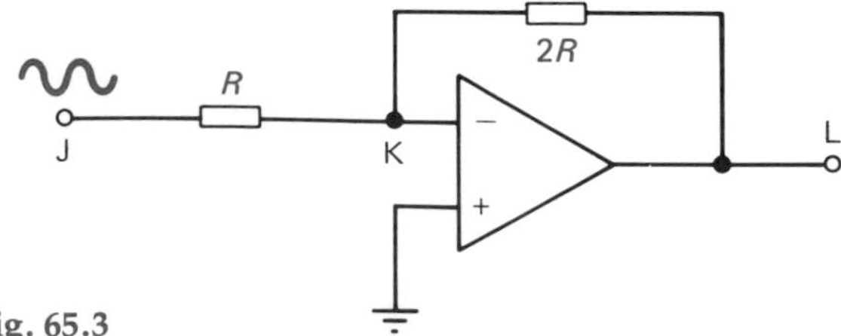

Fig. 65.3

What differences would it make to your graphs if the open-loop gain were somewhat lower, e.g. 1000?

(O. and C.)

3. In the operational amplifier circuit of Fig. 65.4 the power supply voltage is ±15 V and input voltages are applied as shown. Calculate the value of V_o under these conditions. *(A.E.B.)*

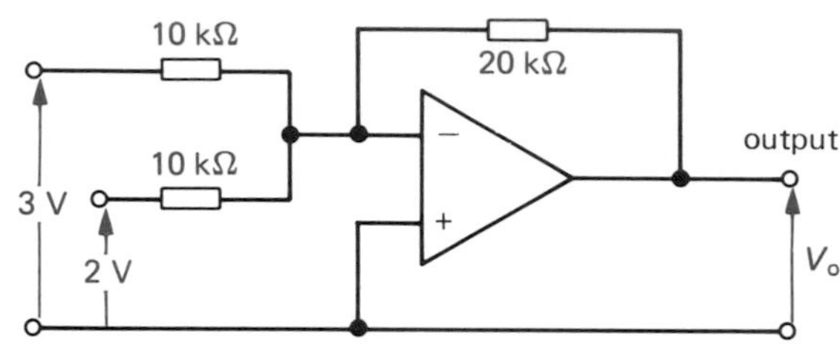

Fig. 65.4

4. What is a differential amplifier?
Explain carefully its main features and how it differs from a simple amplifier.
Given two signal sources A and B, how would you set up a circuit to create (i) B−A, (ii) B + 2A? *(O. and C.)*

5. (a) Draw a diagram of an inverting amplifier with a gain of 50 designed using an operational amplifier (op amp). (You need not show the power supply connections on your diagram.)
(b) Draw a diagram of a non-inverting amplifier with a gain of 50 designed using an op amp. (You need not show the power supply connections on your diagram.)
(c) Draw a sketch graph, with labelled axes, of the output voltage against the input voltage for the circuit of Fig. 65.5. (Consider input voltages between −3 V and +3 V.)
Suggest a use for such a circuit. *(L.)*

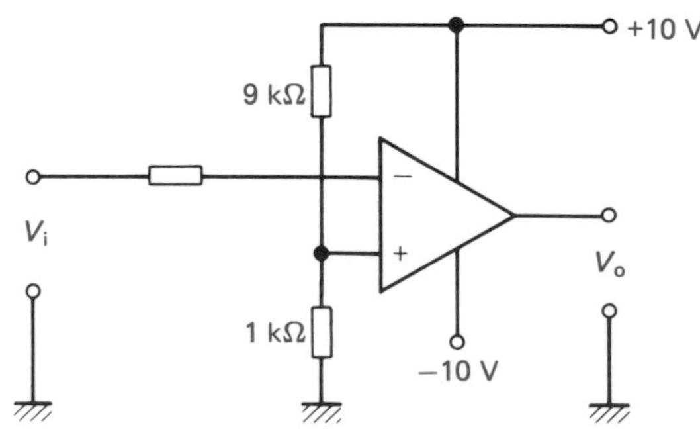

Fig. 65.5

6. This is a question about operational amplifiers and their use.
Explain what is meant by (a) feedback, (b) virtual earth, (c) infinite gain approximation.
Draw a circuit which gives the difference between two input voltages as its output. How is the circuit modified to give an output *K* times the difference of the inputs, where *K* is a constant lying between 1 and 100?
Draw another circuit which gives the integral of an input voltage with respect to time. Account for the output voltage which results when a d.c. voltage is applied at the input to this circuit. (*O. and C.*)

7. A candidate for this examination wished to build a circuit for his project which would operate a relay when the sound level in a room exceeded a particular value. He devised the circuit in Fig. 65.6.

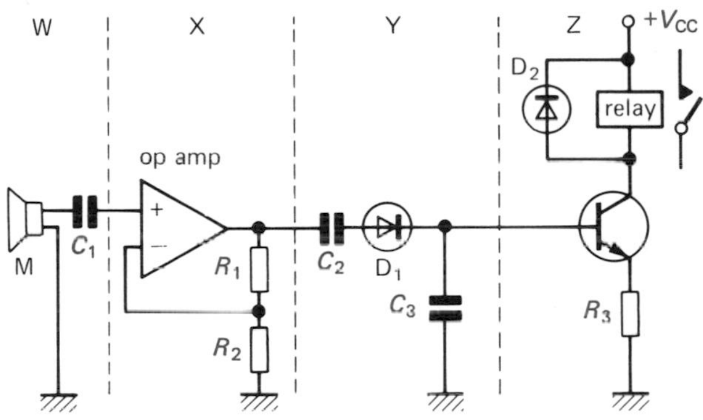

Fig. 65.6

Identify the functions of the sections W, X, Y and Z of the circuit shown.
Explain the purpose of each of the components in section Y.
Explain the purpose of each of the components in section Z.
Explain how a sound above a certain level will cause the relay to operate.
How could the sensitivity of the system be adjusted? (*L.*)

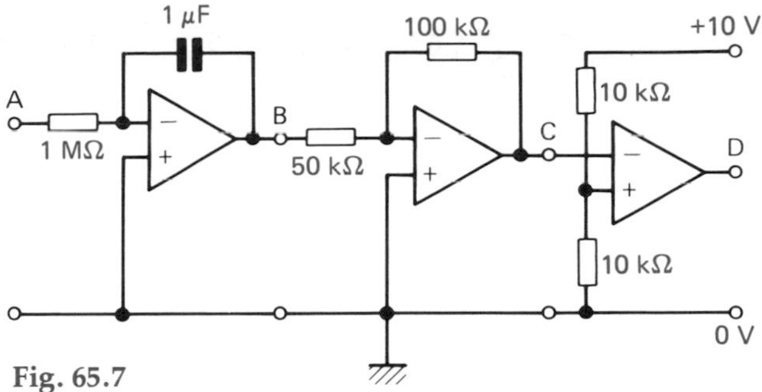

Fig. 65.7

8. The three op amp circuits in Fig. 65.7 are connected as shown. They operate on a ±10 V power supply.
On the same set of axes draw three graphs to show how the voltages at B, C and D vary with time when +1 V is applied to A. Mark clearly the voltage and time scales used.

9. The light-operated switch in Fig. 65.8 uses an op amp as a voltage comparator.

(a) How must V_1 and V_2 compare if the op amp output is to be negative in daylight?

(b) In darkness what happens to (i) the LDR, (ii) V_2 compared with V_1, (iii) the output of the op amp, (iv) Tr, (v) the relay?

(c) How would you alter the circuit to make the relay be off in the dark and switch on in daylight?

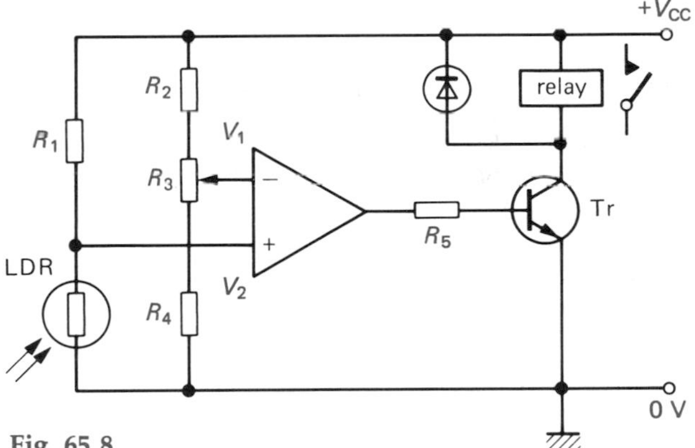

Fig. 65.8

Digital electronics

66 Logic gates

Digital electronics

Digital electronics is concerned with two-state, *switching-type* circuits in which signals are in the form of electrical *pulses*, Fig. 66.1. Their outputs and inputs involve only *two* levels of voltage, referred to as 'high' and 'low'. 'High' is near the supply voltage, e.g. +5 V and 'low' is near 0 V and, in the scheme known as 'positive logic', 'high' is referred to as *logic level 1* (or just 1) and 'low' as *logic level 0* (or just 0).

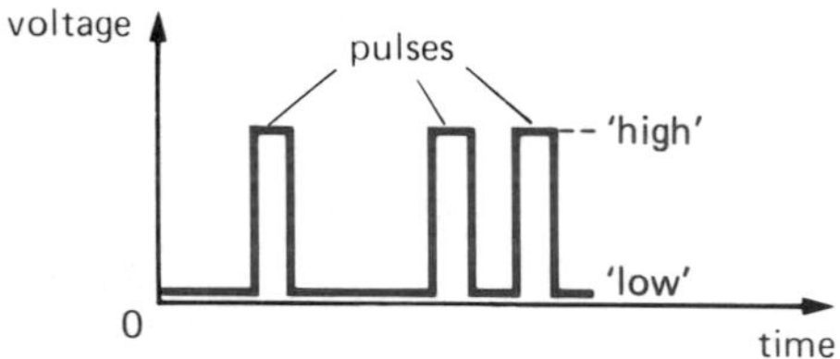

Fig. 66.1

Digital circuits are used in pocket calculators (p. 163), electronic watches (p. 162), digital computers (p. 203), in control systems (e.g. for domestic appliances such as washing machines, or for robots or production processes in industry), in television and other games, and increasingly in telecommunications (p. 166).

Today, digital circuits are made as integrated circuits (ICs, p. 72) and are more complex than the discrete component versions from which they developed.

Types of logic gate

The various kinds of logic gate are switching circuits that 'open' and give a 'high' output depending on the combination of signals at their inputs, of which there is usually more than one. They are decision-making circuits using what is known as *combinational logic*. The behaviour of each kind is summed up in a *truth table* showing in terms of 1s ('high') and 0s ('low') what the output is (1 or 0) for all possible inputs.

(a) NOT gate or inverter. It has one input and one output and the simple transistor switch of Fig. 31.1 behaves as one. The circuit is shown again in Fig. 66.2 along with the British and American symbols and the truth table. It produces a 'high' output (e.g. +6 V) if the input is 'low' (e.g. 0 V) and vice versa. That is, the output is 'high' if the input is *not* 'high' and whatever the input, the gate 'inverts' it.

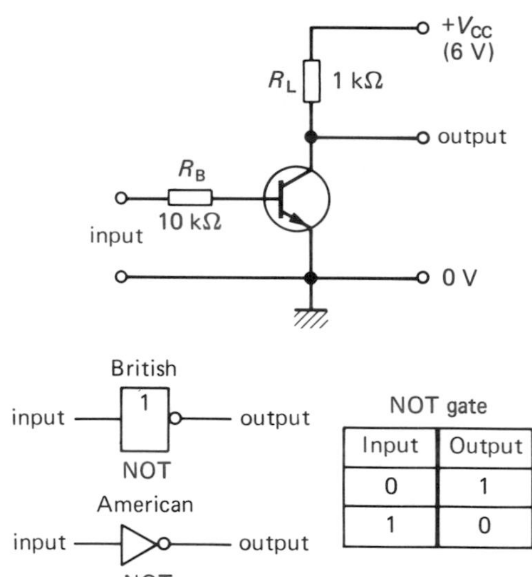

NOT gate

Input	Output
0	1
1	0

Fig. 66.2

The circuit of Fig. 66.2 contains only resistors and a transistor and is said to use Resistor-Transistor-Logic, i.e. RTL.

(b) NOR gate. This is a NOT gate with two or more inputs. The RTL circuit for two inputs A and B is shown in Fig. 66.3, with symbols and truth table. If either or both of A and B are 'high', the transistor is switched on and saturates and the output F is 'low'. When both inputs are 'low', i.e. neither A *nor* B is 'high', F is 'high'.

(c) OR gate. This is a NOR gate followed by a NOT gate; it is shown in symbol form with two inputs in

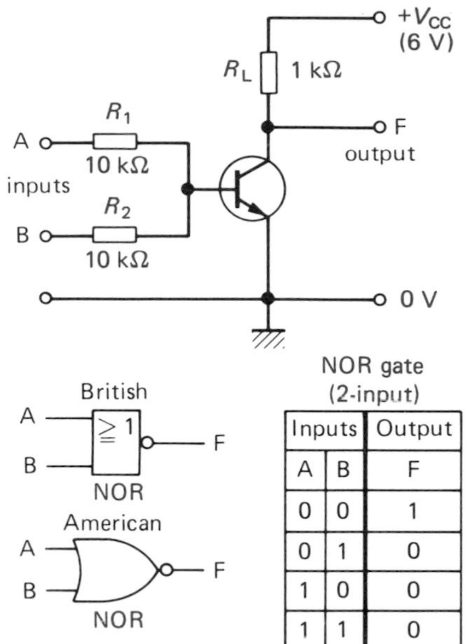

Inputs		Output
A	B	F
0	0	1
0	1	0
1	0	0
1	1	0

Fig. 66.3

Fig. 66.4. The truth table can be worked out from that of the NOR gate by changing 0s to 1s and 1s to 0s in the output. F is 1 when either A *or* B *or* both is a 1, i.e. if any of the inputs is 'high', the output is 'high'. (*Note:* the ≧ sign is sometimes omitted from the British symbols for NOR and OR gates.)

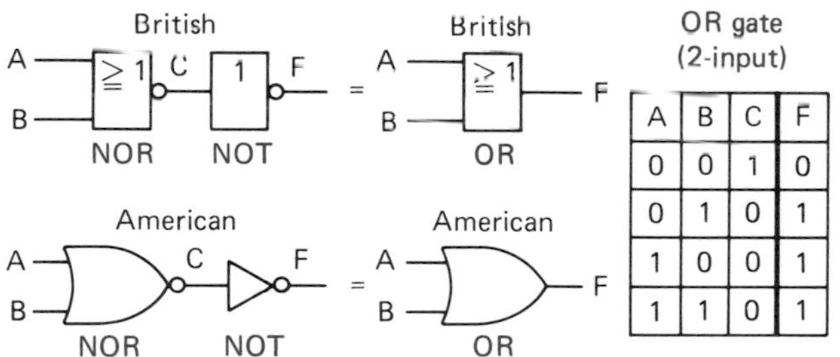

A	B	C	F
0	0	1	0
0	1	0	1
1	0	0	1
1	1	0	1

Fig. 66.4

(d) NAND gate. A circuit for a two-input NAND gate using Diode-Transistor Logic (DTL) is given in Fig. 66.5, along with the symbols and truth table.

When inputs A and B are both 'high', e.g. connected to +6 V (or are left disconnected, i.e. floating), the silicon diodes D_1 and D_2 are reverse biased and behave like high resistors. If V_1 reaches about 1.2 V and V_{BE} about 0.6 V (0.6 V or so is dropped across the forward-biased diode D_3), the transistor turns on. It saturates provided the base current in R_B is large enough, giving a 'low' output.

When either or both A and B are 'low', e.g. connected to 0 V, the current through R_B is by-passed to

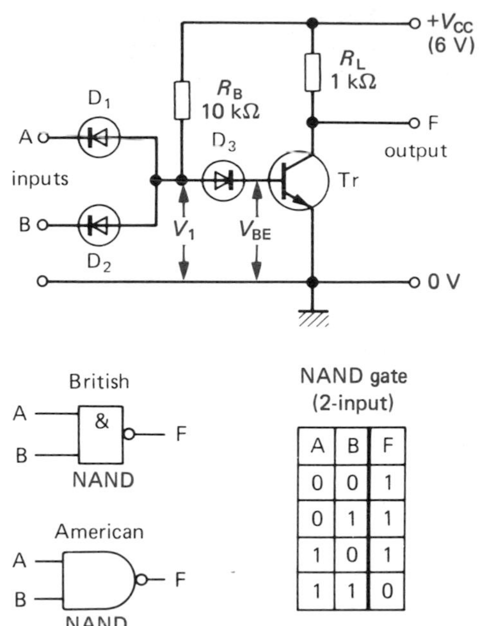

A	B	F
0	0	1
0	1	1
1	0	1
1	1	0

Fig. 66.5

ground via D_1 or D_2 (or both). The transistor is off, because V_1 (the voltage across the conducting diode(s) is now held at +0.6 V or so, giving a 'high' output. (In practice a resistor is needed between the base of Tr and 0 V to provide a path to ground for the minority carriers in the base and so ensure that Tr switches off.)

To sum up, if A *and* B are both 'high', the output is *not* 'high'. (The gate gets its name from this AND NOT behaviour.) Any other input combination gives a 'high' output.

(e) AND gate. This is a NAND gate followed by a NOT gate; it is shown with two inputs in symbol form in Fig. 66.6. The truth table may be obtained from that of the NAND gate by changing 0s to 1s and 1s to 0s in the output. The output is 1 only when A *and* B are 1.

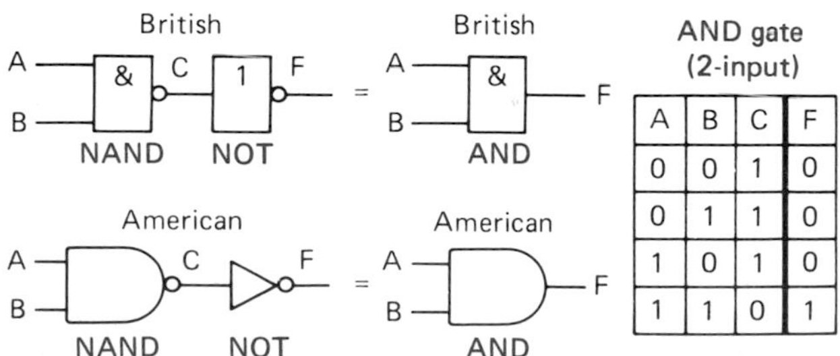

A	B	C	F
0	0	1	0
0	1	1	0
1	0	1	0
1	1	0	1

Fig. 66.6

(f) Exclusive-OR gate. It is an OR gate with only two inputs which gives a 'high' output when either input is 'high' but not when both are 'high'. Unlike the ordinary OR gate (sometimes called the *inclusive*-OR gate) it excludes the case of both inputs being 'high' for a 'high' output. It is also called the *difference* gate because the output is 'high' when the inputs are different.

The symbols and truth table are shown in Fig. 66.7.

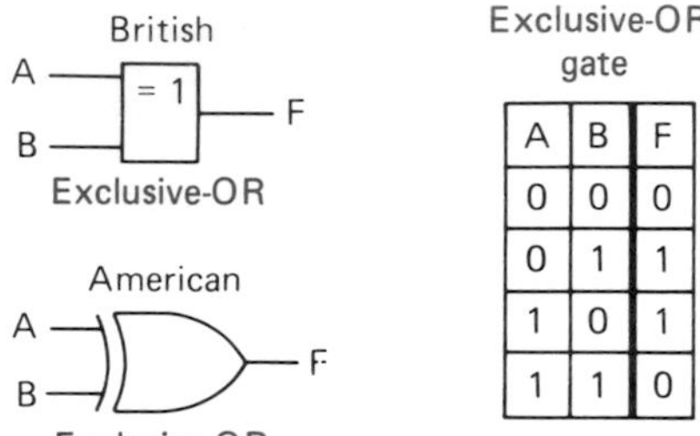

A	B	F
0	0	0
0	1	1
1	0	1
1	1	0

Fig. 66.7

(g) Exclusive-NOR gate. This is a NOR gate with only two inputs which gives a 'high' output when the inputs are equal, i.e. are both 0's or both 1's, Fig. 66.8. For this reason it is also called a *parity* or *equivalence* gate.

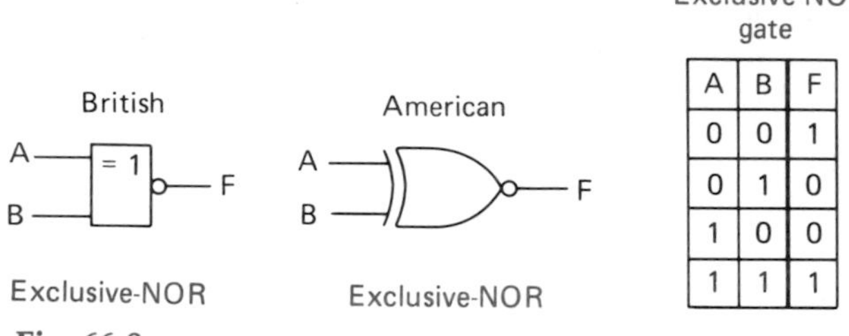

A	B	F
0	0	1
0	1	0
1	0	0
1	1	1

Fig. 66.8

(h) Summary. Try to remember the following:

NOT: output always *opposite* of input
NOR: output 1 *only* when all inputs 0
OR: output 1 *unless* all inputs 0
NAND: output 1 *unless* all inputs 1
AND: output 1 *only* when all inputs 1
excl-OR: output 1 when inputs *different*
excl-NOR: output 1 when inputs *equal*

Other gates from NAND gates

Other logic gates (and circuits) can be made by combining only NAND gates (or only NOR gates), a fact which is often used in logic circuit design, as we will see later (p. 141).

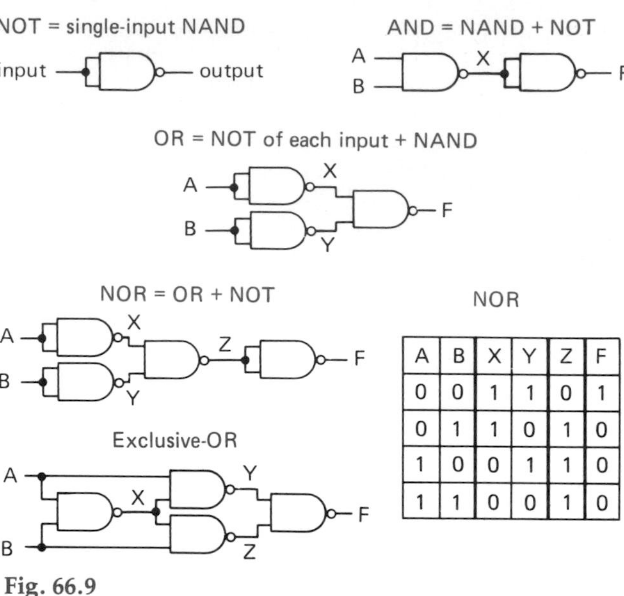

A	B	X	Y	Z	F
0	0	1	1	0	1
0	1	1	0	1	0
1	0	0	1	1	0
1	1	0	0	1	0

Fig. 66.9

The NAND gate equivalents for different gates are shown in Fig. 66.9. You can check that each one gives the correct output for the various input combinations by constructing a stage-by-stage truth table, as has been done for the NOR gate made from four NAND gates. Note:

(a) NOT = single-input NAND (or NOR), made by joining all the inputs together,
(b) AND = NAND followed by NOT,
(c) OR = NOT of each input followed by NAND,
(d) NOR = OR followed by NOT.

Questions

1. The electrical circuits in Fig. 66.10 can be regarded as logic gates. Which type is each one? (Treat A and B as inputs and F as the output.)

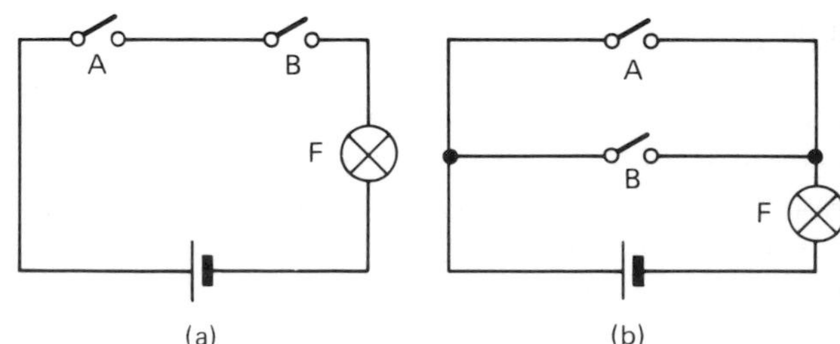

Fig. 66.10

2. Write down the truth table for the circuit (given in British and American symbols) in Fig. 66.11.

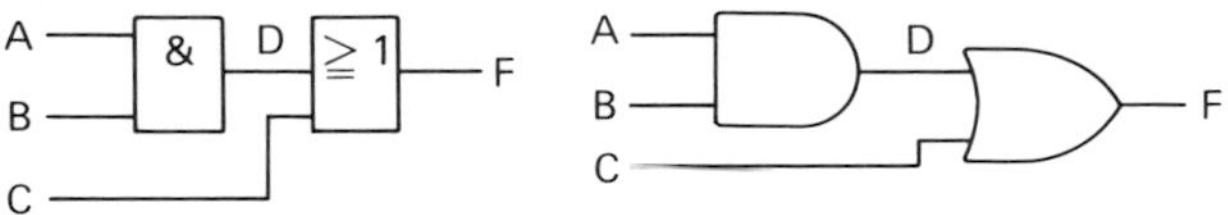

Fig. 66.11

3. Voltages with the waveforms in Fig. 66.12 are applied one to each input of a two-input (a) AND gate, (b) NOR gate. Draw the output waveform for each.

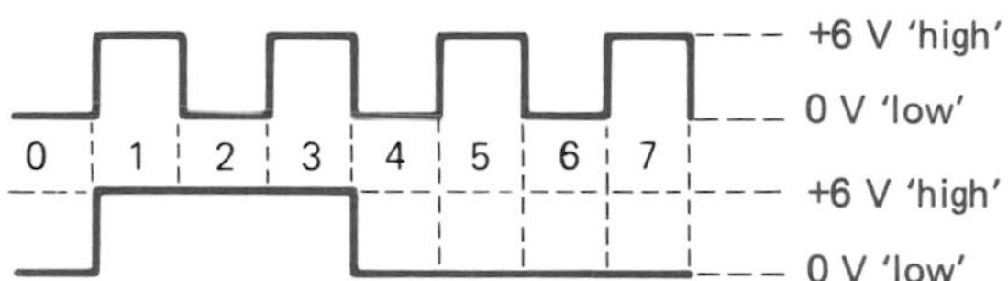

Fig. 66.12

4. The circuit in Fig. 66.13 is of a logic gate using 'diode-resistor' logic. A and B are inputs and F is the output. Write down a truth table and say what kind of gate it is.

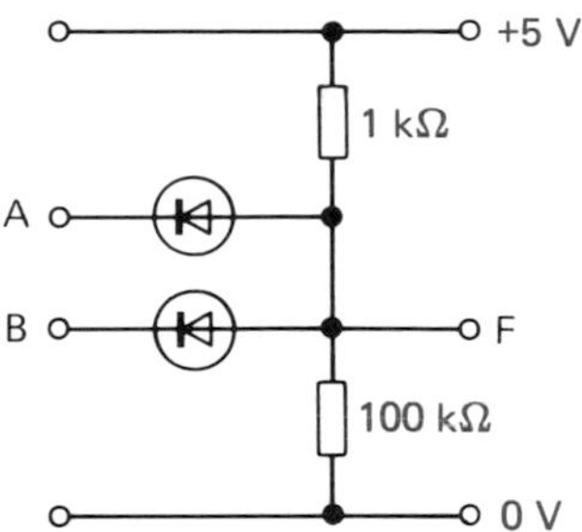

Fig. 66.13

67 Logic families

Introduction

Different types of circuitry or logic can be used to construct logic gates. RTL (p. 132) and DTL (p. 133) were the earliest but are now more or less obsolete. The IC logic gates available today fall into two main groups or families and are based on two transistors to avoid problems that arose with the first simple gates.

One family is TTL (standing for Transistor-Transistor Logic) using bipolar transistors; the other is CMOS (standing for Complementary Metal Oxide Semiconductor and pronounced 'see-moss') based on FETs.

TTL ICs are made as the '7400' series; the figures following '74' indicate the nature of the IC, e.g. the kind of logic gate. CMOS ICs belong to the '4000 B' series; the numbers after '4' depend on what the IC does. Both types are usually in the form of 14 or 16 pin d.i.l. packages (Fig. 34.2, p. 73), often with several gates on the same chip.

For example a quad two-input NAND gate contains four identical NAND gates, with two inputs and one output per gate. Every gate therefore has three pins, making fourteen pins altogether on the package. This includes two for the positive and negative power supply connections which are common to all four gates. Pin connections for the 7400 and 4011B quad two-input NAND gates are given in Fig. 67.1 (using American symbols since these are favoured by component manufacturers and most journals).

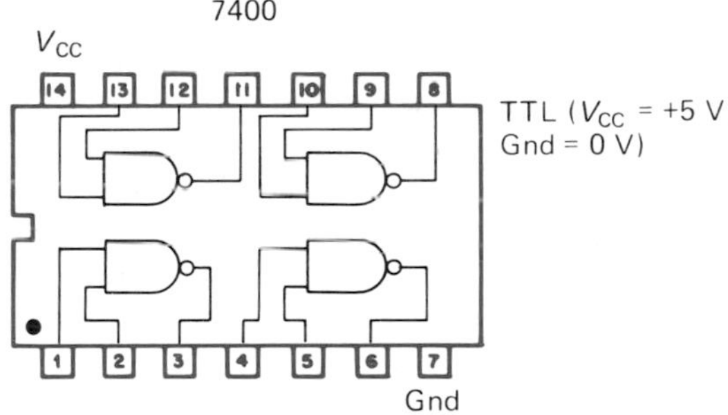

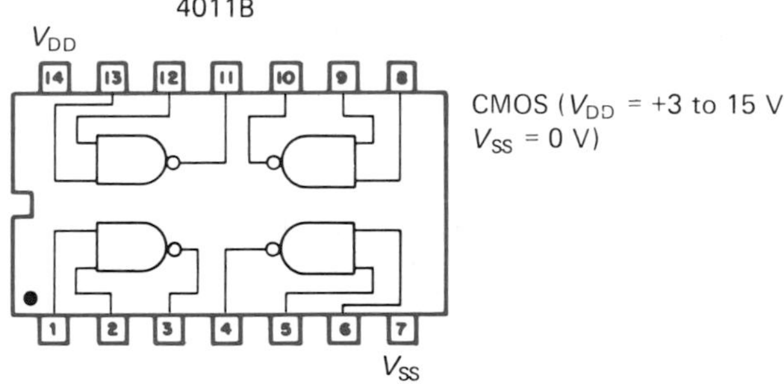

Fig. 67.1

Comparison of TTL and CMOS ICs

Properties of each family are summarized in Table 67.1 and explained below.

Table 67.1

Property	TTL	CMOS
Power supply	5 V ± 0.25 V d.c.	3 V to 15 V d.c.
Current required	Milliamperes	Microamperes
Input impedance	Low	Very high
Switching speed	Fast	'Slow'
Fan-out	Ten	Fifty

(a) Power supply. TTL requires a stabilized 5 V supply while CMOS will work on any unstabilized d.c. voltage between 3 V and 15 V.

(b) Current requirements. In general these are much less (roughly one thousand times) for CMOS than for TTL.

(c) Input impedance. The very high input impedance of a CMOS IC (due to the use of FETs) accounts for its low current consumption but it can allow static electric charges to build up on input pins when for example, they touch insulators, e.g. plastics, clothes, in warm dry conditions. This does not happen with TTL ICs because the low input impedance of bipolar transistors ensures that any charge leaks harmlessly through junctions in the IC.

The static voltages on CMOS ICs can destroy them and so they are supplied in anti-static or conductive carriers. Also, protective circuits are incorporated which operate when the IC is connected to the power supply.

(d) Switching speed. This is much faster for TTL than for CMOS, the switching delay time for TTL being about 10 nanoseconds ($1\,\text{ns} = 10^{-9}\,\text{s}$) compared with 300 ns for CMOS.

(e) Fan-out. For TTL this is ten, which means ten other TTL logic gates can take their inputs from one TTL output and switch reliably, otherwise overloading occurs. For CMOS a fan-out of about fifty is typical for one CMOS IC driving other CMOS ICs (due to their very high input impedance).

(f) Unused inputs. Unused TTL inputs assume logic level 1 unless connected to 0 V; it is good practice to connect them to +5 V. Unused CMOS inputs must be connected to supply positive or negative, depending on the circuit and not left to 'float', otherwise erratic behaviour occurs.

Logic levels and interfacing

(a) Logic levels. Ideally the 'low' and 'high' input and output voltages representing logic levels 0 and 1 respectively should be 0 V and V_{CC} (i.e. +5 V) for TTL. In practice, due to voltage drops across resistors and transistors inside the IC, these values are not achieved.

For example an input voltage of from 0 to 0.8 V behaves as 'low', while one from 2 V to 5 V is 'high' because it can cause an output to change state. Similarly a 'low' output may be 0 to 0.4 V (since the voltage across a saturated transistor is not 0 V) and 'high' is from 2.4 V to 5 V since any voltage in this range can operate other TTL ICs, Fig. 67.2a.

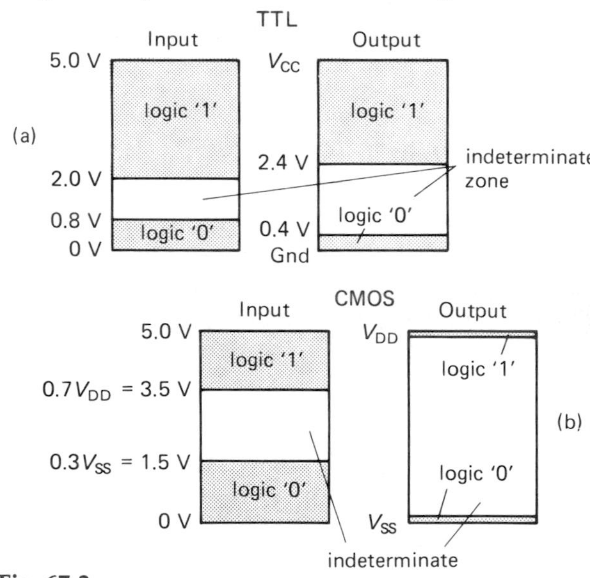

Fig. 67.2

The values for CMOS are given in Fig. 67.2b. The ideal output values are more or less attained in this case and so the output voltage (when unloaded) can swing from zero to near V_{DD}, which is greater than for TTL.

With both families intermediate voltage values (produced for example by overloading outputs) cause faulty switching.

(b) Mixing TTL and CMOS. If operated from a common 5 V supply, the logic levels for TTL and CMOS ICs would be different, i.e. the two families are not compatible even though a TTL IC has the current capability to drive any number of CMOS ICs. For example, the worst-case value of a TTL 'high' output (2.4 V) is lower than the minimum input required by CMOS (3.5 V) to ensure switching. Hence, using TTL and CMOS ICs in the same circuit (to obtain the advantages of each) creates 'interfacing' problems which fortunately can be solved.

Question

1. (a) Possible circuits for a logic level indicator are given in Fig. 67.3a, b, c. In which circuit(s) will a logic level 1 (i.e. +5 V) applied to the input light up the LED?
(b) Table 67.2 for typical TTL input and output sourcing and sinking currents on a 5 V supply, shows that a TTL gate can sink much more current than it can source (e.g. for an output, 16 mA compared with 400 μA).

If when checking TTL logic levels we wish (i) to make a logic 1 light an LED (needing typically 10 mA) and (ii) to avoid overloading a logic 1 output connected to the indicator input (so preventing the logic 1 output voltage from falling into the indeterminate region between 2.4 V and 0.4 V in Fig. 67.2 and producing faulty switching), which circuit should be used for both conditions to be satisfied?
(c) From Table 67.2 work out how many other TTL gate inputs can be driven by one TTL gate output.
(d) The output of a CMOS device can sink or source about 10 mA on a 9 V supply. Which circuit(s) in Fig. 67.3a, b, c, would satisfy the two requirements in (b) if a 9 V supply is used and a 680 Ω resistor replaced the 300 Ω one?

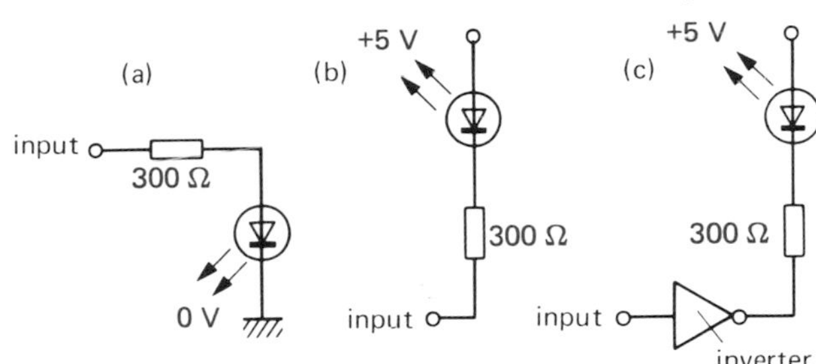

Fig. 67.3

Table 67.2

TTL	**logic 1 (source)**	**logic 0 (sink)**
output	400 μA	16 mA
input	40 μA	1.6 mA

68 Binary adders

Codes

Normally we count on the scale of ten or decimal system using the ten digits 0 to 9. When the count exceeds 9, we place a 1 in a second column to the left of the units column to represent tens. A third column to the left of the tens column gives hundreds and so on. The values of successive columns starting from the right are 1, 10, 100, etc., or in powers of ten, 10^0, 10^1, 10^2, etc.

(a) Binary code. Counting in electronic systems is done by digital circuits on the scale of two or *binary* system. The digits 0 and 1 (called 'bits', from binary digits) are used and are represented electrically by 'low' and 'high' voltages respectively. Many more columns are required since the number after 1 in binary is 10, i.e. 2 in decimal but the digit 2 is not used in the binary system.

Successive columns in this case represent, from the right, powers of 2, i.e. 2^0, 2^1, 2^2, 2^3, etc., or in decimal 1, 2, 4, 8, etc. Table 68.1 shows how the decimal numbers from 0 to 15 are coded in the binary system; a four-bit code is required. Note that the *least significant bit* (l.s.b.), i.e. the 2^0 bit, is on the extreme right in each number and the *most significant bit* (m.s.b.) on the extreme left.

For large numbers, higher powers of two have to be used.

For example:

decimal 29 $= 1 \times 16 + 1 \times 8 + 1 \times 4 + 0 \times 2 + 1 \times 1$
$= 1 \times 2^4 + 1\times 2^3 + 1 \times 2^2 + 0 \times 2^1 + 1 \times 2^0$
= 11101 in binary (a five-bit code)

Table 68.1

Decimal		*Binary*			
10^1 (10)	10^0 (1)	2^3 (8)	2^2 (4)	2^1 (2)	2^0 (1)
0	0	0	0	0	0
0	1	0	0	0	1
0	2	0	0	1	0
0	3	0	0	1	1
0	4	0	1	0	0
0	5	0	1	0	1
0	6	0	1	1	0
0	7	0	1	1	1
0	8	1	0	0	0
0	9	1	0	0	1
1	0	1	0	1	0
1	1	1	0	1	1
1	2	1	1	0	0
1	3	1	1	0	1
1	4	1	1	1	0
1	5	1	1	1	1

(b) Binary coded decimal (BCD). This is a popular, slightly less compact variation of binary but in practice it allows easy conversion back to decimal for display purposes. Each *digit* of a decimal number is coded in binary instead of coding the whole number. For example:

decimal 29 = 0010 1001 in BCD

In binary, decimal 29 requires only five bits. It is obviously essential to know which particular binary code is being used.

Half-adder

Adders are electronic circuits which perform addition in binary and consist of combinations of logic gates.

(a) Adding two bits. A half-adder adds *two bits at a time* and has to deal with four cases (essentially three since two are the same). They are:

$$\begin{array}{cccc} 0 & 0 & 1 & 1 \\ +0 & +1 & +0 & +1 \\ \hline 0 & 1 & 1 & 10 \end{array}$$

In the fourth case, 1 plus 1 equals 2, which in binary is 10, i.e. the right column is 0 and 1 is carried to the next column on the left. The circuit for a half-adder must therefore have two inputs, i.e. one for each bit to be added, and two outputs, i.e. one for the *sum* and one for the *carry*.

(b) Circuit. One way of building a half-adder from logic gates uses an exclusive-OR gate and an AND gate. From their truth tables in Fig. 68.1a and b, you can see that the output of the exclusive-OR gate is always the *sum* of the addition of two bits (being 0 for 1 + 1), while the output of the AND gate equals the *carry* of the two-bit addition (being 1 only for 1 + 1). Therefore if both bits are applied to the inputs of both gates at the same time, binary addition occurs.

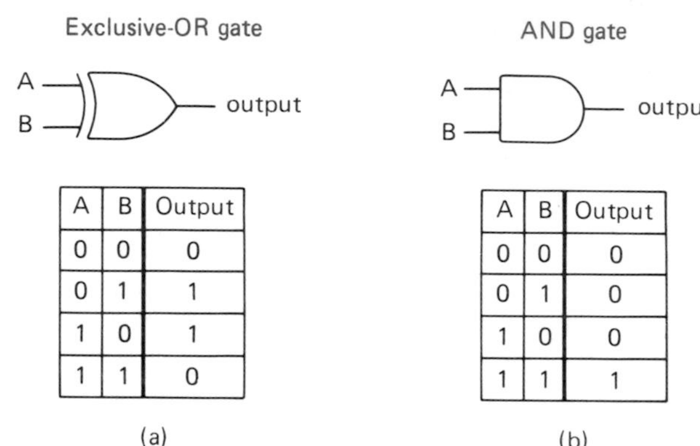

A	B	Output
0	0	0
0	1	1
1	0	1
1	1	0

A	B	Output
0	0	0
0	1	0
1	0	0
1	1	1

Fig. 68.1

The circuit is shown with its two inputs A and B in Fig. 68.2, along with the half-adder truth table.

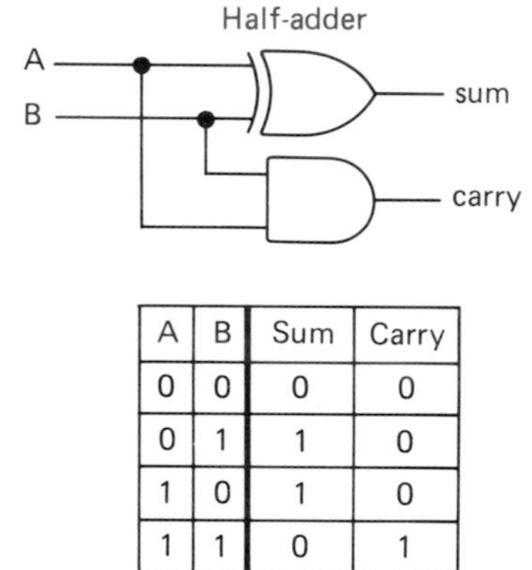

A	B	Sum	Carry
0	0	0	0
0	1	1	0
1	0	1	0
1	1	0	1

Fig. 68.2

Full-adder

(a) Adding three bits. This is a necessary operation when two multi-bit numbers are added. For example, to add 3 (11 in binary) to 3 we write:

$$\begin{array}{r} 11 \\ +11 \\ \hline 110 \end{array}$$

The answer 110 (6 in decimal) is obtained as follows. In the least significant (right-hand) column we have

1 + 1 = sum 0 + carry 1

In the next column three bits have to be added because of the carry from the first column, giving

1 + 1 + 1 = sum 1 + carry 1

(b) Circuit. A full-adder circuit therefore needs three inputs A, B, C (two for the digits and one for the carry) and two outputs (one for the *sum* and the other for the *carry*). It is realized by connecting two half-adders (HA) and an OR gate, as in Fig. 68.3a. We can check that it produces the correct answer by putting A = 1, B = 1 and C = 1, as in Fig. 68.3b.

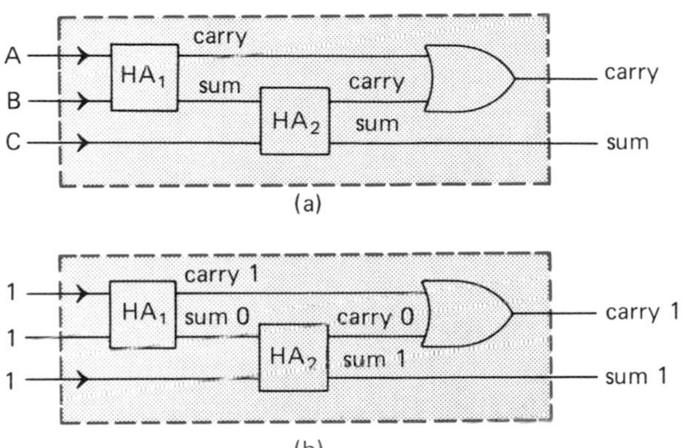

Fig. 68.3

The first half-adder HA_1 has both inputs 1 and so gives a sum of 0 and a carry of 1. HA_2 has inputs 1 and 0 and gives a sum of 1 (i.e. the sum output of the full-adder) and a carry of 0. The inputs to the OR gate are 1 and 0 and, since one of the inputs is 1, the output (i.e. the carry output of the full-adder) is 1. The addition of 1 + 1 + 1 is therefore *sum* 1 and *carry* 1.

The truth table with the other three-bit input combinations is given in Fig. 68.4. You may like to check that the circuit does produce the correct outputs for them.

Full-adder

Inputs			Outputs	
A	B	C	Sum	Carry
0	0	0	0	0
0	0	1	1	0
0	1	0	1	0
1	0	0	1	0
0	1	1	0	1
1	0	1	0	1
1	1	0	0	1
1	1	1	1	1

Fig. 68.4

Multi-bit adder

Two multi-bit binary *numbers* are added by connecting adders in parallel. For example, to add two four-bit numbers, four adders are needed, as shown in Fig. 68.5 for the addition of 1110 (decimal 14) and 0111 (decimal 7). Follow it through to see that the sum is 10101 (decimal 21). Strictly speaking the full-adder FA_1 need only be a half-adder since it only handles two bits.

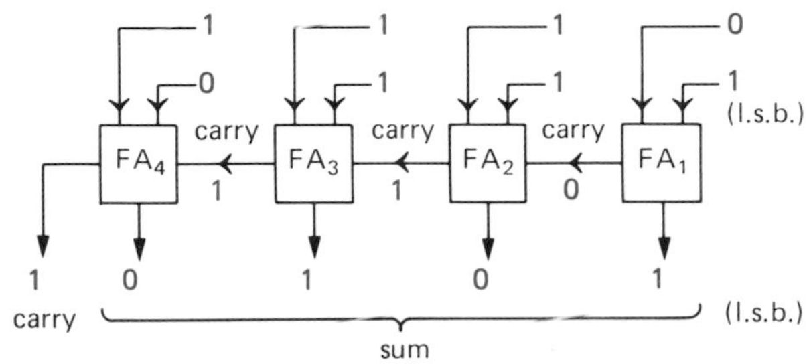

Fig. 68.5

The largest binary numbers that can be added by a four-bit adder are 1111 and 1111, i.e. 15 + 15 = 30. By connecting more full-adders to the left end of the system, the capacity increases.

In the block diagram for a *four-bit parallel adder*, shown in Fig. 68.6, the four-bit number $A_4A_3A_2A_1$ is added to $B_4B_3B_2B_1$, A_1 and B_1 being the least significant bits (l.s.b.). $S_4S_3S_2$ and S_1 represent the sums and C_0, the most significant bit (m.s.b.) of the answer, is the carry of the output.

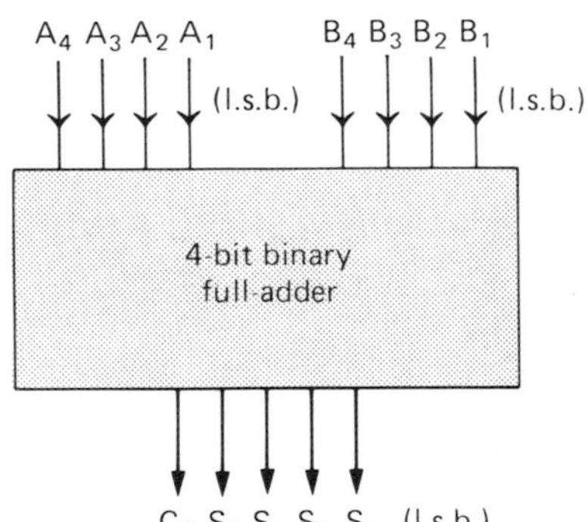

Fig. 68.6

A disadvantage of the adder in Fig. 68.5 is that each stage must wait for the carry from the preceding one before it can decide its own sum and carry, i.e. the carry ripples down from right to left. This is called 'ripple carry addition' and it can be speeded up by including in the adder extra circuits which predict immediately all the carries. In such a 'look-ahead carry adder', all bits are added at the same time giving much faster addition.

Binary subtraction

One method, called the *one's complement method*, is best explained by an example. Suppose we have to subtract 0110 (6) from 1010 (10). We proceed as follows.

(a) Form the one's complement of 0110; this is done by changing 1s to 0s and 0s to 1s giving 1001.

(b) Add the one's complement from (a) to the number from which the subtraction is to be made, i.e. add 1001 and 1010:

```
   1010
 + 1001
  10011
```

(c) If there is a 1 carry in the most significant position of the total in (b), remove it and add it to the remaining four bits to get the answer:

```
[1] [0011] →  0011
 └─────────→ +0001
              0100   (4)
```

The 1 carry that is added, is called the 'end-around carry' (EAC). When there is an EAC, the answer to the subtraction is positive (as above). If there is no EAC, the answer is negative and is in one's complement form. For example, subtracting 0101 (5) from 0011 (3), we get:

```
  0011
 +1010 (one's complement of 0101)
  1101  →  −0010
```

The one's complement of 0101 (i.e. 1010) has been added to 0011 to give 1101. There is no EAC and so to get the final answer we take the one's complement of 1101, i.e. 0010 and put a negative sign in front to give −0010 (−2).

Binary subtraction can thus be done simply by addition and is implemented electronically by an adder since the one's complement of a number is easily obtained by an inverter. A simplified circuit of a four-bit 'subtractor' is shown in Fig. 68.7 where the four-bit number B is subtracted from the four-bit number A. There is an EAC which is applied to the carry-in input of the adder and the answer is S.

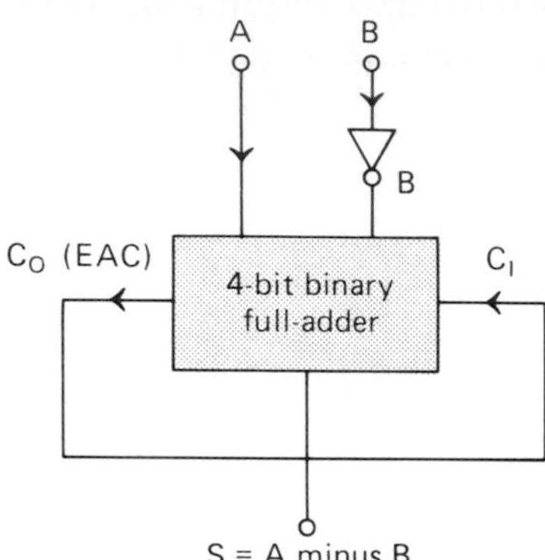

Fig. 68.7

Multiplication (p. 209) and division can be performed by repeated addition and subtraction respectively.

Questions

1. Make the following code conversions:
(a) Decimal numbers 4, 13, 21, 38, 64 to binary.
(b) Binary numbers 111, 11001, 101010, 110010 to decimal.
(c) Decimal numbers 9, 17, 28, 370, 645 to BCD.

2. Add the following pairs of binary numbers and check your answers by converting the binary numbers to decimal:
(a) 10 + 01; (b) 11 + 10; (c) 101 + 011; (d) 1011 + 0111.

3. Subtract the following pairs of numbers by the one's complement method and check your answers by converting the binary numbers to decimal:
(a) 1001 − 0101; (b) 11001 − 00111; (c) 0011 − 0110; (d) 01100 − 10101.

69 Logic circuit design I

Many digital circuits for performing logical operations are built from logic gates. Once it is known what the circuit has to do, a truth table can be drawn up and used as the starting point for the design. The method will be used in this chapter (and the next two) to construct three logic circuits studied previously.

In practice, logic circuits can be made using only NAND (or NOR) gates, as we will see.

Exclusive-OR gate circuit

Lines 2 and 3 of the truth table in Fig. 69.1a show that the output F is 1 only when:

A is 0 *AND* B is 1
OR
A is 1 *AND* B is 0

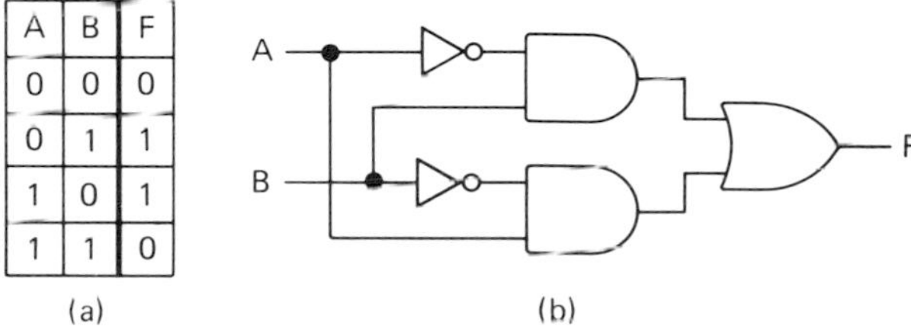

A	B	F
0	0	0
0	1	1
1	0	1
1	1	0

Fig. 69.1

The method requires this statement to be rewritten so that both inputs appear as 1s. Therefore if A is 0, inverting A with a NOT gate makes the input (called 'not A' and written $\bar{A}$) a 1. Similarly instead of writing B as 0, we can say $\bar{B}$ (i.e. not B) is 1. The statement, which is the same as before, becomes F is 1 when:

$\bar{A}$ is 1 *AND* B is 1
OR
A is 1 *AND* $\bar{B}$ is 1

Using the notation of the mathematics of logic circuit design, called *Boolean algebra*, we get

$$F = \bar{A}.B + A.\bar{B} \qquad (1)$$

where a dot (.) represents the AND logic operation and a plus (+) indicates the OR operation.

Using *any* logic gates this means the circuit consists of two two-input AND gates (each with a NOT gate in one input) feeding a two-input OR gate as in Fig. 69.1b.

To implement it with only NAND gates we use the facts stated earlier (p. 134):

NOT = one-input NAND
AND = NAND followed by NOT
OR = NOT of each input followed by NAND

The circuit becomes that of 69.2a but inverting an input twice gives the original input, i.e. the successive NOT gates 6 and 7, also 8 and 9, cancel each other, so only NAND gates 1, 2, 3, 4 and 5 are needed. The equivalent NAND gate circuit is given in Fig. 69.2b; although different from the exclusive-OR circuit in Fig. 66.9, it performs the same logical operation. Note that $\bar{A}$ and $\bar{B}$ are obtained by inverting A and B with one-input NAND gates (1 and 3), i.e. with NOT gates.

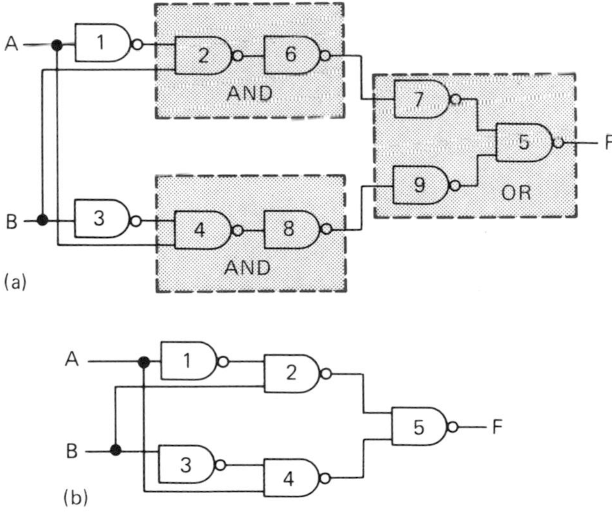

Fig. 69.2

Half-adder circuit

From lines 2 and 3 of the truth table in Fig. 69.3a we see that the SUM output S of the two bits A and B to be added, is 1 only when:

A is 0 *AND* B is 1
OR
A is 1 *AND* B is 0

Fig. 69.3

A	B	S	C
0	0	0	0
0	1	1	0
1	0	1	0
1	1	0	1

(a)

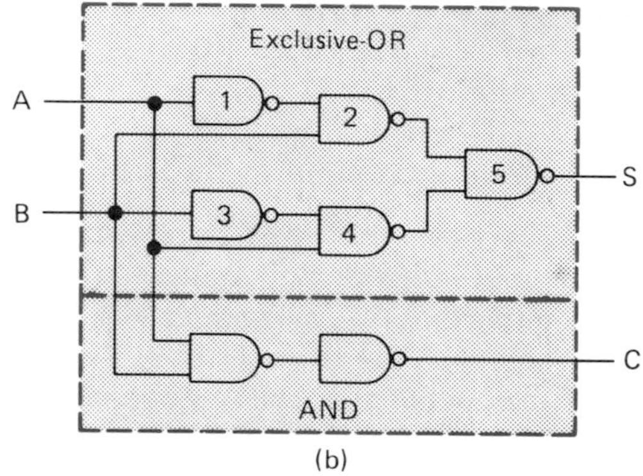

(b)

Following the previous procedure we can therefore write in Boolean notation

$$S = \bar{A}.B + A.\bar{B}$$

This is the same as equation (1), i.e. S is given by an exclusive-OR gate.

Also, from line 4 of the truth table, the CARRY output C is 1 only when:

A is 1 *AND* B is 1

That is, $C = A.B$

This is implemented by a two-input AND gate. The NAND equivalent of the complete half-adder circuit is given in Fig. 69.3b.

Full-adder circuit

A full-adder for adding three bits can be made from two half-adders and an OR gate (as in Fig. 68.3a). It can be designed from first principles using the truth table in Fig. 69.4a.

The Boolean expression for the SUM output is (from lines 2, 3, 4 and 8),

$$SUM = \bar{A}.\bar{B}.C + \bar{A}.B.\bar{C} + A.\bar{B}.\bar{C} + A.B.C$$

The circuit in Fig. 69.4b gives this output. It uses four three-input AND gates (with some inputs inverted) and a four-input OR gate. The simplified equivalent NAND gate circuit is shown on the top half of Fig. 69.5, the cases in which there are two successive NOT gates have been omitted.

Fig. 69.4

A	B	C	Sum	Carry
0	0	0	0	0
0	0	1	1	0
0	1	0	1	0
1	0	0	1	0
0	1	1	0	1
1	1	0	0	1
1	0	1	0	1
1	1	1	1	1

(a)

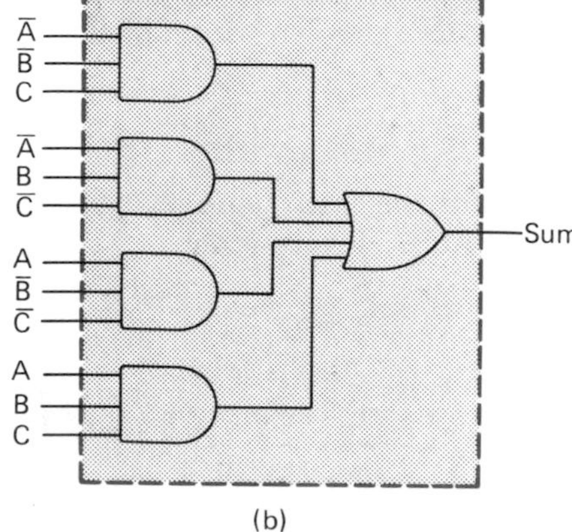

(b)

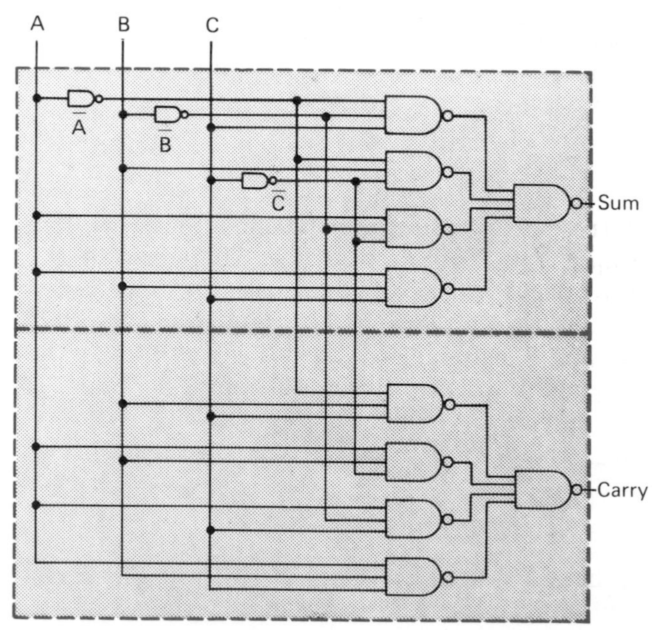

Fig. 69.5

Similarly the Boolean expression for the CARRY output is obtained from lines 5, 6, 7 and 8 of the truth table and is

$$CARRY = \bar{A}.B.C. + A.B.\bar{C} + A.\bar{B}.C + A.B.C$$

It can also be implemented using four AND gates and an OR gate. The lower half of Fig. 69.5 gives the simplified NAND gate equivalent: the whole circuit is for a full-adder using only NAND gates.

Questions

1. Write the truth table for (a) a two-input NAND gate, (b) a two-input NOR gate, (c) the logic circuit in Fig. 69.6.

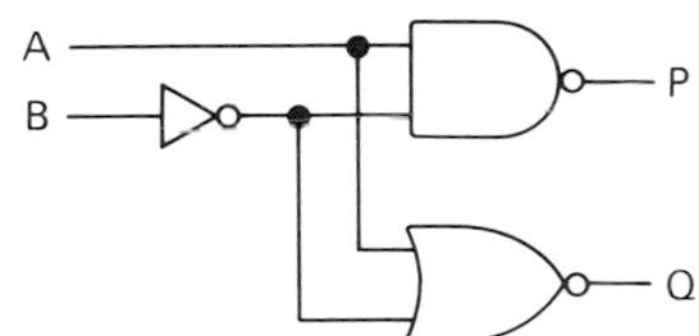

Fig. 69.6

2. Write a truth table for the circuit in Fig. 69.7, including the states at C, D, E, F and G.

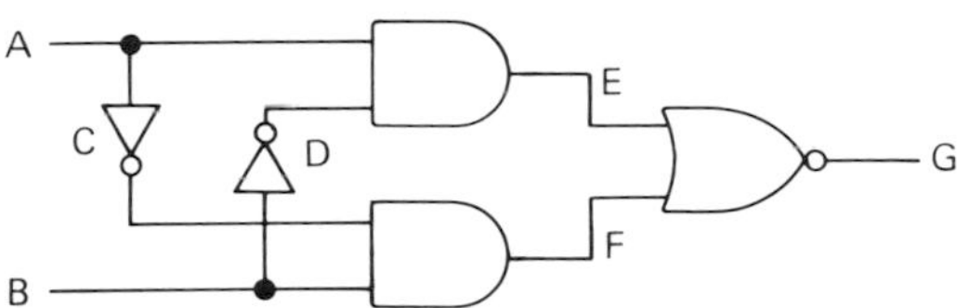

Fig. 69.7

3. Draw a truth table for each of the systems (a) and (b) shown in Fig. 69.8 and identify the logic function which each possesses. (*L.*)

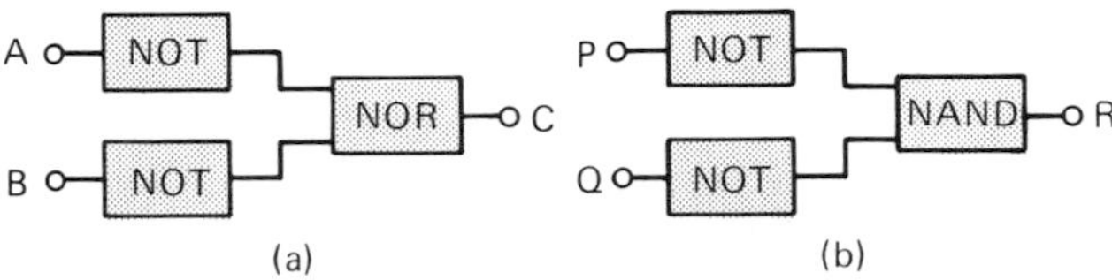

Fig. 69.8

70 Logic circuit design II

The logic circuits to be designed in this chapter are required to produce, for certain input combinations, logic 1 outputs which could be used, for instance, to control some operation or make lamps flash in a particular order.

Circuit 1

The circuit is to have three inputs A, B, C, fed by a three-bit binary code representing the numbers 0 to 7 (C being the l.s.b.). For example, if A = 1 ('high' input), B = 0 ('low' input) and C = 1, the binary number at the input is 101 (5 in decimal).

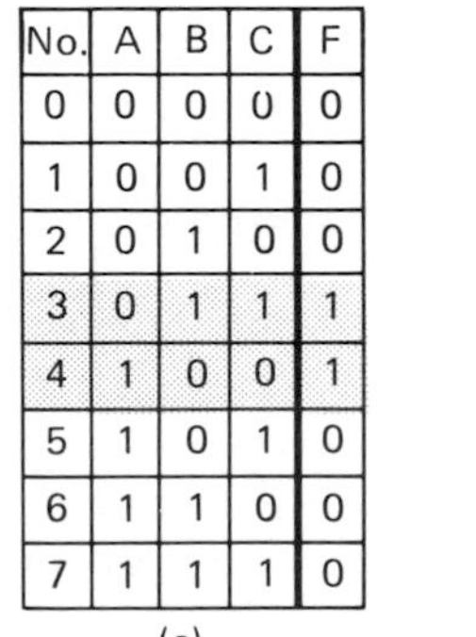

No.	A	B	C	F
0	0	0	0	0
1	0	0	1	0
2	0	1	0	0
3	0	1	1	1
4	1	0	0	1
5	1	0	1	0
6	1	1	0	0
7	1	1	1	0

(a)

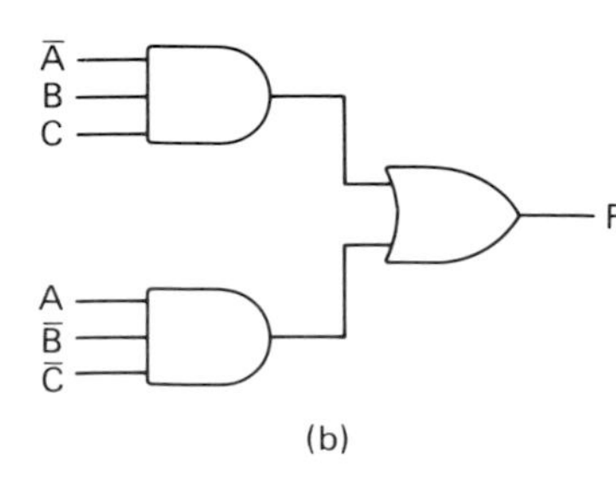

Fig. 70.1

Suppose the function of the circuit is to produce a logic 1 at its output F when numbers 3 and 4 occur. The truth table is given in Fig. 70.1a and lines 3 and 4 show that F is 1 when:

A is 0 *AND* B is 1 *AND* C is 1
OR
A is 1 *AND* B is 0 *AND* C is 0

This statement is the same if it is rewritten so that all inputs appear, as before (p. 141), as 1s. We then get F is 1 when

$\bar{A}$ (not A) is 1 *AND* B is 1 *AND* C is 1
OR
A is 1 *AND* $\bar{B}$ (not B) is 1 *AND* $\bar{C}$ (not C) is 1

In Boolean algebra notation, where a plus (+) indicates the OR logic operation and a dot (.) represents the AND operation, we can write

$$F = \bar{A}.B.C + A.\bar{B}.\bar{C}$$

In terms of logic gates this means the circuit has two three-input AND gates feeding a two-input OR gate as in Fig. 70.1b. To implement it using only NAND gates we use the facts (see p. 134):

AND = NAND followed by NOT
OR = NOT of each input followed by NAND

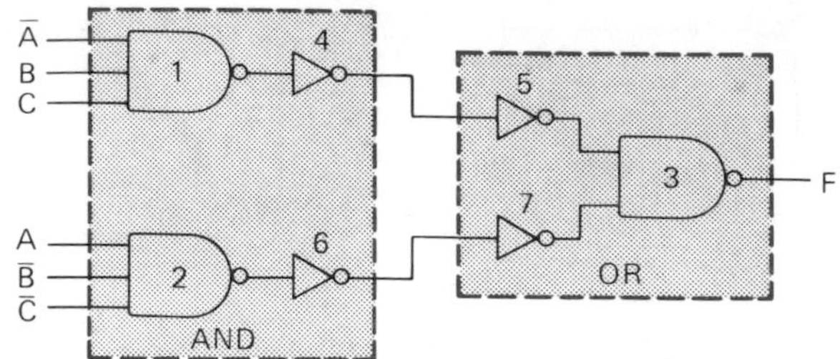

Fig. 70.2

The circuit becomes that of Fig. 70.2 but inverting an input twice gives the original input, i.e. two successive NOT gates cancel each other so only NAND gates 1, 2 and 3 are needed. The equivalent NAND gate circuit is given in Fig. 70.3: note that $\overline{A}$, $\overline{B}$ and $\overline{C}$ are obtained by inverting A, B and C with one-input NAND gates.

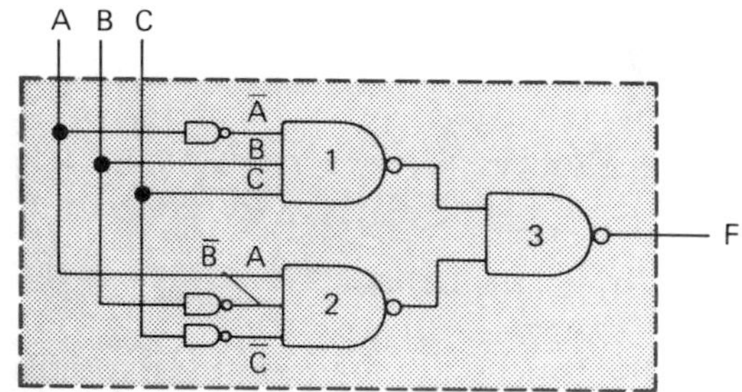

Fig. 70.3

Circuit 2 (traffic lights)

The problem is to design a logic circuit with two inputs A, B and three outputs R, Y, G obeying the truth table in Fig. 70.4. If R, Y and G feed red, yellow and green lights (e.g. LEDs), these would

Fig. 70.4

State	A	B	R	Y	G
0	0	0	1	0	0
1	0	1	1	1	0
2	1	0	0	0	1
3	1	1	0	1	0

flash in the order of British traffic signals when A and B are supplied continuously by a two-bit binary code representing numbers 0 to 3.

From the truth table we can write

(a) R is 1 when (lines 0 and 1):

A is 0 *AND* B is 0 (i.e. $\overline{A}$ is 1 *AND* $\overline{B}$ is 1)
OR
A is 0 *AND* B is 1 (i.e. $\overline{A}$ is 1 *AND* B is 1)

(b) Y is 1 when (lines 1 and 3):

A is 0 *AND* B is 1 (i.e. $\overline{A}$ is 1 *AND* B is 1)
OR
A is 1 *AND* B is 1

(c) G is 1 when (line 2):

A is 1 *AND* B is 0 (i.e. A is 1 *AND* $\overline{B}$ is 1)

Rewriting (a), (b) and (c) in Boolean notation and factorizing as in ordinary algebra:

$$R = \overline{A}.\overline{B} + \overline{A}.B = \overline{A}\,(\overline{B} + B)$$
$$Y = \overline{A}.B + A.B = B\,(\overline{A} + A)$$
$$G = A.\overline{B}$$

Now if the two inputs to an OR gate are an input and its inverse (called its *complement*), one input is a 1, making the output 1 (since for an OR gate, the output is 1 unless all inputs are 0). That is $\overline{A} + A$ is 1 and $\overline{B} + B$ is 1, hence we get

$$R = \overline{A} \quad Y = B \quad G = A.\overline{B}$$

Thus, the logic circuit has to feed R from input A via a NOT gate while Y goes to input B directly and G is supplied by the output of an AND gate having A and the complement of B as its inputs, Fig. 70.5a. Using only NAND gates, the circuit is as in Fig. 70.5b.

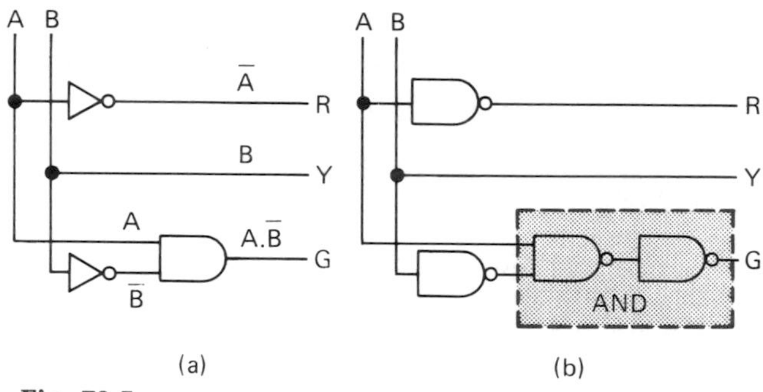

Fig. 70.5

Questions

1. Write the truth table for and, using any gates, design a logic circuit which gives a logic 1 output

(a) with a two-bit binary input when the number is 0, and

(b) with a three-bit binary input when the numbers are 0 and 5.

2. A three-bit binary code is used to represent the numbers 0 to 7 and a logic circuit is required which produces logic 1 outputs when numbers 1 and 3 occur at its inputs.

(a) Write the truth table for this logic function.

(b) Design a logic circuit to implement this function using NAND gates only.

71 Logic circuit design III

The design of three logic circuits which are important in computers and other electronic systems will be considered.

Decoder

While digital electronic systems work in binary code, humans prefer the decimal code. A *decoder* (often at the output end of the system) converts a binary input to a decimal output, frequently for display purposes (e.g. on a seven segment display).

Suppose a two-bit binary decoder is required which will produce a logic 1 output at just one of four outputs (depending on the input), the other three remaining at logic 0. The block diagram and truth table are given in Fig. 71.1. We can regard it as part

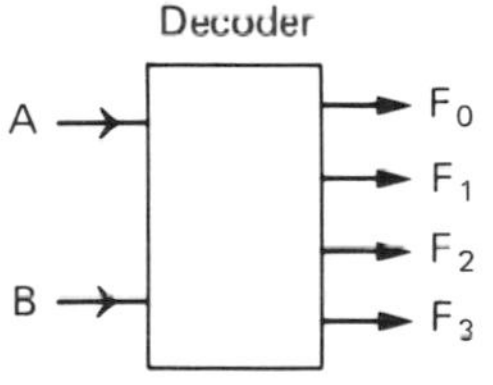

B	A	F_0	F_1	F_2	F_3
0	0	1	0	0	0
0	1	0	1	0	0
1	0	0	0	1	0
1	1	0	0	0	1

Fig. 71.1

of a binary to decimal decoder with F_0, F_1, F_2 and F_3 representing the decimal outputs of 0, 1, 2 and 3 respectively, while A and B are the binary inputs (A being the l.s.b.) causing the appropriate decimal output to go 'high'.

From the truth table we cay say
F_0 is 1 when A is 0 *AND* B is 0
F_1 is 1 when A is 1 *AND* B is 0
F_2 is 1 when A is 0 *AND* B is 1
F_3 is 1 when A is 1 *AND* B is 1

Using Boolean notation and making all inputs 1s, as before (p. 141), we get

$F_0 = \overline{A}.\overline{B}$ $F_1 = A.\overline{B}$ $F_2 = \overline{A}.B$ $F_3 = A.B$

The circuit of the system using AND and NOT gates is shown in Fig. 71.2. The NAND gate only version requires ten gates.
Note. An *encoder* (usually at the input of a system) converts from decimal (e.g. supplied by a numerical

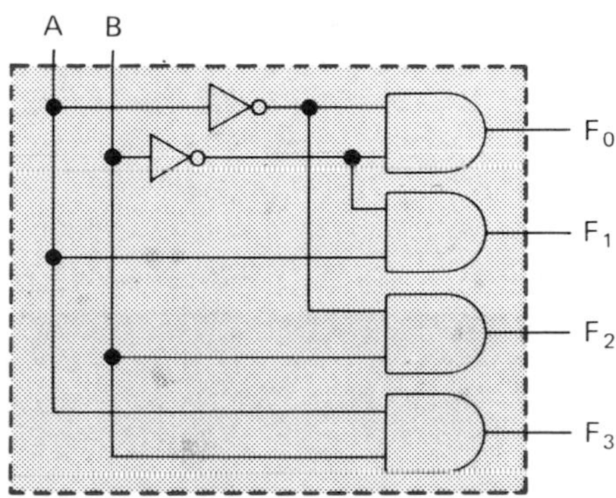

Fig. 71.2

keyboard) into binary and works on similar principles using logic gates. Encoders and decoders are called *code converters*.

Magnitude comparator

A *magnitude comparator* compares two binary numbers. For example, during a counting operation in say a computer, it may be necessary to know when a certain total has been reached before moving on to the next part of the program. This involves the comparison of two numbers.

Suppose the circuit has to compare two one-bit binary numbers A and B and produce a logic 1 output F when they are equal, i.e. both 0s or both 1s. The block diagram and truth table are shown in Fig. 71.3. Proceeding as before we can write

$F = \overline{A}.\overline{B} + A.B$

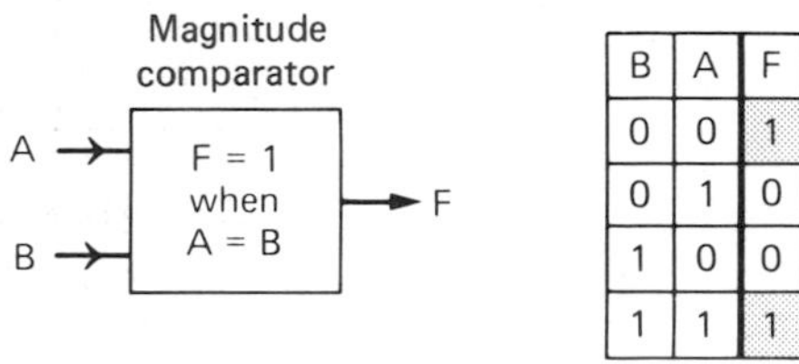

B	A	F
0	0	1
0	1	0
1	0	0
1	1	1

Fig. 71.3

The circuit using AND, OR and NOT gates is given in Fig. 71.4a and using only NAND gates in Fig. 71.4b. It is the same as that for an exclusive-NOR gate (p. 134).

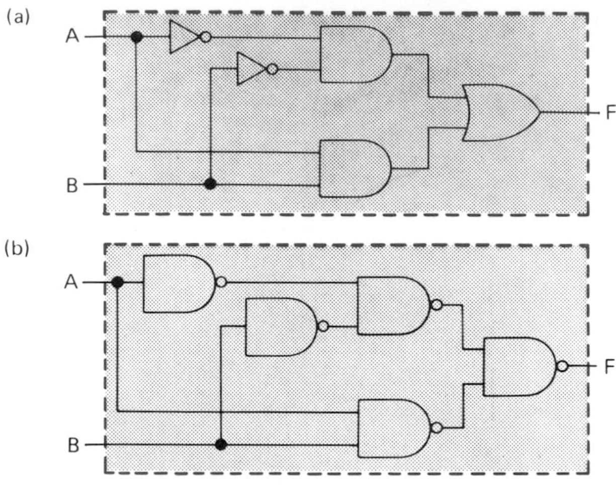

Fig. 71.4

Arithmetic logic unit

An *arithmetic* and *logic unit* (ALU) is at the heart of a digital computer (p. 203). It operates in two alternative ways or modes. In the *arithmetic* mode it performs, in binary, addition, subtraction, multiplication and division, all based on the use of adders (p. 138). In the *logic* mode, logical operations like those done by logic gates are carried out.

Suppose a very simple ALU has two 'ordinary' inputs A and B and a third 'select' input C so that when

(i) C = 0, it is to be in the arithmetic mode and give an output F = 1 when A = B, i.e. it is a magnitude comparator, and
(ii) C = 1, it is to be in the logic mode and perform the AND operation on A and B, i.e. make F = 1 when A = B = 1.

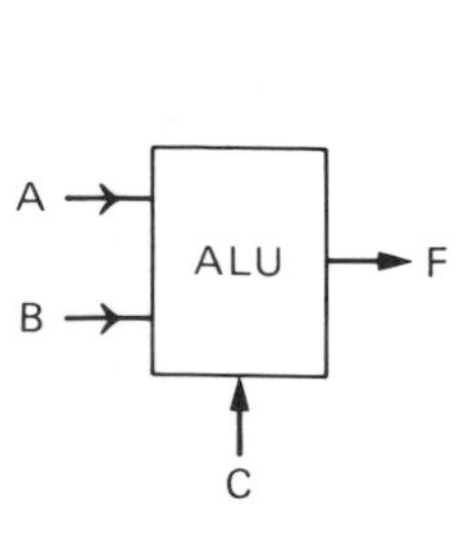

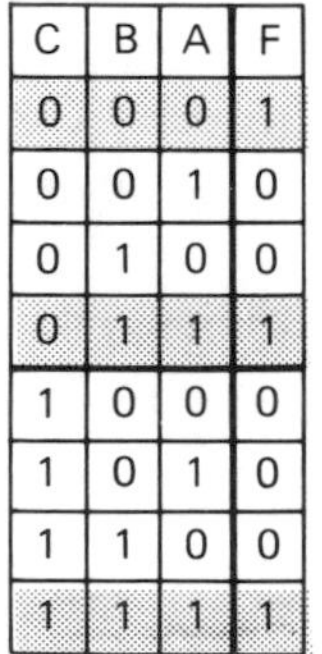

C	B	A	F
0	0	0	1
0	0	1	0
0	1	0	0
0	1	1	1
1	0	0	0
1	0	1	0
1	1	0	0
1	1	1	1

Fig. 71.5

The block diagram and truth table are given in Fig. 71.5. We can write

$$
\begin{aligned}
F &= \bar{A}.\bar{B}.\bar{C} + A.B.\bar{C} + A.B.C \\
&= \bar{A}.\bar{B}.\bar{C} + A.B\,(\bar{C} + C) \\
&= \bar{A}.\bar{B}.\bar{C} + A.B \quad \text{since } \bar{C} + C = 1 \text{ (p. 144)}
\end{aligned}
$$

The circuit using AND, NOT and OR gates is given in Fig. 71.6a. Note how the need to use a three-input AND gate is avoided by 'ANDing' $\bar{A}$ and $\bar{B}$ first, then 'ANDing' their output $\bar{A}.\bar{B}$ with $\bar{C}$. The simplified equivalent NAND gate circuit is shown in Fig. 71.6b.

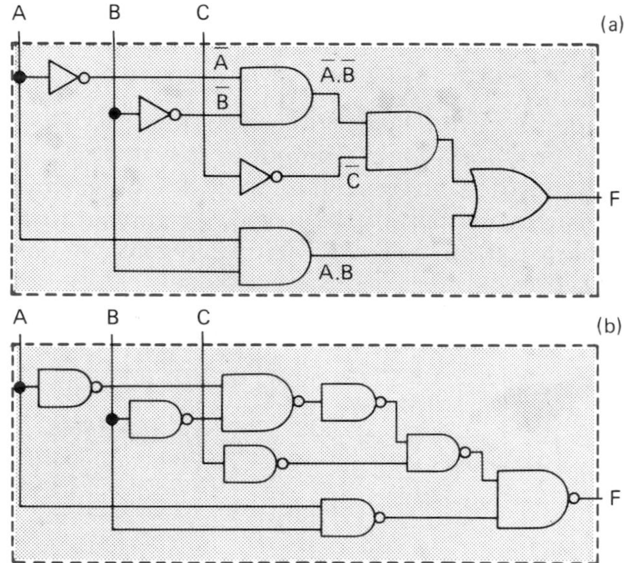

Fig. 71.6

Questions

1. Design a magnitude comparator with two inputs A and B which gives a logic 1 at its output F only when A is greater than B.

2. Design a two-way electronic switch (called a *data selector* or a 2-to-1 line *multiplexer*) which connects its output F to input A when the 'command' input C is at logic 0 and to input B when C is at logic 1, Fig. 71.7.

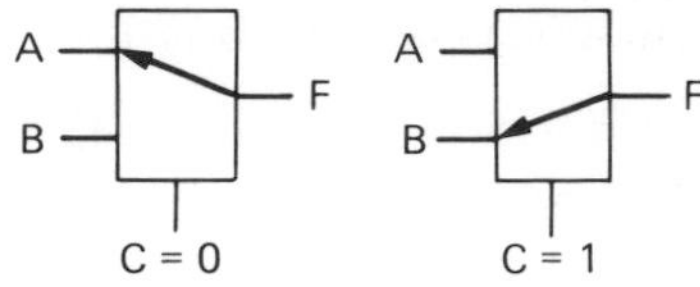

Fig. 71.7

72 Progress questions

1. (a) Draw the symbols for the following gates: (i) NOT, (ii) AND, (iii) NAND, (iv) NOR, (v) OR, (vi) exclusive-OR and (vii) exclusive-NOR.
(b) State what the input conditions must be for each of the gates in (a) to give a 'high' (logic 1) output.
(c) Draw *six* diagrams to show the NAND gate equivalents of the other six gates listed in (a).

2. The gas central heating system represented by the block diagram in Fig. 72.1 has digital electronic control. The gas valve turns on the gas supply to the boiler when the output from the logic gate is 'high'. This is so only if *both* of the following conditions hold:
(i) the output from the *thermostat* is 'high', indicating that the room temperature has fallen below that desired and in effect saying 'yes', the room needs more heat, and
(ii) the output from the *pilot flame sensor* is 'high', meaning 'yes', the pilot is lit.
What type of logic gate is required?

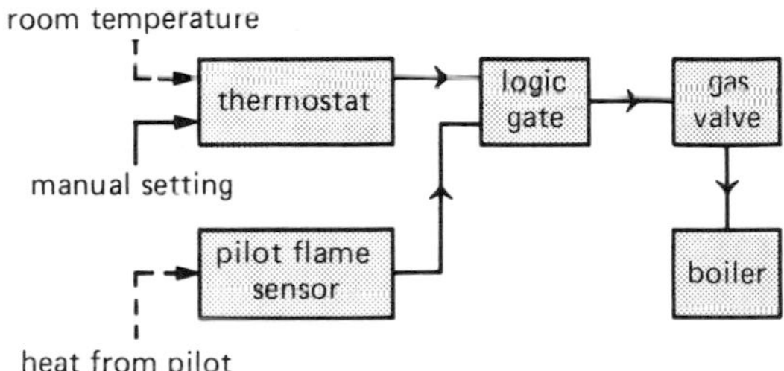

Fig. 72.1

3. In the RTL logic gate of Fig. 72.2, the inputs and outputs are at logic 1 when they are at +6 V and they are at logic 0 when at 0 V.
(a) If input B is kept at 0 V, write the truth table showing output F when input A is changed from 0 to 1. What kind of logic gate is it?
(b) If both A and B can be changed, write the truth table showing how F behaves. What kind of gate is it?
(c) Use a combination of these two gates to draw the circuit for an OR gate.

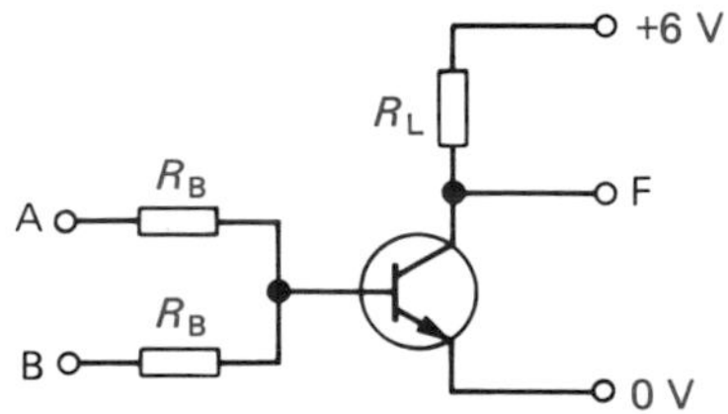

Fig. 72.2

4. (a) Explain the meaning of the term *saturation* applied to a transistor.
(b) Explain why in the circuit of Fig. 72.3 $V_o \approx 0$ when $V_i \approx V_{CC}$. What condition must R_B satisfy for this to be true?
Explain why $V_o = V_{CC}$ when $V_i = 0$.

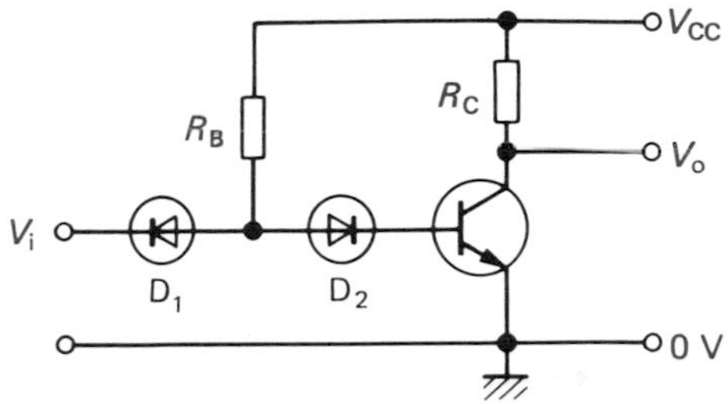

Fig. 72.3

(c) Draw a truth table relating the inputs A and B to the output C for the circuit in Fig. 72.4. The circuit is to be used in a logic system in which logical 1 is represented by a voltage $\approx V_{CC}$ and logical 0 is represented by a voltage ≈ 0.
Identify the logic function represented by this circuit. Explain why in this circuit the voltage at C in normal use will be less than V_{CC}. (*L.*)

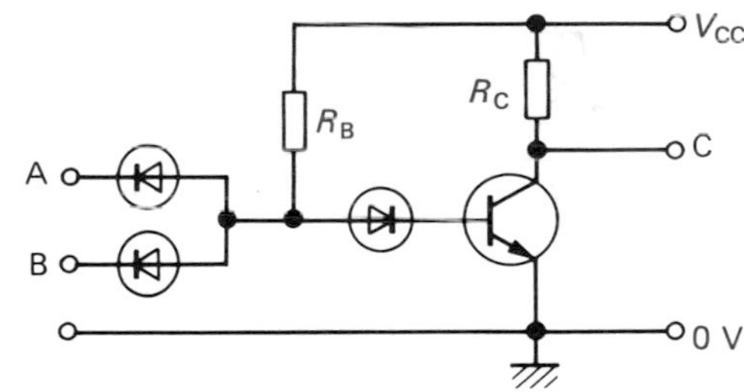

Fig. 72.4

5. Describe the action of an AND gate.
Construct a truth table for a two-input AND gate.
State the output from this AND gate in Boolean form. (*C.*)

6. Draw a truth table for the logic system shown in Fig. 72.5, including the states at C, D and E. (*L.*)

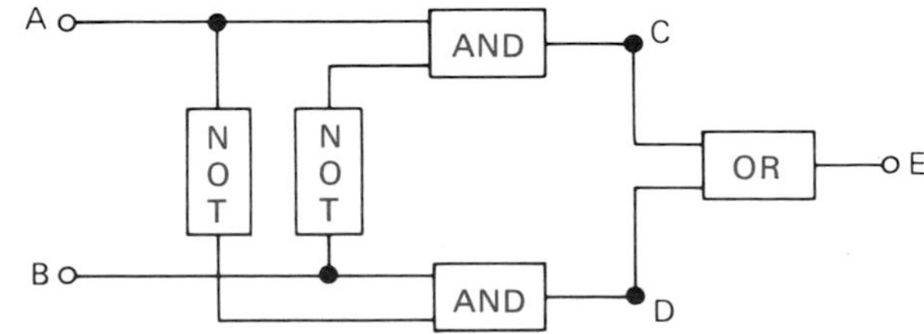

Fig. 72.5

7. Complete a truth table giving the logic states at A, B, C, D and E for the circuit in Fig. 72.6. (*O. and C.*)

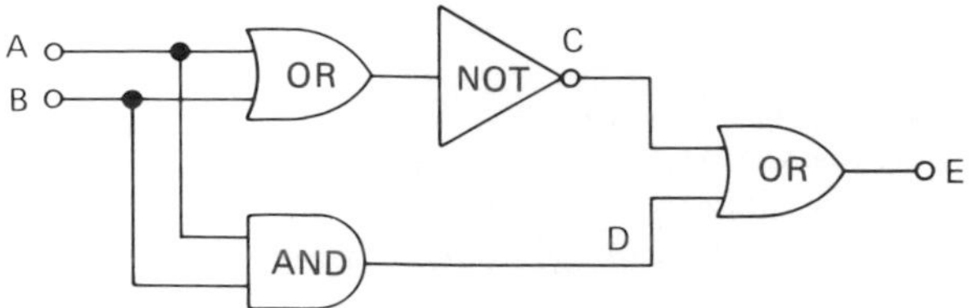

Fig. 72.6

8. Two numbers are written in binary as 11011 and 10010. These numbers are represented as digital signals by two pulse trains. They are then fed into an OR logic gate.

(a) Copy and complete Fig. 72.7 to show the pulse train fed into input Y and the output from Z and the binary number that this output represents.

(b) What is the binary sum of 11011 and 10010?

(c) Explain why the answers to (a) and (b) are not the same. (*A.E.B. 1982 Electronics*)

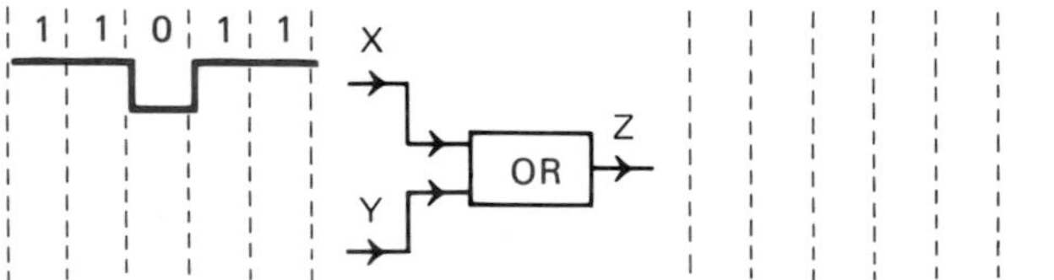

Fig. 72.7

9. (a) Explain in simple terms what is meant by an OR circuit and by a NAND circuit. In each case draw an appropriate circuit diagram and explain how the circuit works.

How are the situations changed if there are three inputs rather than two?

(b) Draw a circuit with three inputs A, B and C, which would give a high output if input C were high together with a high voltage at input A or input B, or if the inputs A, B and C were all low.

If the same performance were claimed for a different circuit, how could you test the claim? (*O. and C.*)

10. A game is devised in which rings are thrown to land on hooks A, B, C, D and E mounted on a board as in Fig. 72.8. Each hook is connected to a switch which closes and gives a logic level 1 output when a ring lands on it. The aim of the game is to score exactly 4 with two rings on separate hooks. Hooks A and B each score 1, hook C scores 3 and hooks D and E each score 2. When a player throws two rings so as to score 4 the light is illuminated.

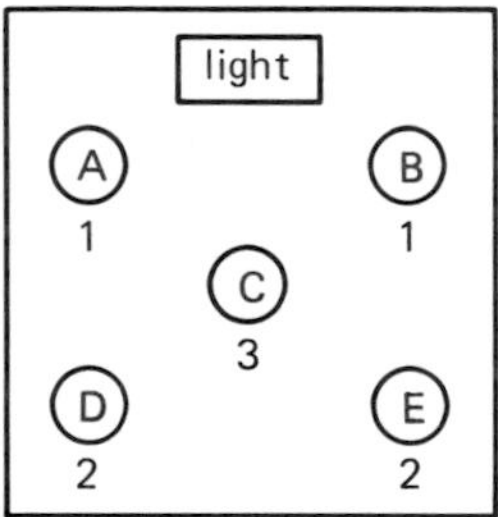

Fig. 72.8

(i) If logic level 1 will turn on the light, write a logic statement of the function which is required to activate the light, considering all possibilities, or list the separate conditions which will activate the light.

(ii) Draw a block diagram of a logic system to fulfil the function given in your answer to (i).

(iii) If you only have NAND gates available, explain how you would construct the other gates you need. (*L. part qn.*)

73 Bistable multivibrators I

Multivibrators

Multivibrators are two-stage switching circuits in which the output of the first stage is fed to the input of the second and vice versa. When one output is 'high', the other is 'low', i.e. their outputs are *complementary*. Switching between the two logic levels is so rapid that the output voltage waveforms are 'square'. The term 'multivibrator' arises from this since a square wave consists of a large number of sine waves with frequencies that are odd multiples of the fundamental.

Multivibrators are of three types. *Bistables* or *flip-flops*, of which there are several varieties, are used in counters (p. 158), shift registers (p. 160) and memories (p. 160) and will be considered first; *astables* and *monostables* are treated later.

Transistor SR bistable

The basic circuit is shown in Fig. 73.1. The collector of each transistor is coupled to the base of the other by a resistor R_1 or R_2. The output of Tr_1 is thus fed to the input of Tr_2 and vice versa.

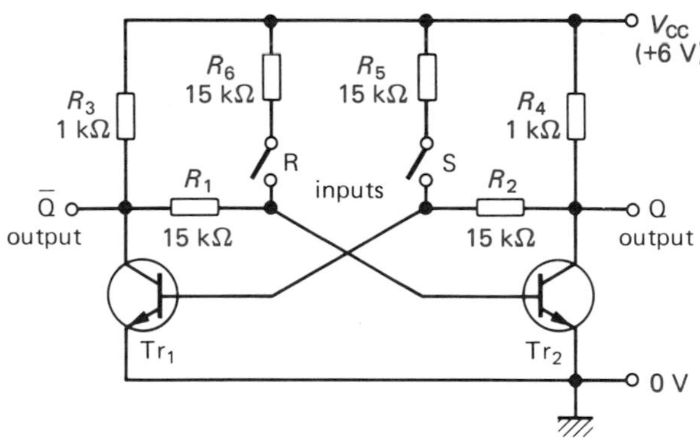

Fig. 73.1

When the supply is connected, Tr_1 and Tr_2 both draw base current but, because of slight differences (e.g. in h_{FE}), one, say Tr_1, has a larger collector current and conducts more rapidly than the other. It quickly saturates while Tr_2 is driven to cut-off.

The collector voltage of Tr_1 is therefore 'low' (most of V_{CC} being dropped across R_3) and so no current flows through R_1 into the base of Tr_2. Tr_2 remains off, its collector voltage is thus 'high' (V_{CC}) and causes current to flow via R_2 into the base of Tr_1, reinforcing Tr_1's saturation condition. The *feedback is positive* and the circuit is in a stable state which it can keep indefinitely with output Q = 1 (since Tr_2 is cut-off) and its complementary output $\overline{Q}$ (not Q) = 0 (since Tr_1 is saturated).

The state can be changed by applying a positive voltage (> 0.6 V for silicon transistors) to the base of the 'off' transistor, i.e. to Tr_2 via R_6 (e.g. by temporarily connecting the *reset* input R to V_{CC}). Tr_2 then draws base current (through R_6), which is large enough to drive it into saturation. Its collector voltage falls from V_{CC} to near zero, so cutting off the base current to Tr_1. Tr_1 switches off, its collector voltage rises from near zero to V_{CC} and is fed via R_1 to the base of Tr_2 to keep it saturated. The circuit is now in its second stable state (hence bistable) but with Tr_1 off ($\overline{Q}$ = 1) and Tr_2 saturated (Q = 0).

To return to the first state a positive voltage must be applied to the base of the 'off' transistor (Tr_1) at the *set* input S. The output of each transistor can thus be made to 'flip' to V_{CC} or 'flop' to 0 V, according to the truth table in Fig. 73.2 which also gives the SR (set reset) bistable symbol. Taking the output of the

State	S	R	Q	$\overline{Q}$
set	1	0	1	0
reset	0	1	0	1

S Q
R $\overline{Q}$

Fig. 73.2

circuit as the collector voltage of Tr_2, each state in effect stores (remembers) one bit of 'information', i.e. a 1 when Q = 1 or a 0 when Q = 0 and the bistable acts as a *one-bit memory*.

NAND gate SR bistable

The basic circuit is shown in Fig. 73.3 with its truth table, feedback being from each output to one of the inputs of the other two-input NAND gate.

(a) Set state. If S = 0 and R = 1, NAND gate X has at least one of its inputs at logic 0, therefore its output Q must be at logic 1 (since the output of a NAND is always 1 unless all inputs are 1s). Q is fed back to input B and so both inputs to NAND gate Y are 1s;

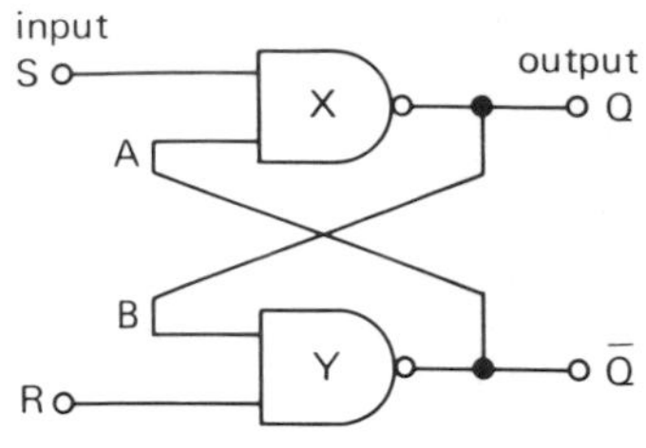

State	S	R	Q	$\bar{Q}$
set	0	1	1	0
	1	1	1	0
reset	1	0	0	1
	1	1	0	1
indeterminate	0	0	1	1
	1	1	?	?

Fig. 73.3

hence output $\bar{Q} = 0$. This is a stable state, called the *set state*, with Q = 1 and $\bar{Q} = 0$ given by S = 0 and R = 1. (In the transistor SR bistable S = 1 and R = 0 gives this state.)

If S becomes 1 with R still 1, gate X inputs are now S = 1 and A = 0 (since $\bar{Q} = 0$), i.e. one input is a 0 and so Q stays at 1. The circuit has thus 'remembered' or 'latched' the state Q = 1.

(b) Reset state. In this second stable state Q = 0 and $\bar{Q} = 1$. It is given by S = 1 and R = 0, which you can check (see question 2). (In the transistor version the reset state is obtained when S = 0 and R = 1.)

If R becomes 1 with S still 1, Q remains at 0, showing that the reset state has been latched.

Note. When S = 1 and R = 1, Q (and $\bar{Q}$) can be either 1 or 0, depending on the state before this input condition existed. The previous output state is retained as shown by the second and fourth rows of the truth table. Hence the logic levels of Q and $\bar{Q}$ depend on the *sequence* which makes both inputs 1.

(c) Indeterminate state. When S = 0 and R = 0, we get Q = 1 and $\bar{Q} = 1$. If both inputs are then made 1 *simultaneously*, we cannot predict whether the bistable will return to the 'set' or the 'reset' state. This undesirable situation is avoided by changing the inputs *alternately*.

Antibounce circuit

If a mechanical switch, for example on a keyboard, is used to change the state of a bistable (or any other logic system), more than one electrical pulse may be produced due to the metal contacts of the switch not staying together at first but bouncing against each other rapidly and creating extra unwanted pulses, Fig. 73.4a. The effect of this 'contact bounce' can be eliminated by using an SR bistable in the 'anti-bounce' circuit of Fig. 73.4b to 'clean up' the switch action.

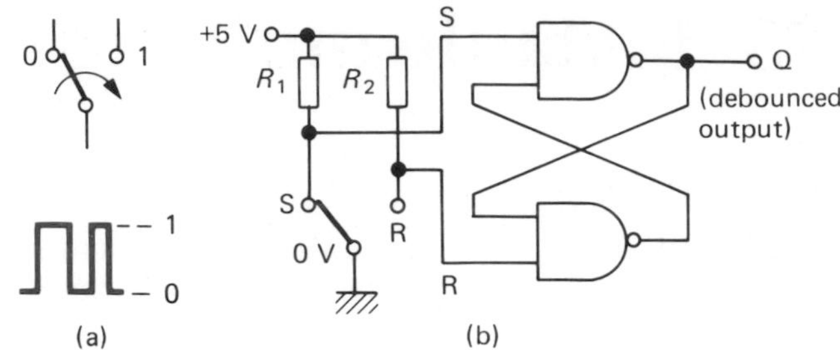

Fig. 73.4

When the switch is in the position shown, S is connected to 0 V and is at logic 0 while R is at logic 1 due to its connection via R_2 to supply positive. Hence Q = 1 (from the truth table in Fig. 73.3). If the switch is then operated to make the circuit at the other contact and 'bounces' once, the logic levels at S and R for different positions are given in Fig. 73.5. The last three lines show that Q stays at logic 0 despite the bounce.

Switch position		S	R	Q
S R	making contact at S	0	1	1
S R	moving to make contact at R	1	1	1
S R	making contact at R	1	0	0
S R	bounces back from contact at R	1	1	0
S R	remakes contact at R	1	0	0

Fig. 73.5

Triggered (T-type) bistable

A T-type bistable is a modified SR type with extra components that enable successive pulses, applied to an input called the *trigger* T, Fig. 73.6a, to make the bistable switch to and fro (or 'toggle') from one stable state (e.g. Q = 0 and $\bar{Q} = 1$) to the other (e.g. Q = 1 and $\bar{Q} = 0$).

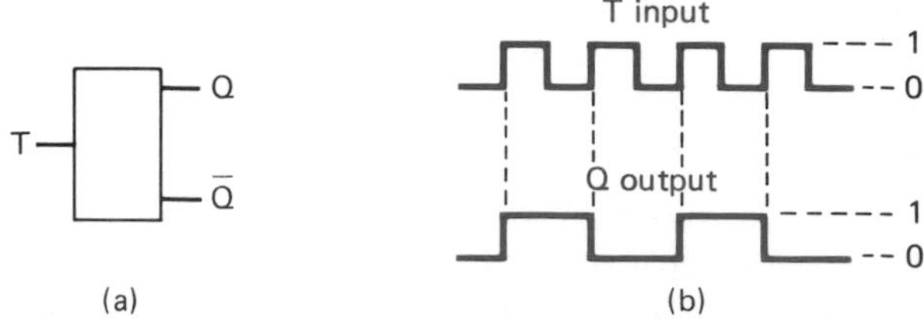

Fig. 73.6

Two trigger input pulses are required to give one output pulse at either Q or $\bar{Q}$. The frequency of the output pulses is therefore half that of the input pulses as the waveforms in Fig. 73.6b show. A T-type flip-flop *divides the input frequency by two* and forms the basis of binary counters (p. 158). It can be built from transistors or NAND gates but is not available as an IC since other types can be made to toggle, as we will see shortly.

In the transistor version of Fig. 73.7 the pulse 'steering' is done by connecting between the collector and base of each transistor, a diode (D_1 or D_2) and a resistor (R_5 or R_6), their junctions going via capacitors (C_1 or C_2) to the trigger input. The action is as follows. Suppose Tr_1 is on (saturated) and Tr_2 off. The base of Tr_1 (point N) is therefore at +0.6 V and its collector (point O) near 0 V, so D_1 is on the verge of conducting since it is almost forward biased (the voltage between points N and M being close to 0.6 V). D_2 on the other hand is heavily reverse biased because the collector of Tr_2 (point Z) is at +6 V (and so also is point X because no current flows through R_6) and its base (point Y) is at 0 V.

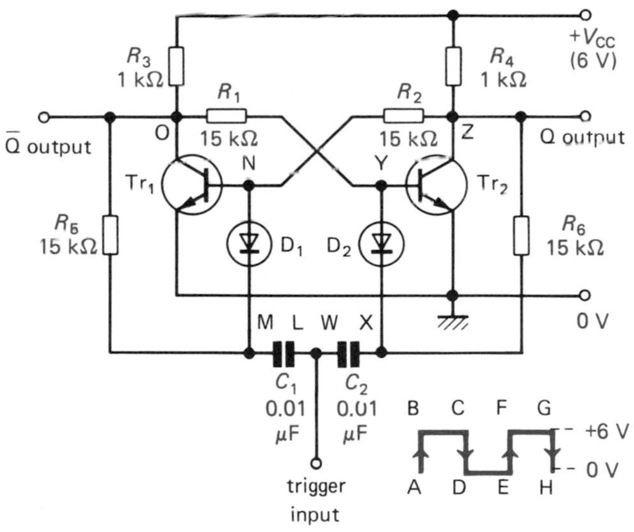

Fig. 73.7

When square-wave pulses (6 V amplitude) are applied to the trigger input, the rapidly rising, positive-going edge of the first pulse (i.e. AB) raises the voltage of both L and M to +6 V and of W and X to +6 V and +12 V respectively since the charges on C_1 and C_2 and therefore the voltages across them, cannot change instantly. As a result the voltage of the lower ends of D_1 and D_2 (points M and X) are raised by 6 V, forcing D_1 into reverse bias and making D_2 even more reverse biased than previously. The state of the bistable stays the same (Tr_1 on and Tr_2 off) and the charges on C_1 and C_2 quickly adjust to restore the voltage at M and X to 0 V and 6 V respectively.

The rapidly falling negative-going edge of the first pulse (i.e. CD) suddenly lowers the voltage of L from 6 V to 0 V, of M from 0 V to −6 V and of both W and X from 6 V to 0 V. This means that while D_2 is now on the point of conducting, D_1 becomes forward biased, conducts heavily and in effect short-circuits the base-emitter of Tr_1, diverting the current into C_1 which becomes charged. Tr_1 is cut off, its collector current falls, thereby causing its collector voltage to rise and turn on Tr_2. This switches the bistable into its second state (Tr_1 off and Tr_2 on).

The switching process is repeated and the bistable returns to its first state (Tr_1 on and Tr_2 off) when the negative-going edge (GH) of the second input pulse arrives and so on.

Questions

1. The circuit in Fig. 73.8 is to be used in a quiz game where there are two competitors. It operates a system whereby a lamp indicates which of them has pressed his switch first. Explain why if switch A is closed first, lamp L_B will not light if the other competitor then closes switch B. *(L. part qn.)*

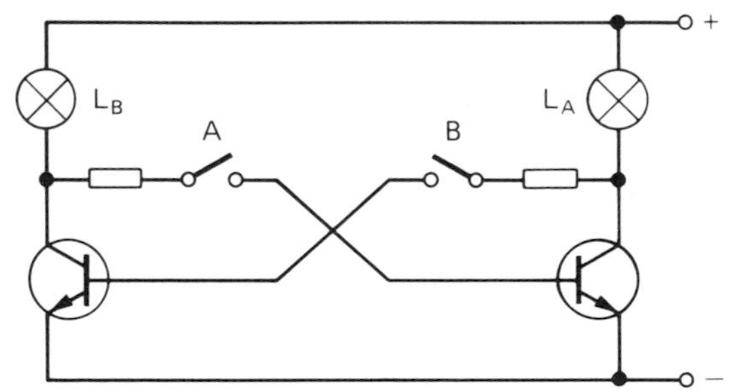

Fig. 73.8

2. In the circuit of Fig. 73.3 if S = 1 and R = 0 what is the logic level of input (a) A, (b) B?

3. (a) A T-type flip-flop 'toggles'. Explain this statement.
(b) Copy Fig. 73.6b and draw under it the $\bar{Q}$ output.

74 Bistable multivibrators II

The bistables described in this chapter are more versatile than the SR type from which they have been developed. While they can also be made from discrete transistors or NAND gates, it is much more convenient to use them in IC form. Their internal circuits will not be considered. For our purposes it is enough to know what they do and how to connect them, i.e. to treat them as 'black boxes'.

Clocked bistables

(a) Clocked logic. In large digital systems containing hundreds of interconnected bistables, outputs do not respond immediately to input changes but wait until a *clock* pulse is received (also called a *trigger* or *enabling* pulse), If the same clock pulse is applied to all bistables simultaneously, they change together, i.e. they are synchronized. Otherwise timing sequences can be upset by a 'race condition' in which the output from the system is decided by the speed of operation of individual gates rather than by the logic rules they should obey. In a clocked logic system changes occur in an orderly way, a step at a time, as commanded by the clock pulses.

A clocked SR bistable can be made by adding a two-input AND gate before each of the bistable inputs S_B and R_B, as in Fig. 74.1a. Then, if the clock input CK is at logic 1, S_B is 1 if the upper AND gate input S_A is 1 and R_B is 0 if the lower AND gate input R_A is 0. Data as 1's and 0's thus passes from S_A and R_A to the bistable only when CK is 1. If CK is 0, S_B and R_B cannot change even when S_A and R_A do.

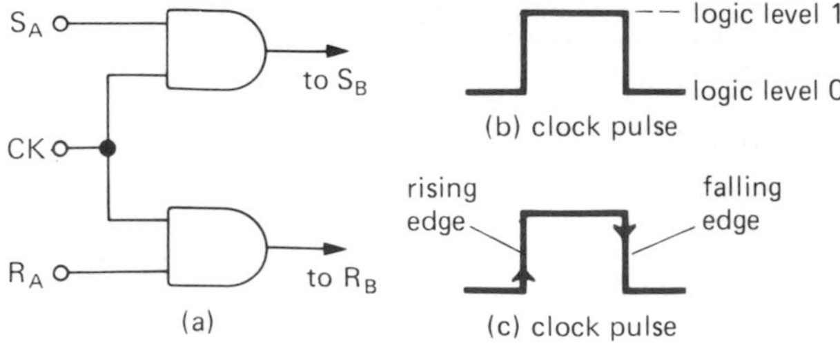

Fig. 74.1

(b) 'Clocks'. Clock pulses are supplied by some form of pulse generator. It may be a crystal controlled oscillator with a very steady repetition frequency, e.g. 10 MHz, or an astable multivibrator (p. 155) or a mechanical switch turning a d.c. supply on and off. The pulses should have fast rise and fall times, i.e. be good 'square' waves, and any switches should have antibounce circuits (p. 150).

(c) Level and edge triggering. There are two types of clocking or triggering operation. In *level* triggering, bistables change state when the logic *level* of the clock pulse is 1 or 0, Fig. 74.1b. The SR bistables described in chapter 73 are level triggered as is the clocked SR bistable just considered.

In *edge* triggering, a *change* in voltage level causes switching. If it occurs during the rise of the clock pulse from logic level 0 to 1 it is rising—or positive-edge triggering and most modern clocked logic is of this type. In falling—or negative-edge triggering, switching occurs when the clock pulse falls from 1 to 0, Fig. 74.1c. The T-type bistable (p. 150) is edge triggered.

In general edge triggering is more satisfactory than level triggering because in the former, output changes occur at an *exact* instant during the clock pulse and any further input changes do not affect the output until the next clock pulse rise (or fall). In level triggering, output changes can occur at any time while CK is 1.

D-type bistable

(a) As a data latch. The symbol for a clocked D-type bistable is given in Fig. 74.2a. CK is the input for *clock* pulses. D is the input to which a bit of *data* (a 0 or a 1) is applied for 'processing' by the bistable, while Q and $\overline{Q}$ are the two outputs, one always being the complement of the other (i.e. if Q = 1, $\overline{Q}$ =

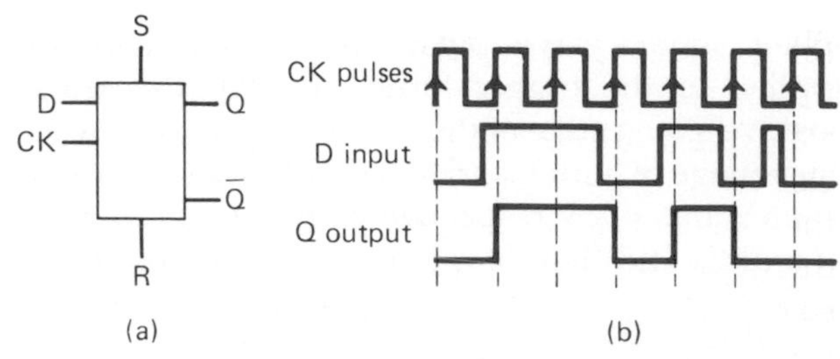

Fig. 74.2

0 and vice versa). S is the *set* input allowing Q to be set to 1 and R is the *reset* input by which Q can be made 0.

Suppose the flip-flop is rising-edge triggered, then the logic level of the D input is transferred to the Q output on the rising edge of a clock pulse, as shown by the *timing diagram* in Fig. 74.2b which you should study carefully. It shows that the Q output 'latches' on to the D input and stores it *at the instant a clock pulse changes its logic level from 0 to 1*. During the rest of the clock pulse D can change without affecting Q. The flip-flop is a *data latch* which acts as a *buffer* between input and output and does not 'communicate' with the D-input again until the next rising edge arrives.

Since a D-type bistable has only one data input, the indeterminate state (of two inputs at logic level 0: p. 150) cannot rise.

Quad data latches (i.e. four D-type bistables) are used to obtain a reading on a rapidly changing numerical display that is counting pulses (which would otherwise be seen as a blur) coming at a high rate from a counter. The latches are connected between the counter and the decoder (e.g. binary to decimal, p. 145) feeding the display, as in Fig. 74.3. When the latch is enabled by receiving an appropriate signal, it stores the count (as a four-bit binary number) and holds it for display until the next count enters the latch. In the meantime, the counter can carry on counting.

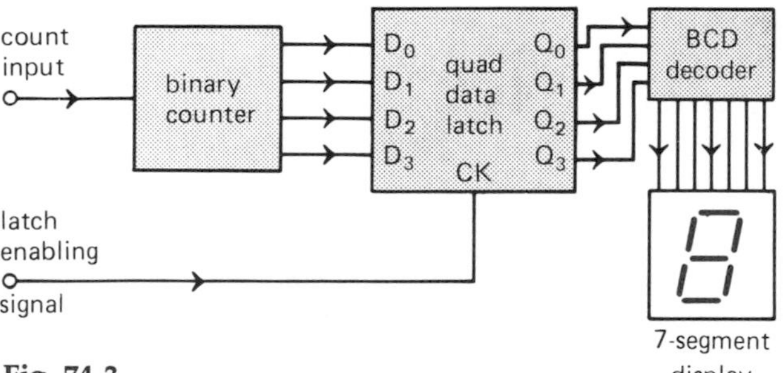

Fig. 74.3

(b) As a T-type bistable. If the $\bar{Q}$ output is connected to the D input as in Fig. 74.4a, successive clock pulses make the flip-flop 'toggle'. If the first clock pulse leaves Q = 1 and $\bar{Q}$ = 0 (because D = 1), then feedback (from $\bar{Q}$ to D) makes D = 0 and during the second clock pulse D is transferred to Q, so now Q = 0 and $\bar{Q}$ = 1. D now becomes 1 and the third clock pulse makes Q = 1 and $\bar{Q}$ = 0 again and so on.

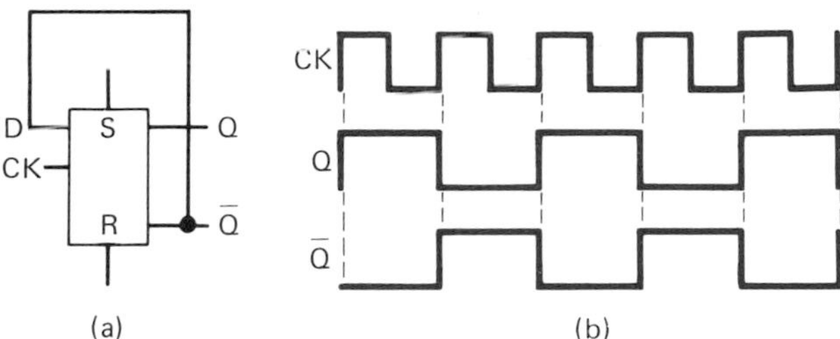

Fig. 74.4

Hence, Q = 1 *once* every *two* clock pulses, that is, the output has *half* the frequency of the clock pulses, Fig. 74.4b, and is dividing the clock frequency by two. It is used for this purpose in binary counters (p. 158).

J–K bistable

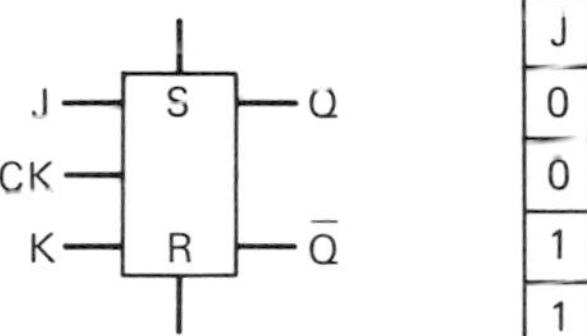

J	K	Q after clock pulse
0	0	stays at 0 or 1
0	1	resets to 0
1	0	sets to 1
1	1	toggles

Fig. 74.5

The clocked J–K bistable is as versatile as a bistable can be and has many uses. Its symbol and truth table are given in Fig. 74.5. There are two inputs, called J and K (for no obvious reason), a clock input CK, two outputs Q and $\bar{Q}$ and set S and reset R inputs. All four input combinations can be used on J and K and there is no disallowed, indeterminate state. When clock pulses are applied to CK it:

(i) retains its present state if J = K = 0,
(ii) acts as a D-type flip-flop if J and K are different,
(iii) acts as a T-type flip-flop if J = K = 1.

Schmitt trigger

The Schmitt trigger is a type of bistable whose output switches very rapidly from one state to the other; it is available as an IC. Among other things, it can be used as a 'rise time improver' to convert a slowly changing input into one with a fast rise time, Fig. 74.6a, as is required for reliable operation of many digital circuits.

The action is shown by the characteristic curve in Fig. 74.6b. The value of V_i which triggers the circuit

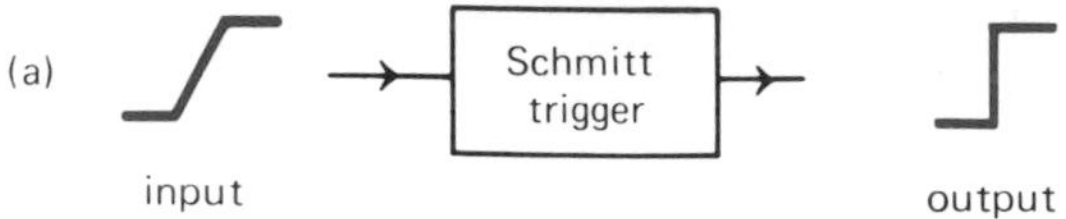

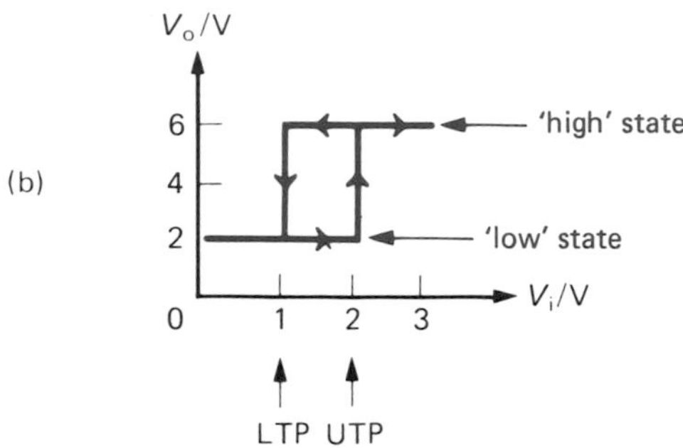

Fig. 74.6

and makes the output V_o jump from the 'low' to the 'high' state is called the *upper trip point* (UTP), here 2 V. The change from the 'high' to the 'low' state occurs at a lower value of V_i, called the *lower trip point* (LTP), here 1 V. A sine wave input would give a good square wave output.

The basic circuit is shown in Fig. 74.7. Positive feedback is obtained via a resistor R_2 that is common to the emitter circuits of both transistors (rather than via resistors connected from the collector of one to the base of the other as in an ordinary bistable).

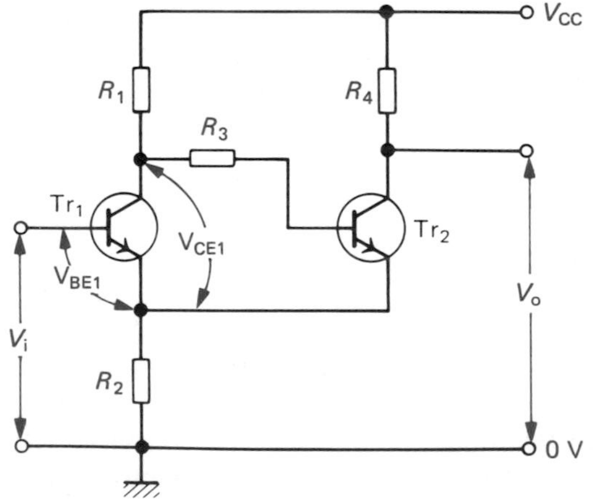

Fig. 74.7

Suppose the input voltage V_i is 'low'. Then Tr_1 is off, Tr_2 is on and the output voltage V_o is 'low'. Current from Tr_2 flows through R_2. If V_i is made 'high' (i.e. ≥ UTP), Tr_1 starts on switch on, its collector-emitter voltage V_{CE1} falls and so starts to switch off Tr_2. The current through R_2 and therefore the voltage across it decreases, so increasing the base-emitter voltage V_{BE1} and reducing the time it takes for Tr_1 to reach its turn-on voltage. This speeds up the switching on of Tr_1 and off of Tr_2 and makes V_o go 'high' rapidly.

Sequential logic

Bistables are 'memory-type' circuits which use *sequential* logic, i.e. their outputs depend not only on *present* inputs (as was the case with combinational logic circuits, p. 132), but on *previous* ones as well. The order or sequence in which inputs are applied is important (p. 150).

The chief requirement of a sequential logic circuit is that it should 'remember' the earlier inputs. This is achieved by feedback connections which ensure that, when the input changes, the effect of the previous input is not lost.

Questions

1. Copy Fig. 74.8 which refers to a D-type bistable operating on the *rising edges* of clock pulses. Draw under it the waveform for the Q output.

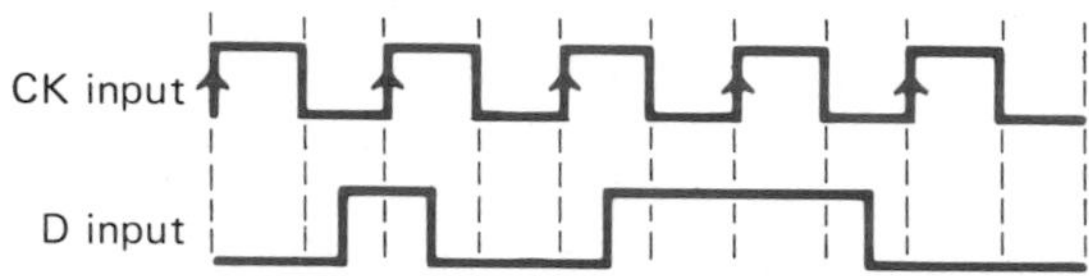

Fig. 74.8

2. Clock pulses of frequency 2 Hz are applied to the CK input of a D-type flip-flop connected to toggle. What is the frequency of the output pulses?

3. (a) Draw the characteristic of a Schmitt trigger which has a UTP = 3 V, a LTP = 2 V, a 'high' state = 9 V and a 'low' state = 3 V.

(b) Sketch the output voltage waveform from the Schmitt trigger of (a) if the input is a sine wave of peak value 5 V.

4. The circuit in Fig. 74.9 is suggested for a level triggered D-type flip-flop. Copy the truth table and by completing it (using the truth table for an SR flip-flop in Fig. 73.3), decide if it is suitable.

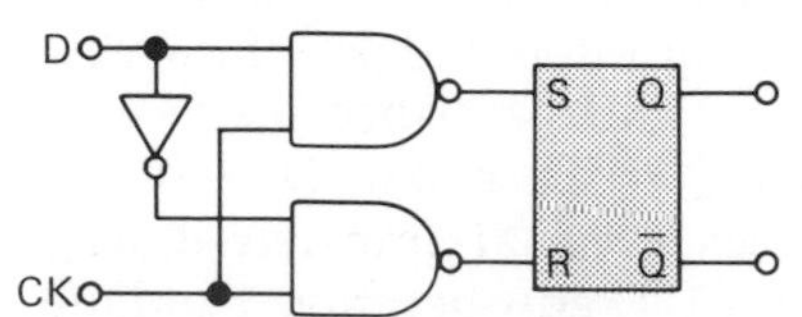

D	CK	S	R	Q
0	1			
0	0			
1	1			
1	0			

Fig. 74.9

75 Astables and monostables

Introduction

(a) Astable or free-running multivibrator. This has no stable states. It switches from the 'high' state to the 'low' state and vice versa automatically at a rate determined by the circuit components. Consequently it generates a continuous stream of almost square wave pulses, i.e. it is a square wave oscillator and belongs to the family of relaxation oscillators (p. 117).

The op amp version was considered earlier (p. 128); here the transistor and NAND gate versions will be described.

(b) Monostable or 'one-shot' multivibrator. It has one stable state and one unstable state. Normally it rests in its stable state but can be switched to the other state by applying an external trigger pulse, where it stays for a certain time before returning to the stable state. It will convert a pulse of unpredictable length (time) from a switch into a 'square' pulse of predictable length and height (voltage) for the input to another circuit.

It can be built from transistors, NAND gates or an op amp but only the first two will be considered.

Transistor astable

The circuit, Fig. 75.1, is similar to that of the transistor bistable (Fig. 73.1) but feedback is via capacitors.

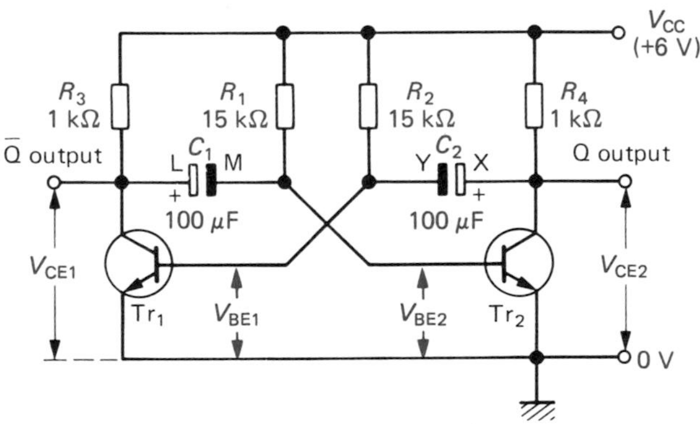

Fig. 75.1

(a) Action. When the supply is connected, one transistor quickly saturates and the other cuts off (as with the bistable). Each then switches to its other state, then back to its first state and so on. As a result, the output voltage, which can be taken from the collector of either transistor, is alternately 'high' (+6 V) and 'low' (near 0 V) and is a series of almost square pulses.

To see how these are produced, suppose that Tr_2 was saturated (i.e. on) and has just cut off, while Tr_1 was off and has just saturated. Plate L of C_1 was at +6 V, i.e. the collector voltage of Tr_1 when it was off; plate M was at +0.6 V, i.e. the base voltage of Tr_2 when it was saturated. C_1 was therefore charged with a p.d. between its plates of (6 V − 0.6 V) = +5.4 V.

At the instant when Tr_1 suddenly saturates, the voltage of the collector of Tr_1 and so also of plate L, falls to 0 V (nearly). But since C_1 has not had time to discharge, there is still +5.4 V between its plates and therefore the potential of plate M must fall to −5.4 V (p. 34). This negative potential is applied to the base of Tr_2 and turns it off.

C_1 starts to charge up via R_1 (the p.d. across it is now 11.4 V), aiming to get to +6 V, but, when it reaches +0.6 V, it turns on Tr_2. Meanwhile plate X of C_2 has been at +6 V and plate Y at +0.6 V (i.e. the p.d. across it is 5.4 V). Therefore, when Tr_2 turns on, X falls to 0 V and Y to −5.4 V and so turns Tr_1 off. C_2 now starts to charge through R_2 and when Y reaches +0.6 V, Tr_1 turns on again. The circuit thus switches continuously between its two states.

(b) Voltage waveforms. The voltage changes at the base and collector of each transistor are shown graphically in Fig. 75.2. We see that V_{CE1} (collector voltage of Tr_1) and V_{CE2} (collector voltage of Tr_2) are complementary (i.e. when one is 'high' the other is 'low'). Also, they are not quite square but have a rounded rising edge. This happens because as each transistor switches off and V_{CE} rises from near 0 V to V_{CC}, the capacitor (C_1 or C_2) has to be charged (via the collector load R_3 or R_4). In doing so it draws current and causes a small temporary voltage drop across the collector load which prevents V_{CE} rising 'vertically'.

(c) Frequency of the square wave. The time t_1 for which Tr_1 is on (i.e. saturated with $V_{CE1} \approx 0$) depends on how long C_1 takes to charge up through R_1 from −5.4 V to +0.6 V (i.e. 6.0 V = V_{CC}) and

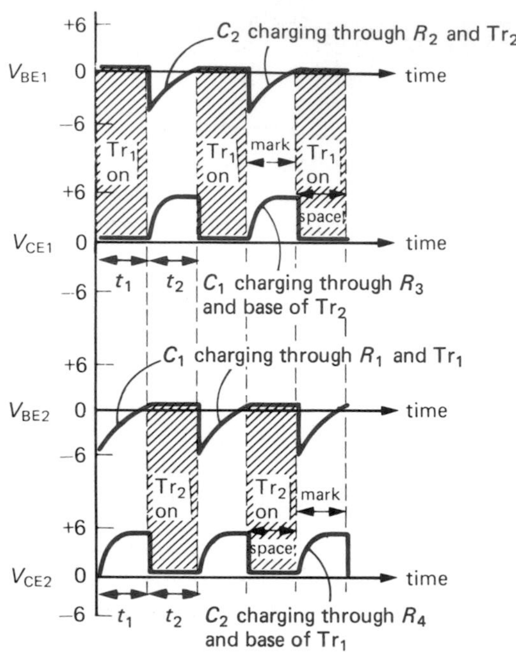

Fig. 75.2

switch on Tr_2. It can be shown that it depends on the time constant C_1R_1 (p. 33) and is given by

$$t_1 = 0.7\ C_1R_1$$

Similarly the time t_2 for which Tr_2 is on (i.e. the time for which Tr_1 is off) is given by

$$t_2 = 0.7\ C_2R_2$$

The frequency f of the square wave is therefore:

$$f = \frac{1}{t_1 + t_2} = \frac{1}{0.7\ (C_1R_1 + C_2R_2)}$$

If $C_1 = C_2$ and $R_1 = R_2$ then $t_1 = t_2$, i.e. the transistors are on and off for equal times and their *mark-to-space ratio* (p. 19) is 1. We then have

$$f = \frac{1}{1.4\ C_1R_1} = \frac{0.7}{C_1R_1}\ \text{Hz}$$

if C_1 is in farads (F) and R_1 in ohms (Ω). For example, if $C_1 = C_2 = 100\,\mu\text{F} = 100 \times 10^{-6}\,\text{F} = 10^{-4}\,\text{F}$ and $R_1 = R_2 = 15\,\text{k}\Omega = 1.5 \times 10^4\,\Omega$, then $f = 0.7/(10^{-4} \times 1.5 \times 10^4) = 0.7/1.5 \approx 0.5\,\text{Hz}$.

The mark-to-space ratio of V_{CE1} and V_{CE2} can be varied by choosing different values for C_1R_1 and C_2R_2. However the base resistors R_1 and R_2 should be low enough to saturate the transistors. (The condition for satisfactory saturation and cut-off is (as for the bistable) R_1/R_3 and R_2/R_4 both less than h_{FE}. In Fig. 75.2 this means the transistors must have $h_{FE} > 15$.)

Transistor monostable

The circuit is shown in Fig. 75.3; one stage is coupled by a capacitor C_1 and the other by a resistor R_2. When the supply is connected, the circuit settles in its one stable state with Tr_1 off and Tr_2 held on (saturated) by resistor R_1. The output voltage is therefore zero, i.e. Q = 0.

A positive voltage pulse applied via R_5 from the 6 V

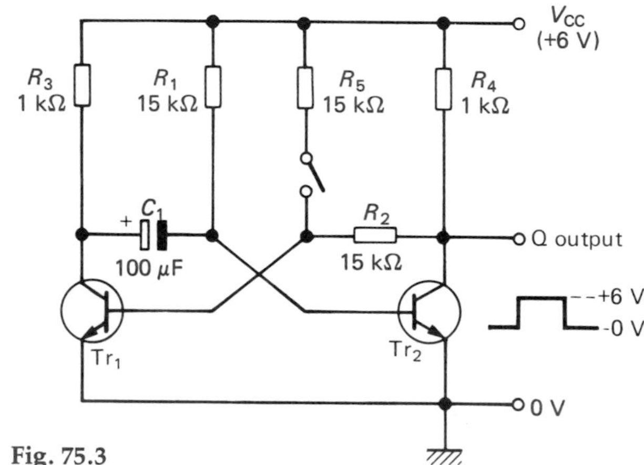

Fig. 75.3

supply to the base of Tr_1 switches it on. The potential of the right hand plate of C_1 therefore falls rapidly from +0.6 V to −5.4 V (as explained for the bistable, p. 151) and switches off Tr_2, making Q go 'high' (+6 V). The monostable is now in its second state but only until the right hand plate of C_1 charges up through R_1 to +0.6 V. Then Tr_2 is switched on again and Q goes 'low' (0 V).

The time T of the 'square' output pulse is the time for which Q is 'high' and is given approximately by

$$T = 0.7\ C_1R_1$$

NAND gate astable and monostable

(a) Astable. The circuit using two CMOS inverters A and B (e.g. two one-input NANDs) is shown in Fig. 75.4. Output B is the input to A and, if at switch-on it is 'high', output A, and so also the astable output V_o, will be 'low'. Since input B must be 'low' (if its output is 'high'), C will start to charge up through R.

When the right-hand plate of C (in the diagram) is

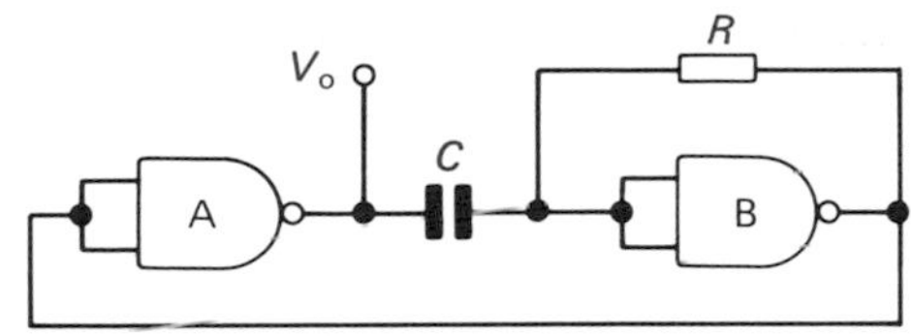

Fig. 75.4

positive enough, i.e. 'high', output B begins to go 'low' driving input A 'low' and making V_o 'high'. As a result the potential of the right plate (as well as the left) of C rises suddenly because the charge on a capacitor cannot be changed in zero time. The resulting positive feedback reinforces the direction input B was going and confirms the 'high' state of V_o.

Because the input to B is now 'high' and its output 'low', C starts to discharge through R and when the input voltage to B becomes 'low', B switches back to its first state with a 'high' output. This in turn makes V_o 'low'. The process is repeated to give a square wave output V_o, at a frequency determined by the values of C and R.

(b) Monostable. The circuit of Fig. 75.5 is in its stable state with output B (V_o) 'low' because its input is 'high' via R_1. Closing S briefly takes output A from 'high' (due to its 'low' input from output B) to 'low' and so forces input B 'low' and V_o 'high'. C then charges through R_1 until input B is 'high' enough to switch its output back to the stable state with V_o 'low'. The 'length' of the pulse so produced depends on the values of C and R_1.

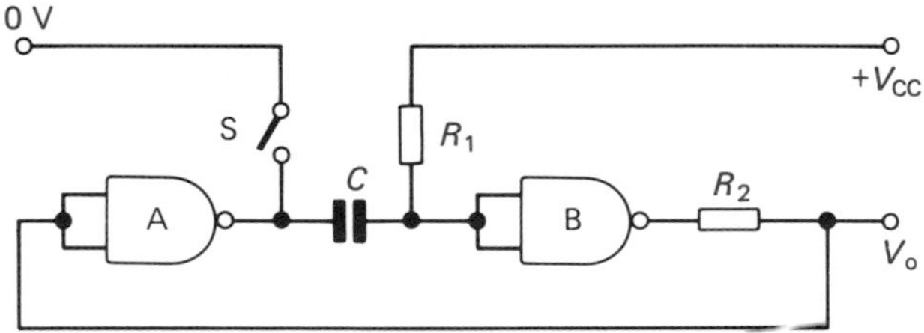

Fig. 75.5

555 Timer IC

In many electronic systems timing operations are controlled by an astable or a monostable. The popular eight-pin d.i.l. 555 timer IC, Fig. 75.6, can be used as either. It works on any d.c. supply from 3 to 15 V and has the following properties.

(i) When the voltage between 'trigger' (pin 2) and 'ground' (pin 1) falls below $\frac{1}{3}V_{CC}$, the IC is triggered, the 'output' (pin 3) rises to near V_{CC} and 'discharge' (pin 7) is disconnected from 'ground'.

Fig. 75.6

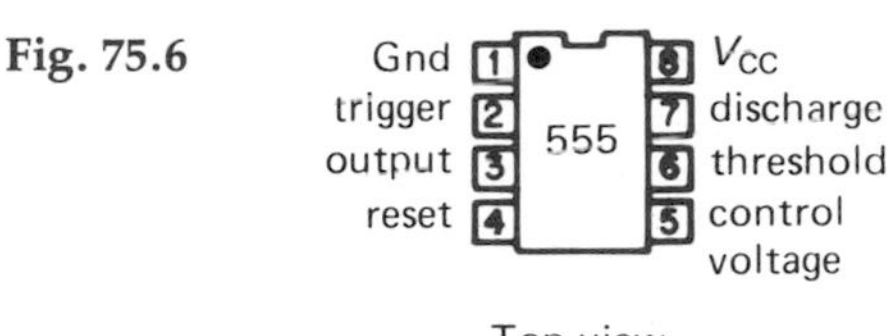

(ii) When the voltage between 'threshold' (pin 6) and ground exceeds $\frac{2}{3}V_{CC}$, the 'output' falls to 0 V and 'discharge' is connected to 'ground'.

(a) Monostable operation. The basic connections are shown in Fig. 75.7. R_1 and C_1 are external components whose values determine the time T (given in seconds, by $1.1\,R_1C_1$ if R_1 is in Ω and C_1 in F, or R_1 in MΩ and C_1 in μF) of the single 'square' output pulse produced when S_1 is switched from X to Y and back to X again. This sends the output 'high' (see (i)) and allows C_1 to charge up through R_1 (since 'discharge' is now on open circuit).

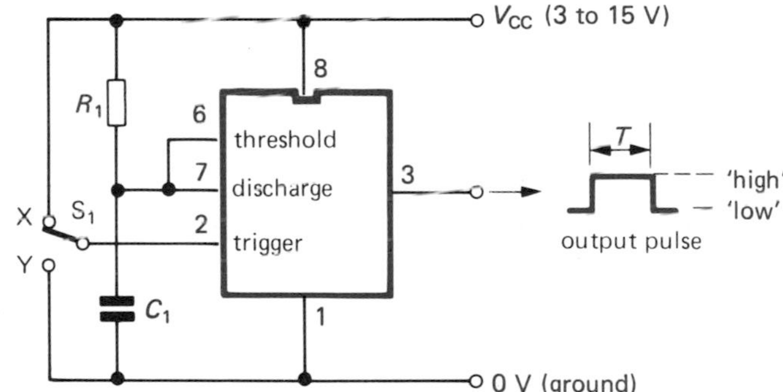

Fig. 75.7

When the voltage across C_1 reaches $\frac{2}{3}V_{CC}$, the output goes 'low' (see (ii)) and C_1 discharges (since 'discharge' is now connected to ground), returning the circuit to its stable state to await the next trigger pulse.

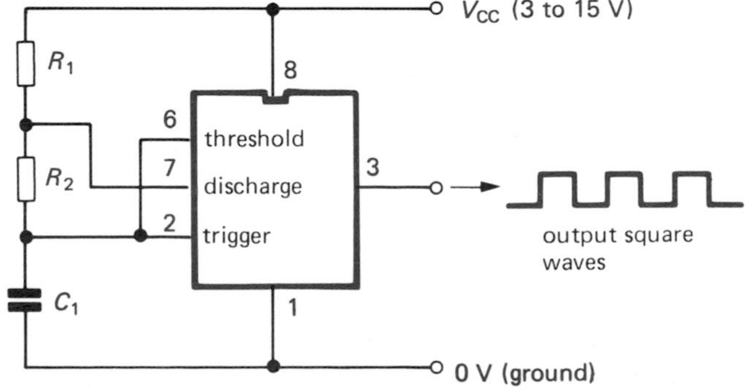

Fig. 75.8

(b) Astable operation. In the simplified circuit of Fig. 75.8, R_1, R_2 and C_1 are external components; their values decide the frequency f (given in Hz by $f = 1.4/(R_1 + 2R_2)C_1 \approx 0.7/(R_2C_1)$ if $R_2 \gg R_1$) of the square wave oscillations produced automatically at the output. These occur because C_1 charges up through R_1 and R_2 and when the p.d. across it reaches $\frac{2}{3}V_{CC}$, the output goes 'low'. C_1 then starts to discharge (via 'discharge' which is now connected to ground) and when the voltage across it falls below $\frac{1}{3}V_{CC}$, the output goes 'high' and C_1 recharges (since 'discharge' is again an open circuit). This sequence is repeated continuously.

Questions

1. Using two transistors, two resistors and two capacitors draw a circuit which will switch two small lamps L_1 and L_2 on and off in turn for equal times.

(a) What is the circuit called?

(b) State two ways the circuit could be altered so that in each case L_1 would stay on for twice as long as L_2.

(c) If car headlamps rated at 12 V 48 W were to be used for L_1 and L_2, what extra components would you add to the circuit?

2. In the transistor monostable circuit of Fig. 75.3 Tr_1 may come on first and Tr_2 be off. Explain why the situation is soon reversed and the circuit settles in its stable state with Tr_1 off and Tr_2 on.

76 Binary counters

Counters consist of bistables connected so that they toggle when the pulses to be counted are applied to their clock input. Counting is done in binary code, the bits 1 and 0 being represented by the 'high' and 'low' states of the bistable's Q output.

Binary up-counter

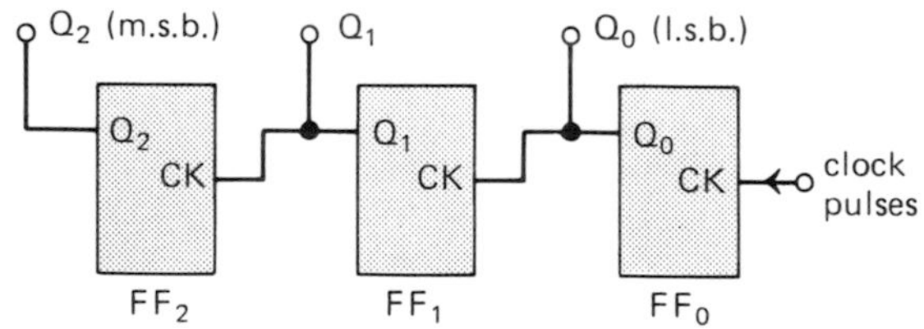

Fig. 76.1

A simple three-bit binary up-counter is shown in Fig. 76.1 consisting of cascaded (in series) toggling flip-flops FF_0, FF_1 and FF_2 (i.e. with $\bar{Q}$ joined to D if they are D-types and J = K = 1 if they are J–K types) with the Q output of each feeding the clock input (CK) of the next. The total count is given at any time by the states of Q_0 (the l.s.b.), Q_1 and Q_2 (the m.s.b.) and it progresses upwards from 000 to 111 (7 in decimal) as shown in Table 76.1, before resetting to 000.

(a) Action. Suppose the flip-flops are triggered on the *falling* edge of a pulse and that initially Q_0, Q_1 and Q_2 are all reset to zero.

Table 76.1

Number of clock pulse	Outputs		
	Q_2	Q_1	Q_0
0	0	0	0
1	0	0	1
2	0	1	0
3	0	1	1
4	1	0	0
5	1	0	1
6	1	1	0
7	1	1	1

On the falling edge ab of the first clock pulse, shown in Fig. 76.2, Q_0 switches from 0 to 1. The resulting rising edge AB of Q_0 is applied to CK of FF_1, which does not change state because AB is not a falling edge. Hence the output states are $Q_2 = 0$, $Q_1 = 0$ and $Q_0 = 1$, giving a binary count of 001.

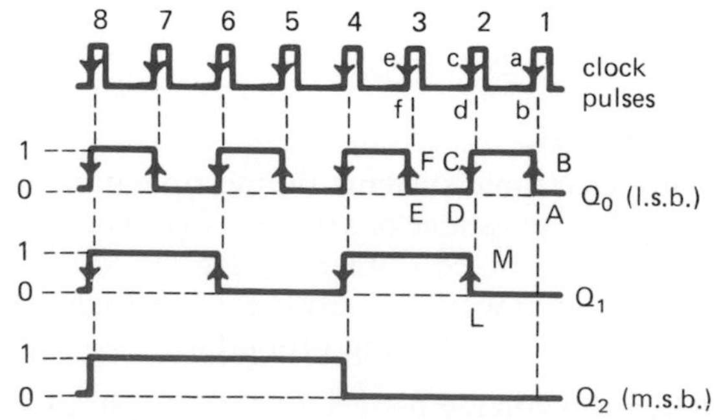

Fig. 76.2

The falling edge cd of the second clock pulse makes FF_0 change state again and Q_0 goes from 1 to 0. The falling edge CD of Q_0 switches FF_1 this time, making $Q_1 = 1$. The rising edge LM of Q_1 leaves FF_2 unchanged. The count is now $Q_2 = 0$, $Q_1 = 1$ and $Q_0 = 0$, i.e. 010.

The falling edge ef of the third clock pulse to FF_0 changes Q_0 from 0 to 1 again but the rising edge EF does not switch FF_1 leaving $Q_1 = 1$, $Q_2 = 0$ and the count at 011. The action thus ripples along the flip-flops, each one waiting for the previous flip-flop to supply a falling edge at its clock input before changing state.

(b) Modulo. The modulo of a counter is the number of output states it goes through before resetting to zero. A counter with three flip-flops counts from 0 to $(2^3 - 1) = 8 - 1 = 7$; it has eight different output states representing the decimal number 0 to 7 and is a modulo-8 counter.

(c) Dividers. In a modulo-8 counter one output pulse appears at Q_2 every eighth clock pulse. That is, if f is the frequency of a regular train of clock pulses, the frequency of the pulses from Q_2 is $f/8$ (i.e. Q_2 is 1 for four clock pulses and 0 for the next four pulses). A modulo-8 counter is a 'divide-by-8' circuit as well. Counters are used as dividers in digital watches (p. 162).

Binary down-counter

In a down-counter the count decreases by one for each clock pulse. To convert the up-counter in Fig. 76.1 into a down-counter, the $\bar{Q}$ output (instead of the Q) of each flip-flop is coupled to the CK input of the next, as in Fig. 76.3. The count is still given by the Q outputs.

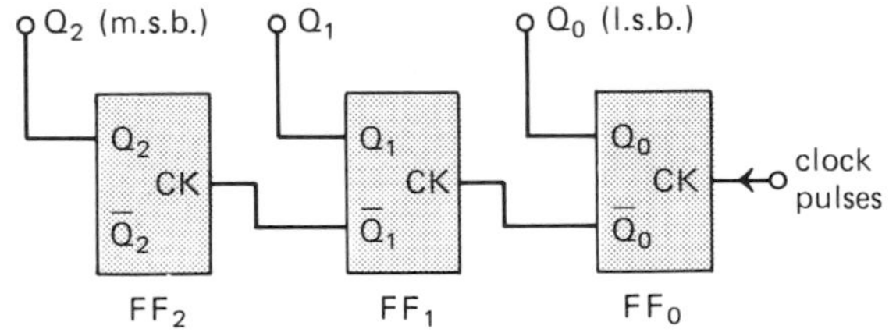

Fig. 76.3

Decade counter

A four-bit binary up-counter, modified as in Fig. 76.4, acts as a modulo-10 counter and counts up from 0 to 9 before resetting. When the count is 1010

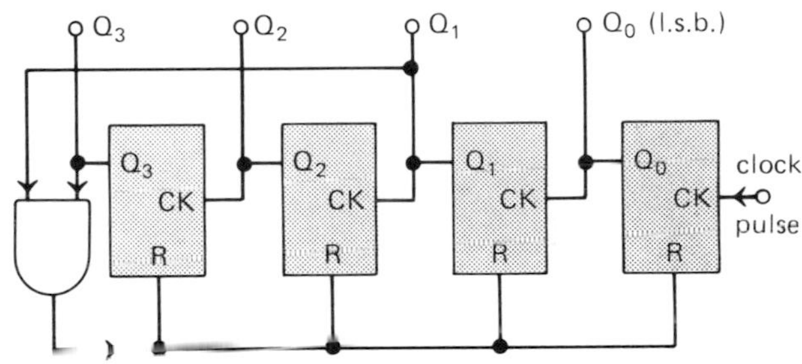

Fig. 76.4

(decimal 10), $Q_3 = 1$, $Q_2 = 0$, $Q_1 = 1$ and $Q_0 = 0$ and since both inputs to the AND gate are 1's (i.e. Q_3 and Q_1), its output is 1 and this resets all the flip-flops to 0 (otherwise it would be a modulo-16 counter counting from 0 to 15).

Questions

1. (a) How many output states are there in a counter which has (i) 1, (ii) 2, (iii) 3, (iv) 4 and (v) 5 bistables?
(b) What is the highest decimal number each counter in (a) can count before resetting?

2. (a) What is meant by the modulo of a counter?
(b) How many flip-flops are needed to build counters of modulo- (i) 2, (ii) 5, (iii) 7, (iv) 10 and (v) 30?
(c) If f is the frequency of the clock pulses applied to a modulo-10 counter, what is the frequency of the output pulses from the last flip-flop?

3. Draw the block diagram for a modulo-6 binary up-counter which resets every sixth clock pulse.

77 Registers and memories

Shift registers

A shift register is a memory which stores a binary number and shifts it out when required. It consists of several D-type or J–K flip-flops, one for each bit (0 or 1) in the number. The bits may be fed in and out serially, i.e. one after the other, or in parallel, i.e. all together. Shift registers are used, for example, in calculators to store two binary numbers before they are added.

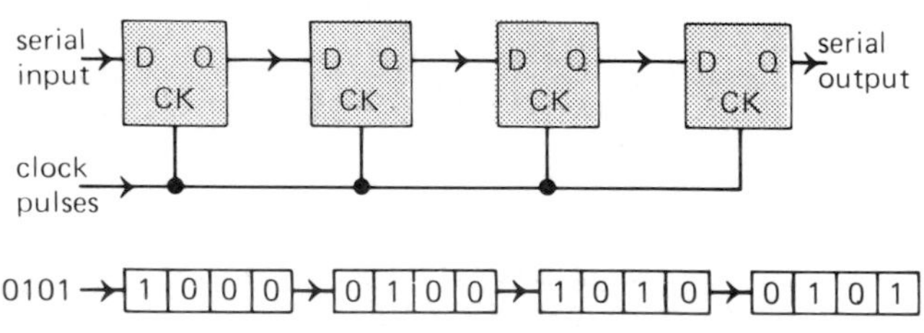

(a) Serial-input-serial-output

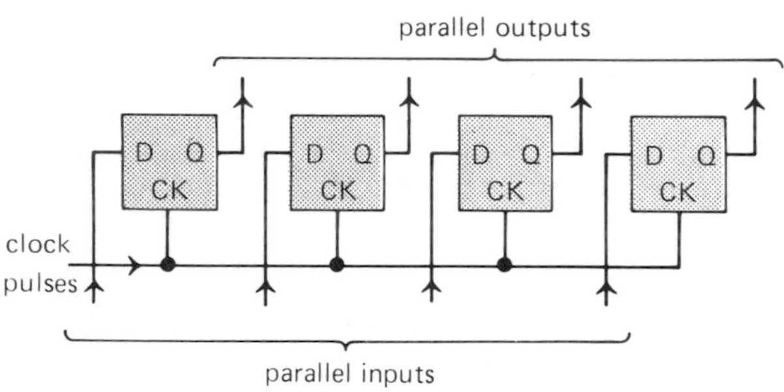

(b) Parallel-input-parallel-output

Fig. 77.1

In the four-bit serial-input-serial-output (SISO) type of Fig. 77.1a, clocked D-type flip-flops are used, the Q output of each one being applied to the D-input of the next. The bits are loaded one at a time, usually from the left and move one flip-flop to the right every clock pulse. Four pulses are needed to enter a four-bit number such as 0101 and another four to move it out serially.

In the parallel-input-parallel-output (PIPO) type of Fig. 77.1b, all bits enter their D inputs simultaneously and are transferred together to their respective Q outputs (where they are stored) by the same clock pulse. They can then be shifted out in parallel.

Two other types are serial-input-parallel-output (SIPO) and parallel-input-serial-output (PISO).

Memories

(a) Organization. A memory stores *data*, i.e. the information to be processed and the *program* (note the spelling) *of instructions* to be carried out. An IC semiconductor memory consists of an array of flip-flops, called memory cells, each storing one bit of data. The array is organized so that the bits are in groups or 'words' of typically, 1, 4, 8 or 16 bits.

Every word has its own location or *address* in the memory which is identified by a certain binary number. The first word is at address zero, the second at one, the third at two and so on. The table in Fig. 77.2 shows part of the contents of a sixteen word, four-bit memory, i.e. it has sixteen addresses each storing a four-bit word. For example in the location with address 1110 (decimal 14), the data stored is the four-bit word 0011 (decimal 3).

Address					Data				
Decimal	Binary m.s.b.			l.s.b.	Binary m.s.b.			l.s.b.	Decimal
0	0	0	0	0	0	1	0	1	5
1	0	0	0	1	1	1	0	0	12
2	0	0	1	0	0	1	1	0	6
3	0	0	1	1	1	0	0	1	9
⋮	⋮	⋮	⋮	⋮	⋮	⋮	⋮	⋮	⋮
14	1	1	1	0	0	0	1	1	3
15	1	1	1	1	0	1	1	1	7

Fig. 77.2

In a *random access memory* all words can be located equally quickly, i.e. access is random and it is not necessary to start at address zero.

(b) Types. There are two main types—Read Only Memories, i.e. ROMs, and Read and Write Memories which are confusingly called RAMs (because they allow random access, as ROMs also do). A RAM not only lets the data at any address be 'read' but it can also have new data 'written' in. Whereas a RAM loses the data stored almost as soon as the power to it is switched off, i.e. it is a *volatile* memory, a ROM does not, i.e. it is *non-volatile*. ROMs are used for permanent storage of fixed data such as the program in a computer. RAMs are required when data has to be changed.

The programmable ROM or PROM lets the user 'burn' the pattern of bits, i.e. the program, into a ROM by applying a high voltage which fuses a link in the circuit. The disadvantage of this type is that it does not allow changes or corrections to be made later. When alterations are necessary, for instance during the development of a program, an erasable PROM or EPROM is used in which the program is stored electrically and is erased by exposure to ultraviolet radiation before reprogramming.

(c) Structure. The simplified structure of a sixteen word four-bit RAM is shown by the block diagram of Fig. 77.3; that for a ROM is similar but it has no 'write' provision.

To 'write' a word into a particular address the four-bit binary number of the address is applied to the address inputs and the word (also in binary) is set up at the data inputs. When *write enable* is at the appropriate logic level (say 'high'), the word is stored automatically at the correct address in the memory array, as located by the address decoder.

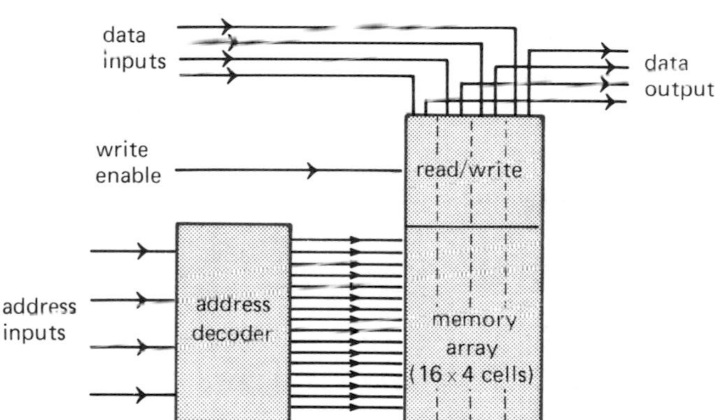

Fig. 77.3

To 'read' a word stored at a certain address, the address code is applied as before and the word appears at the data outputs if write enable is at the required logic level (say 'low').

(d) Storage capacity. A sixteen word four-bit memory has a storage capacity of $16 \times 4 = 64$ bits (it has 64 memory cells) and is limited to four-bit words.

An eight-bit word is called a *byte*. In computer language the symbol K (capital K) is used to represent 1024 (2^{10}). For example, a memory with a capacity of 4 K bits stores $4 \times 1024 = 4096$ bits, i.e. 512 words if it is organized in bytes or 1024 four-bit words. Do not confuse K with k (small k) which stands for kilo, i.e. 1000.

Questions

1. (a) What does a shift register do?
(b) How many clock pulses are needed to shift an eight-bit binary number into an eight flip-flop serial shift register?

2. (a) Explain: data, bit, word, address, write, read, random access, volatile, byte.
(b) Distinguish between (i) a ROM and a RAM, (ii) a PROM and an EPROM.
(c) What is the storage capacity of a sixty-four word eight-bit memory in (i) bits, (ii) bytes?

78 Some digital systems

Microelectronic options

When designing an electronic system using large-scale integrated (LSI) circuits, two broad approaches are possible. In the first, the job done by the system is controlled by its circuit and cannot be changed without altering the circuit. This is the *circuit-controlled* option. In the second, the job done is decided by a program of instructions (software) which tells the system exactly how to perform the task. To change the job done, the program is changed. This is the *program-controlled* option.

The first approach may be best when the system has just one task and the second when it has several. Within the circuit-controlled option there are three choices, to be outlined now. The program-controlled option uses a microprocessor, to be discussed later (p. 210).

(a) Wired logic. The system is wired together on a suitable board using standard, off-the-shelf SSI and MSI chips and, perhaps discrete components. The chips might range from logic gates to an ALU. This is often the simplest and cheapest method but it is inflexible.

(b) Custom chips. In this case all functions demanded of the system are done by one 'dedicated' LSI chip containing the exact number of components for the job. It gives maximum efficiency, minimum size and minimum power consumption.

(c) Uncommitted logic array (ULA). This approach is half-way between **(a)** and **(b)**. It uses a chip containing an array of standard logic 'cells', each complete in itself (consisting of logic gates, adders, etc.) but not interconnected into any circuit pattern. Connections are made by the chip manufacturer once a logic diagram has been constructed for the customer's application. ULAs are cheaper than custom chips and can be designed in about one-third of the time. They are also more versatile, within limits.

Digital watch

The block diagram of Fig. 78.1 shows the main parts of a digital watch. When activated by a battery, the quartz crystal oscillator produces very stable electrical pulses with a frequency of exactly 32 768 Hz (2^{15} Hz). The dividers reduce this to one pulse per second and after counting and decoding they are applied to the appropriate electrodes of the seven-segment LCD or LED decimal displays to show the time or date.

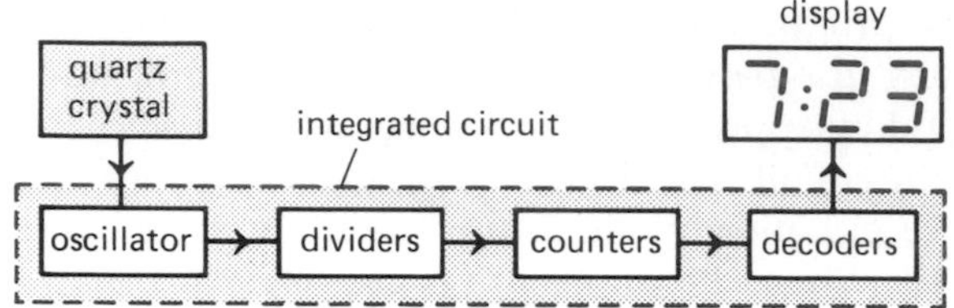

Fig. 78.1

The IC is a custom chip containing over 2000 transistors and incorporates all the system's major circuits.

Digital voltmeter

The main parts of a digital voltmeter (p. 97) are shown in the simplified block diagram of Fig. 78.2.

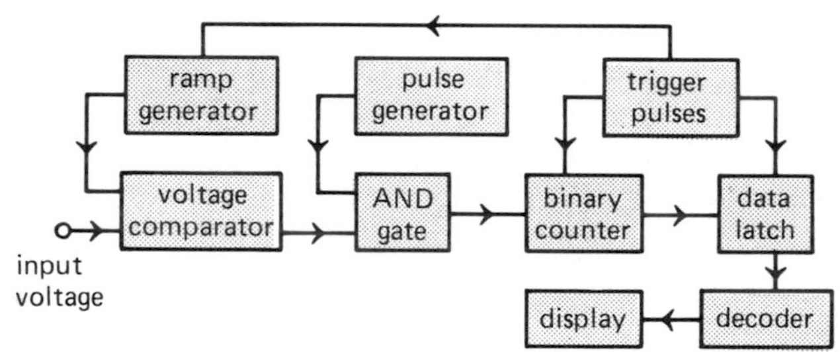

Fig. 78.2

The action is:

(i) a *trigger pulse* sets the *binary counter* to zero and starts the *ramp generator* (p. 208) which produces a repeating sawtooth waveform,

(ii) the *voltage comparator* (p. 124) output changes from 'high' to 'low' when the ramp voltage equals the input voltage,

(iii) the *AND gate* inputs are supplied by the *comparator* and a steady train of pulses from the *pulse generator*; the latter pass through the gate until the *comparator* output goes 'low' and the number which does so depends on the time taken by the ramp voltage to equal the input voltage, i.e. it is proportional to the input voltage (if the ramp is linear),

(iv) the *counter* records the number of output pulses from the *AND gate*,

(v) the *data latch* passes this number to the *decoder* for conversion to decimal before they reach the *display*, where it is held until the next count enters the latch.

The system produces the digital equivalent of the analogue input voltage, i.e. it is an analogue-to-digital (A/D) converter (p. 208) and the waveforms in Fig. 100.3 for a steady d.c. input should also be studied.

Electronic calculator

In many calculators a microprocessor is used as the 'brain' but the principles involved can be shown for the addition of two numbers using the very simple system of Fig. 78.3. It has a four-bit adder (p. 139) as its ALU (p. 146).

Suppose the two numbers to be added are 3 and 5. When switch '3' is pressed on the *keyboard*, the four-bit binary form of decimal 3 is produced by the *encoder* at its four outputs Q_0, Q_1, Q_2 and Q_3. In this case the output values would be Q_3 (m.s.b.) = 0, Q_2 = 0, Q_1 = 1 and Q_0 (l.s.b.) = 1. The binary number 0011 is thus applied to the four inputs of *shift register B*. If the 'store' switch (STO) on the *keyboard* is pressed next, a 'clock' pulse causes *shift register B* to shift 0011 from its inputs to its outputs, where it is stored and becomes the input to *shift register A*.

The second number can now be entered and if switch '5' is pressed, the binary number applied to the inputs of *shift register B* becomes 0101, i.e. Q_3 = 0, Q_2 = 1, Q_1 = 0 and Q_0 = 1. On applying a second 'clock' pulse, the first number (0011 = 3) is shifted from the inputs of *shift register A* to its outputs where it becomes the $A_3A_2A_1A_0$ input to the *adder*. At the same time, the second number (0101 = 5) is shifted from the inputs of *shift register B* to its outputs and becomes the $B_3B_2B_1B_0$ input to the *adder*.

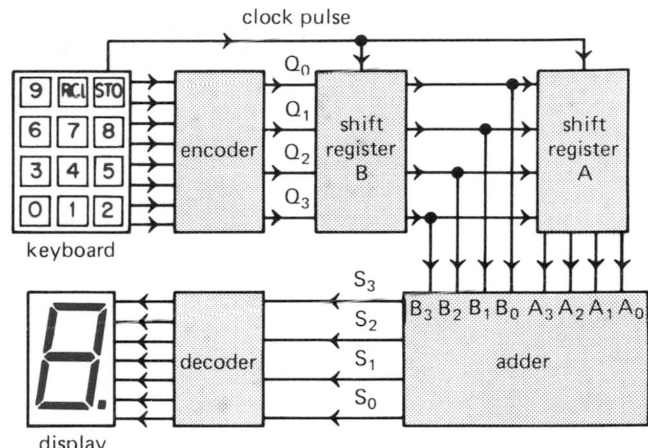

Fig. 78.3

The *adder* adds the two binary numbers $A_3A_2A_1A_0$ and $B_3B_2B_1B_0$ immediately and produces their binary sum $S_3S_2S_1S_0$ at its four outputs. These outputs are applied to the four inputs of the *decoder* which, if it was driving a seven-segment decimal *display*, would create seven outputs each capable of driving one segment. In this example all seven segments would light up and give 8 as the sum of 3 + 5.

79 Progress questions

1. 'One of the characteristics of the bistable circuit is that it has memory'. Explain this statement.
What do you consider are the other important characteristics of the bistable circuit?
How are the properties of the bistable (regarded as a basic module) applied in (a) a scaling (i.e. counting) system, (b) a shift register? *(O. and C.)*

2. (a) Fig. 79.1 represents three bistables connected in series. When a bistable is in the state logical 1 the light-emitting diode (LED) is lit, and when it is in the state logical 0 the LED is off. If a bistable changes state from logical 1 to logical 0, it causes the next bistable along to change state.

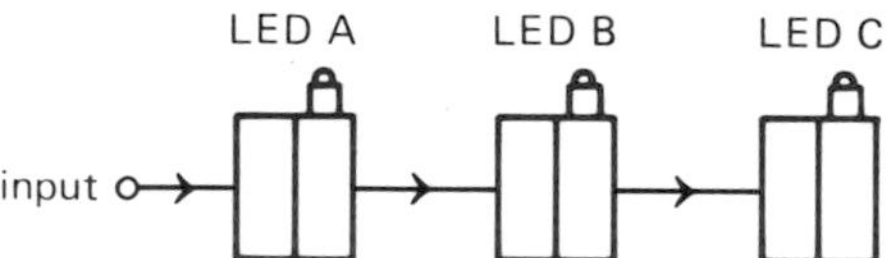

Fig. 79.1

At the start, all the LEDs are off. A series of switching pulses is then fed to the input.
(i) Copy and complete the table to show the states (1 or 0) of the LEDs after each of 7 pulses.

Pulse number	LED A	LED B	LED C
1			
2			
3			
4			
5			
6			
7			

(ii) Explain what happens for the first 3 pulses.
(iii) Describe a use for such an arrangement of bistables.
(b) Fig. 79.2 shows a bistable circuit. The lamp A is on after the circuit is first connected up.

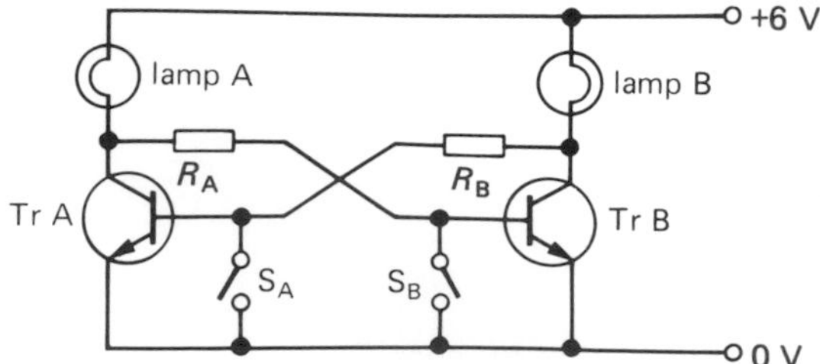

Fig. 79.2

(i) In which transistor is there a large collector current flowing?
(ii) Which switch must be closed momentarily to turn off lamp A? Explain.
(iii) What effect does the closing of this switch have on the base voltage of the transistor Tr B?
(iv) What happens as a result to lamp B? Explain.
(v) How can the initial condition with lamp A be restored? Explain. *(O.L.E.)*

3. (a) Draw the truth table for a two-input NAND gate, Fig. 79.3a.

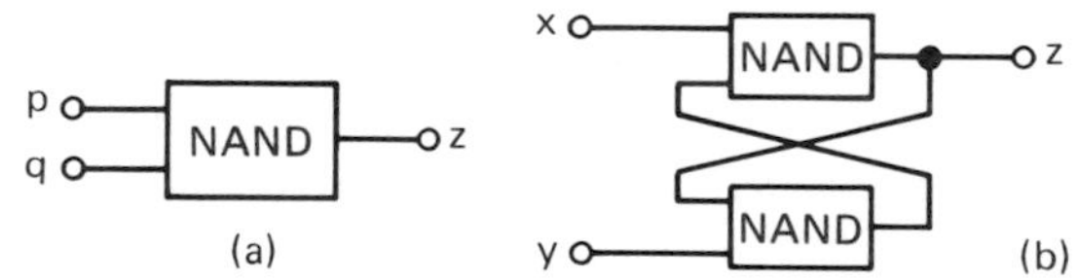

Fig. 79.3

(b) Suppose two such NAND gates are joined as in Fig. 79.3b so that the output of each gate is connected to one of the inputs of the other gate. Considering x and y as inputs of this system, and z as its output, copy and complete the following truth table, making sure you make each change in sequence.

Sequence	x	y	z
1st	1	0	0
2nd	1	1	
3rd	0	1	
4th	1	1	
5th	1	0	

A circuit which uses this logic combination is that of a burglar alarm, Fig. 79.4. When all the switches are closed the input will be connected to +6 V (logic 1). If one or more switches are open the input is connected through the 100 Ω resistor, to 0 V (logic 0). The lower input is normally at logic 1 unless the reset button is pressed. The

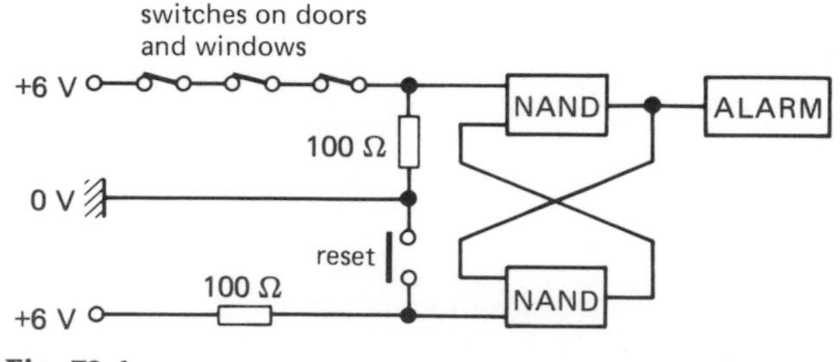

Fig. 79.4

alarm rings when it receives a logic 1 signal from the NAND gate output.

(c) Describe the sequence of steps to 'set' the alarm.
(d) Describe and explain why the alarm rings when an intruder opens a door or window.
(e) Why can an intruder not stop the alarm ringing by just shutting the window or door?
(f) How can the alarm be switched off?
(g) It would obviously help to locate the intruder if some indication of which window or door is opened could be given. Design an addition to the burglar alarm circuit that would do this.

(A.E.B. 1982 Electronics)

4. What is (a) a data latch, (b) code conversion? Describe how a digital measuring unit holds, displays and refreshes a reading (as for example in a digital voltmeter). In your answer, be sure to quote all the stages of the process and discuss each as fully as you can.

(O. and C.)

5. (a) With the aid of an example, explain the reason for using a Schmitt trigger.
(b) A certain Schmitt trigger switches on when the input voltage rises to 2.0 V and then off when it falls to 1.6 V. Draw the output waveform for the input shown in Fig. 79.5, given that the output voltage is 0 V with trigger off and 10 V when it is on.
What is the duration of each output pulse?

(O.L.E. part qn.)

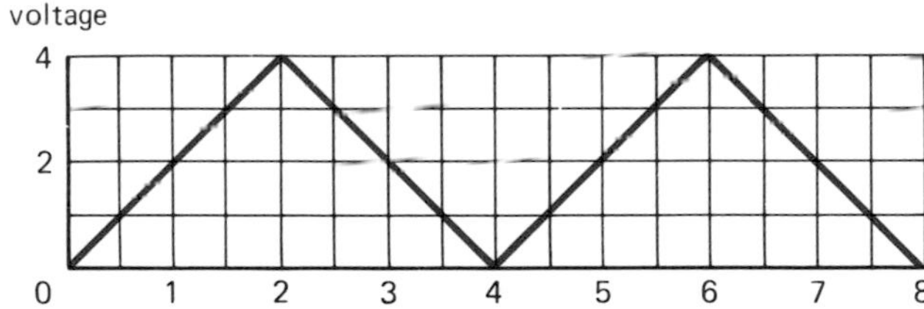

Fig. 79.5

6. Fig. 79.6 shows the circuit of an astable multivibrator.

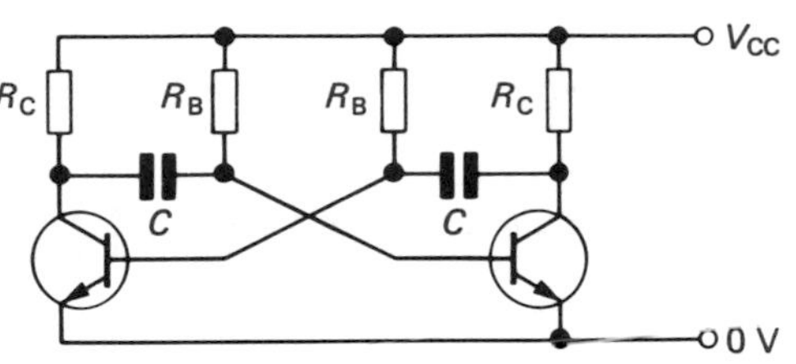

Fig. 79.6

(a) Explain the role of the resistors R_B in the operation of the circuit. What factors limit the maximum value which R_B can have? What effect would an increase in the value of R_B have?
(b) Explain the role of the capacitors C. What effect would a decrease in the value of C have?
(c) Explain the part played by the resistors R_C in the operation of the circuit.
(d) Sketch graphs of the variation with time of (i) the p.d. between collector and emitter, and (ii) the p.d. between the base and emitter, for one of the transistors.
(e) What effect would a variation in V_{CC} have on the operation of the circuit?
(f) How would you change the circuit so that one transistor was on for longer than the other? *(L.)*

7. How is electronics used to convert measurement of an electrical quantity (e.g. a d.c. voltage) into digital form? Discuss in simple terms (a) how a linear relationship is obtained between digital reading and input signal, (b) how 'zero error' is avoided.
What part does a latch play in the process of making and recording a digital measurement? *(O. and C.)*

Information and electronics

80 Information technology

Introduction

Information technology (IT) is concerned with the ability of *computers* to store and process vast amounts of information in split seconds and of *telecommunications* to transmit it almost instantaneously. *Microelectronics* enables the equipment required to be made incredibly small and cheaply. IT is cropping up in every aspect of our lives, bringing changes, as profound as those of the Industrial Revolution, that will affect our homes, shops, offices, factories and schools as well as our health and leisure.

Links are being established allowing information to be sent anywhere in much the same way as telephones enable us to speak to each other now. In the broadest sense, information includes the messages, programmes and other 'traffic' carried by telephones, radio, television, as well as by computer signals and the various data processing devices used by banks, insurance companies and other business and government organizations. Today, information is seen, like minerals and energy, as a basic resource which is becoming more easily and widely accessible. Its control, however, raises important issues.

Some scientists are wondering whether with the possible development of ultra-intelligent machines, we may be seeing the next major step in the evolution of the universe. They argue that just as over three billion years ago, some very exotic molecules appeared on Earth which produced life by entering into a new combination of matter, energy and information, so may life evolve into something else in the future, heralded by the advances now occurring in IT.

Representing information

Information or data can be represented electrically in two ways.

(a) Digital method. In this method electricity is switched on and off and the information is in the form of electrical pulses. For example, in the simple circuit of Fig. 80.1a, data can be sent by the 'dots' and 'dashes' of the Morse code by closing the switch for a short or a longer time. In Fig. 80.1b the letter A (·–) is shown.

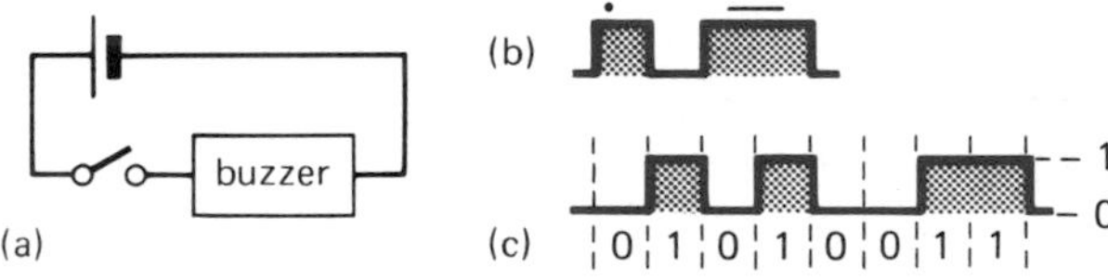

Fig. 80.1

Computers use the simpler binary code with 1 and 0 represented by 'high' and 'low' voltages respectively. They can only handle numbers (i.e. 1 and 0) and so a pattern of 1s and 0s has to be agreed for each of the 26 letters of the alphabet if words are also to be processed.

A five-bit code has 2^5 (32) variations, i.e. from 00000 to 11111, which would be enough. However, many digital systems used eight-bit words, i.e. bytes. The *American Standard Code for Information Interchange* (ASCII) is an eight-bit code that allows $2^8 = 256$ characters to be coded in binary. This is adequate for all letters of the alphabet (capitals and small), the numbers 0 to 9, punctuation marks and other symbols. Fig. 80.1c shows one eight-bit pulse, it represents the letter capital S in ASCII code.

(b) Analogue method. In this case the flow of electricity is *regulated* (not switched) and a continuous range of voltages (or currents) between 0 and some maximum, is possible, Fig. 80.2a. The actual value at any instant stands for a number.

A crystal microphone (p. 43) produces a voltage which is the analogue of information (in the form of sound). A carbon microphone (p. 43) gives current as the analogue of information.

Advantages of digital signals

Information in digital form has certain advantages over that in analogue form despite the fact that most transducers produce analogue signals which can be amplified readily. There are two main reasons for this.

First, with digital signals it is only necessary to detect the presence or absence of a pulse (i.e. whether it is 1 or 0). Should they pick up 'noise' (i.e. stray, unwanted voltages or currents) at any stage of processing or transmission, it does not matter as it would with an analogue signal whose waveform would be distorted.

Second, digital signals fit in with modern technology and can be used with both telecommunications and data processing equipment.

Analogue-to-digital conversion

With the increasing popularity of digital signals this operation is frequently necessary. It is performed by an analogue-to-digital converter (p. 208) in a process called *pulse code modulation* (PCM) which involves 'sampling' its value regularly.

Suppose the analogue voltage has the waveform shown in Fig. 80.2a. It is divided into a number of equally-spaced voltage levels (six in this case) and measurements taken at equal intervals to find the level at each time. Every level is represented in binary code by a number which has a characteristic series of on–off electrical pulses, i.e. a particular

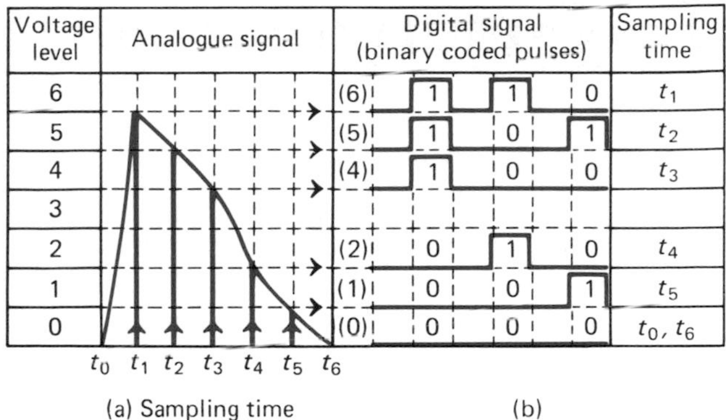

Fig. 80.2

digital bit-pattern. A three-bit code can represent up to eight levels (0 to 7), as in Fig. 80.2b; four bits will allow sixteen levels to be coded.

The accuracy of the representation increases with the number of voltage levels and the sampling frequency. For speech the frequency has to be about 8000 times per second, i.e. 125 μs between samples; for music it must be more and for television signals very much more.

The analogue voltage shown would be represented in digital form by the train of pulses in Fig. 80.3 using a three-bit code.

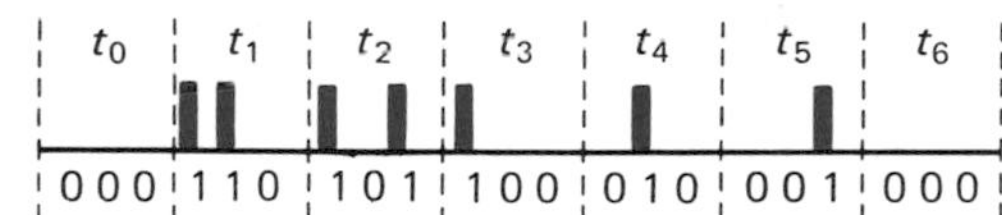

Fig. 80.3

81 Communication systems

Transmission of information

Electrical signals representing 'information' from a microphone, a TV camera, a computer, etc., can be sent from place to place using either cables or radio waves. Information in the form of audio frequency (a.f.) signals may be transmitted directly by a cable but in general, and certainly in radio and TV, they require a 'carrier'. This has a higher frequency than the information signal, its amplitude is constant and its waveform sinusoidal.

The general plan of *any* communication system is shown in Fig. 81.1. Signals from the *information source* are added to the carrier in the *modulator* by the process of 'modulation'. The modulated signal is sent along a 'channel' in the 'propagating medium' (i.e. cable or radio wave) by the *transmitter*.

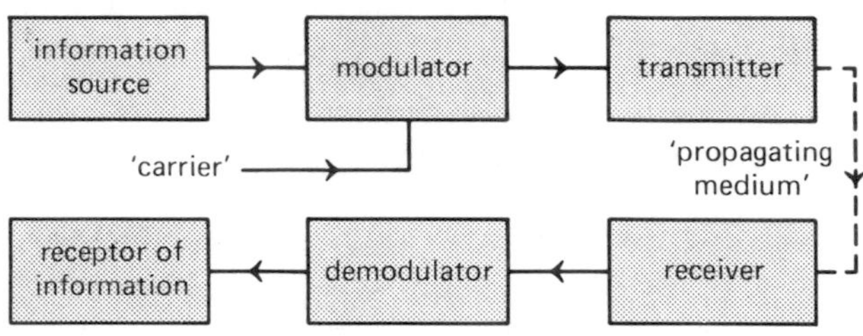

Fig. 81.1

At the receiving end, the *receiver* may have to select (and perhaps amplify) the modulated signal before the *demodulator* extracts from it the information signal for delivery to the *receptor of information*.

Types of modulation

(a) Amplitude modulation (AM). The information signal from, for example, a microphone, is used to vary the *amplitude* of the carrier so that it follows the wave shape of the information signal, Fig. 81.2.

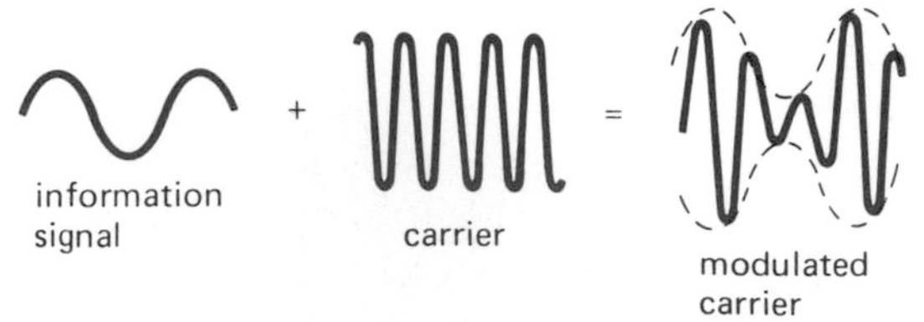

Fig. 81.2

(b) Frequency modulation (FM). In this case the information signal varies the *frequency* of the carrier, which increases if the signal is positive and decreases if it is negative. The effect, much exaggerated, is shown in Fig. 81.3.

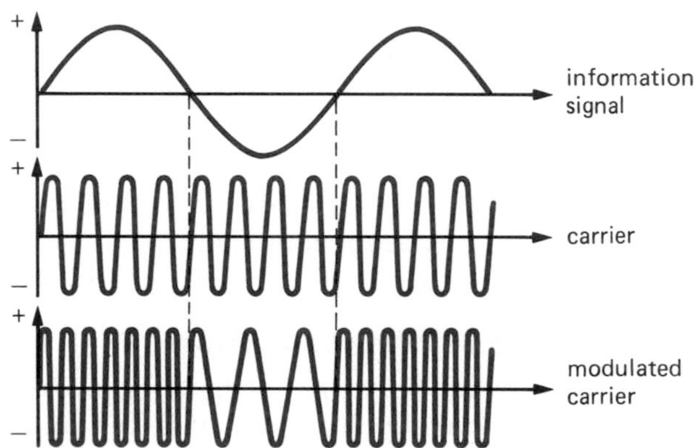

Fig. 81.3

(c) Pulse code modulation (PCM). The carrier is not transmitted continuously but is modulated to form a pattern of *pulses* which, as explained on p. 166, represents in binary code, regular samples of the amplitude of the information signal.

Bandwidth

The term is used in two ways.

(a) Bandwidth of a signal. This is the range of frequencies a signal occupies. For intelligible speech it is about 3 kHz (e.g. 300 to 3400 Hz as in the telephone system), for high quality music it is 20 kHz or so and for television signals about 8 MHz.

The bandwidth of a modulated signal is due to the fact that when modulated, other frequencies, called *side frequencies*, are created on either side of the carrier (which is a single frequency). In AM, if the carrier frequency is f_c and a modulating frequency f_m, two new frequencies of $f_c - f_m$ and $f_c + f_m$ are produced, one below f_c and the other above it, Fig. 81.4a.

If, as usually occurs in practice, the carrier is modulated by a range of a.fs., each a.f. gives rise to a pair

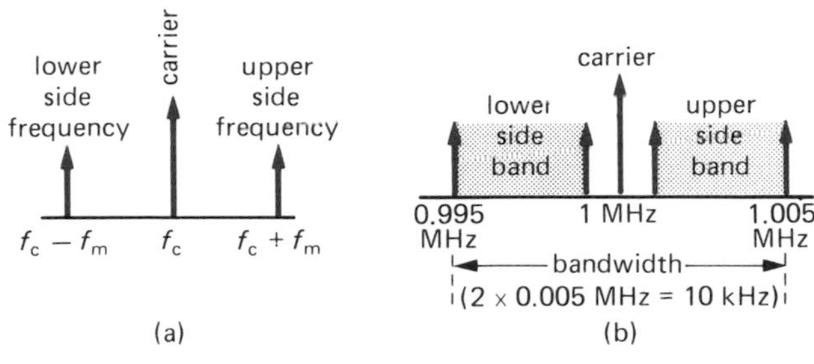

Fig. 81.4

of side frequencies. The result is a band of frequencies, called the *lower* and *upper sidebands*, stretching below and above the carrier by the value of the highest modulating frequency. For example if f_c = 1 MHz and the highest f_m = 5 kHz = 0.005 MHz, then $f_c - f_m$ = 0.995 MHz and $f_c + f_m$ = 1.005 MHz, Fig. 81.4b. The bandwidth of a carrier modulated by a.f. signals up to 5 kHz is thus 10 kHz. Sidebands also arise in FM.

(b) Bandwidth of a channel. This is the range of frequencies a communication channel can accommodate. High frequency transmission channels have greater bandwidths than lower frequency ones, i.e. their information-carrying capacity is greater. For instance, the v.h.f. radio band, which extends from 30 MHz to 300 MHz has 'space' for 2700 10 kHz wide signals. The medium waveband, 300 kHz to 3 MHz, can carry only 270 such signals.

Cables also have different bandwidths.

Multiplexing

Multiplexing involves sending several different information signals along the same communication channel so that they do not interfere. Two methods are outlined.

(a) Frequency division. The signals (e.g. speech in analogue form) modulate carriers of different frequencies which are then transmitted, a greater bandwidth being required.

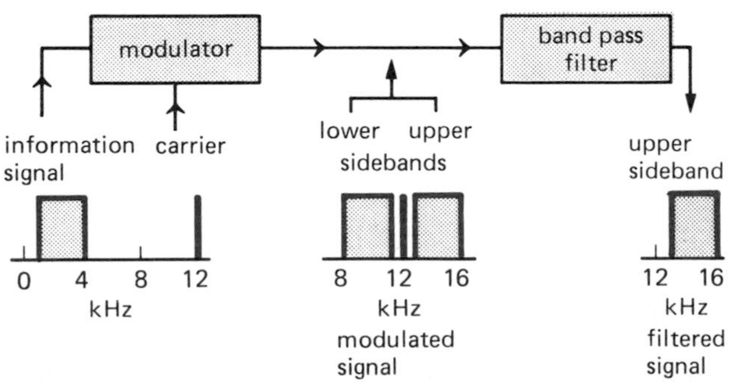

Fig. 81.5

The principle is shown in Fig. 81.5 for cable transmission. The information signal contains frequencies up to 4 kHz and the carrier frequency is 12 kHz. *Each* sideband contains *all* the modulating frequencies, i.e. all the information and so only one is really necessary. Here, a 'bandpass filter' allows just the upper sideband to pass.

The multiplexing of three 4 kHz wide information signals 1, 2 and 3, using carriers of 12, 16 and 20 kHz is shown in Fig. 81.6.

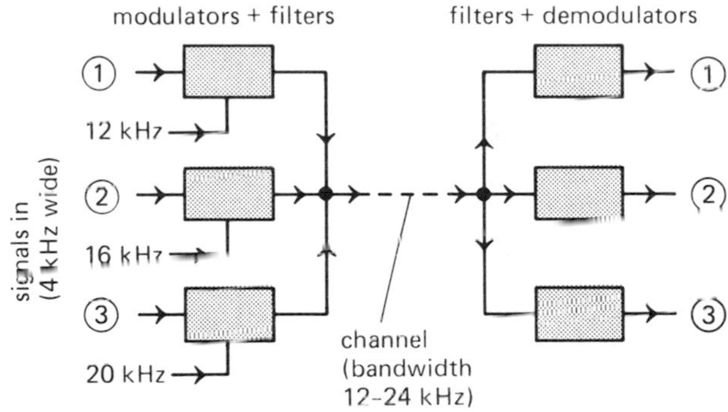

Fig. 81.6

(b) Time division. The method is shown in Fig. 81.7 for three signals. An electronic switch, i.e. a multiplexer (p. 146 and here 3 : 1 line), samples each signal in turn (for speech 8000 times per second).

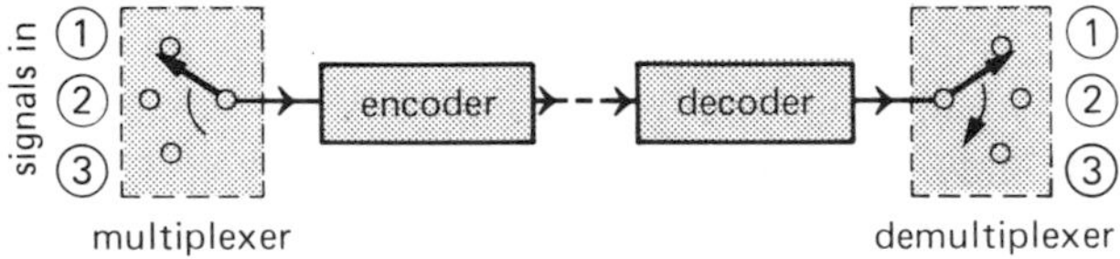

Fig. 81.7

The encoder converts the samples into a stream of pulses, representing, in binary, the level of each signal (as in PCM). The decoder and a demultiplexer (1 : 3 line), which are synchronized with the multiplexer and encoder, reverse the operation at the receiving end.

Questions

1. What is the bandwidth when frequencies in the range 500 Hz to 3500 Hz amplitude modulate a carrier?

2. A carrier of frequency 800 kHz is amplitude modulated by frequencies ranging from 1 kHz to 10 kHz. What frequency range does each sideband cover?

Audio systems

82 Sound recording

Tape recording

(a) Recording. Sound can be stored by magnetizing plastic tape coated with finely powdered iron oxide or chromium oxide. The *recording head* is an electromagnet having a tiny gap filled with a strip (a 'shim') of non-magnetic material between its poles, Fig. 82.1. When a.c. from the recording microphone passes through the windings, it causes an alternating magnetic field in and just outside the gap. This produces in the tape a chain of permanent magnets, in a magnetic pattern which represents the original sound.

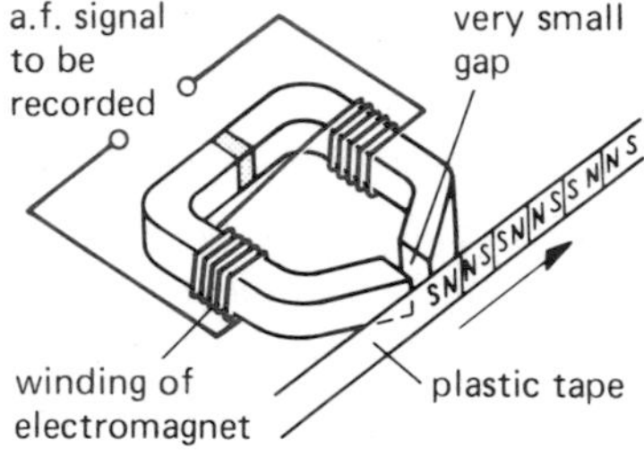

Fig. 82.1

The *strength* of the magnet produced at any part of the tape depends on the value of the a.c. when that part passes the gap and this in turn depends on the loudness of sound being recorded. The *length* of a magnet depends on the frequency of the a.c. (which equals that of the sound) and on the speed of the tape. The effect of high and low frequency a.c. is shown in Fig. 82.2.

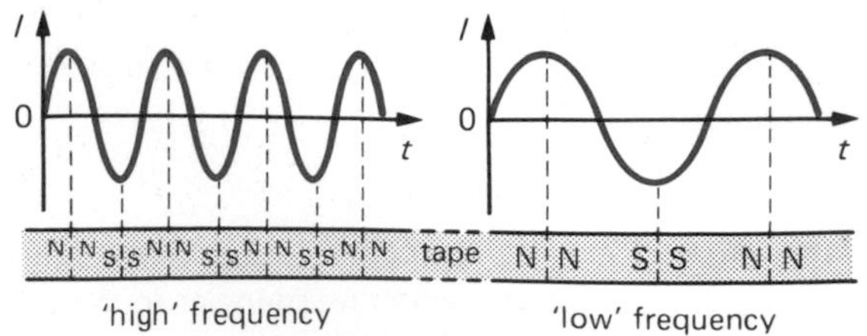

Fig. 82.2

High frequencies and low speeds create shorter magnets and very short ones tend to lose their magnetism immediately. Each part of the tape being magnetized must have moved on past the gap in the recording head before the magnetic field reverses. High frequencies and good quality recording require high tape speeds.

Chromium oxide tapes record a wider range of frequencies and their output is greater on playback.

Two-track and stereo tapes are shown in Fig. 82.3. In the former, half the tape width is used each time the tape is turned over. In the latter, the recording head has two electromagnets, each using one-quarter tape-width.

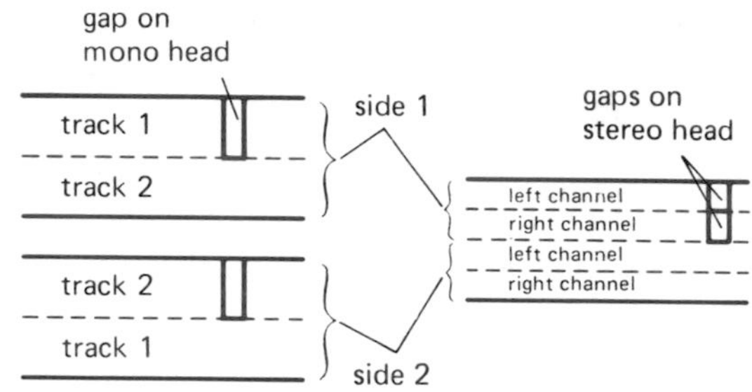

Fig. 82.3

(b) Bias. Magnetic materials are only magnetized if the magnetizing field exceeds a certain value. When it does, doubling the field does not double the magnetization. That is, the magnetization is not directly proportional to the magnetizing field, as the graph in Fig. 82.4a shows. Therefore, signals fed to the recording head would be distorted when played back (as they also would if the saturation value of the magnetic material was exceeded by too strong a recording signal).

To prevent this, *a.c. bias* is used. A sine wave a.c. with a frequency around 60 kHz is fed into the recording head along with the signal. The a.c. bias, being supersonic, is not recorded but it enables the signal to work on linear parts of the graph, Fig. 82.4b.

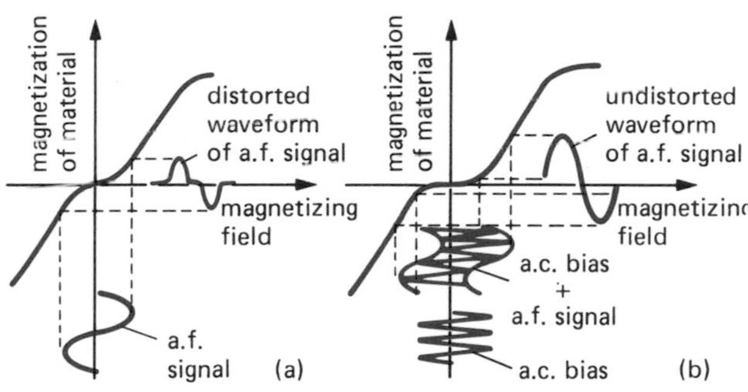

Fig. 82.4

(c) Playback. On playback, the tape runs past another head, the *playback head*, which is similar to the recording head (sometimes the same head records and plays back). As a result, the varying magnetization of the tape induces a small a.f. voltage in the windings of the playback head. After amplification, this voltage is used to produce the original sound in a loudspeaker.

During playback the tape must move at the same constant speed as when recorded. Any variation changes the pitch of the sound produced; it is called *wow* if the frequency wobbles slowly and *flutter* if it is fast. The driving electric motor is usually d.c. and more than one may be used.

(d) Erasing. The a.c. bias is also fed during recording to the *erase head*, which is another electromagnet but with a larger gap so that its magnetic field covers more tape. It is placed before the recording head and removes any previous recording by subjecting the passing tape to a decreasing, alternating magnetic field which demagnetizes it.

A simplified block diagram of a tape recorder is given in Fig. 82.5.

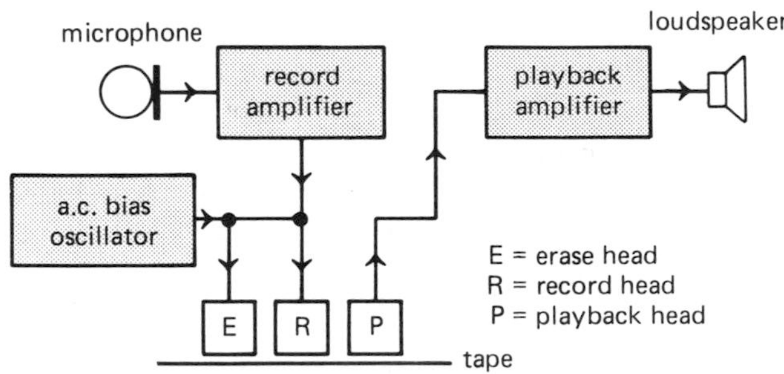

Fig. 82.5

(e) Cassette recorders. The audio cassette recorder is now rated as a hi-fi system. At the quite low tape speed of 4.75 cm/s, it can handle frequencies up to 17 kHz. The improvement is due to better head design and tape quality and to the use of noise reduction circuits, e.g. *Dolby*, which improve higher frequency recording by reducing the level of 'hiss' that occurs during quiet playback spells.

The video cassette recorder (VCR) works on the same principle but plays back on a TV receiver, the high frequency video signals stored on the tape. In this case it is more practical for the playback head to be rotated rapidly (25 revolutions per second) while the tape moves slowly past it (since a much greater relative speed is necessary between tape and head at these frequencies).

Disc recording

(a) Making the disc. Previously, a master disc was cut during a live recording but it is much more usual today to make a master tape first (from which tapes can also be copied). After the master tape is edited, the first disc is made from it on acetate. The wavy groove which stores the sound, Fig. 82.6a, is produced by a heated cutter with a fine point, driven towards the centre of the rotating acetate disc, Fig. 82.6b. At the same time it is moved a little from side to side by the a.f. currents being recorded.

Once cut, the acetate disc is electroplated with nickel to form a 'negative' copy, i.e. one which has wavy 'humps' instead of wavy grooves. This is used to make a 'positive' copy which is then chromium plated to give a tough 'negative' for use as a die in the hot press producing the plastic copies sold to the public.

(b) Pick-up. The operation of mono and stereo pick-ups for producing electrical signals from the

Fig. 82.6

disc during playback was discussed earlier (p. 50). For the stylus to follow the groove accurately, the pick-up arm must be light with near-frictionless bearings. The force required to keep the stylus in the groove depends on the type of cartridge, being greater for the crystal type than the magnetic. It can be adjusted by movable weights on the other end of the arm, Fig. 82.7.

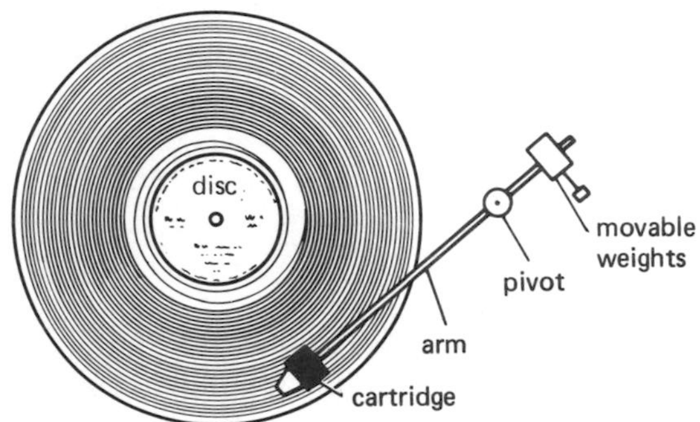

Fig. 82.7

(c) Turntable. The disc must be replayed at exactly the same speed as during recording (to avoid wow and flutter) and this is achieved by using an electric motor to drive a heavy turntable either directly or via a rubber idler wheel as in Fig. 82.8 or by a belt, at $33\frac{1}{3}$ rpm or 45 rpm. Low frequency 'noise', called *rumble*, can be transmitted to the pick-up from the bearing on which the turntable rotates, as can motor

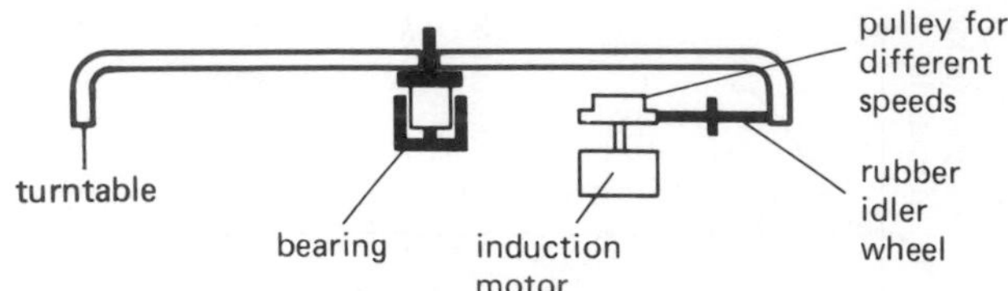

Fig. 82.8

vibration. Care is needed in the mechanical design to avoid such problems.

(d) Digital disc recorders. These play back audio (sound) and/or video (picture) signals which are in digital form in a series of reflecting pits of different lengths and spacing in a disc (5 inch for audio and 12 inch for video). Both types use laser pick-ups (p. 50). Fig. 82.9 shows a digital audio disc (DAD) player.

Fig. 82.9

83 Audio amplifiers

Audio frequency (a.f.) amplifiers fall into two broad groups—preamplifiers and power amplifiers.

Preamplifier

A preamplifier is a voltage (or small-signal) amplifier which amplifies the signal supplied to its input by a transducer, e.g. microphone, pick-up, or tape playback head. To obtain maximum voltage transfer, the input impedance of the amplifier should be high compared with that of the transducer (p. 113).

A preamplifier has also to provide equalization and tone control, both are achieved using frequency-dependent negative feedback (n.f.b.).

(a) Equalization. This ensures that the output is a faithful reproduction of the original sound, i.e. that the same frequencies are present in both in the same relative proportions. The need for equalization arises because when, for example, sound is recorded on tape or disc, the low (bass) frequencies have to be attenuated (weakened) and the high (treble) frequencies boosted.

In disc recording this is necessary because low frequencies cause greater sideways motion of the cutter and if it were not restricted one part of the groove might break through to adjacent parts. Conversely high frequencies have to be boosted to increase the amplitude of the groove displacements.

When discs and tapes are played the preamplifier should amplify low frequencies *more* than high frequencies to compensate for what happens during recording.

(b) Tone control. This permits the overall frequency response of the amplifier to be, to some extent, under the listener's control so that allowance can be made for different loudspeaker performances, room acoustics, etc. Tone control lets the bass and treble frequencies be cut or boosted as required.

(c) Frequency-dependent n.f.b. If the n.f.b. path in an amplifier contains a capacitor C and a resistor R, its impedance varies with frequency f. For example, in the inverting op amp circuit of Fig. 83.1, the reactance X_C of C ($= 1/(2\pi fC)$) is large at low frequencies since f is small, giving a small amount of n.f.b. and therefore a small reduction in the gain A (p. 112).

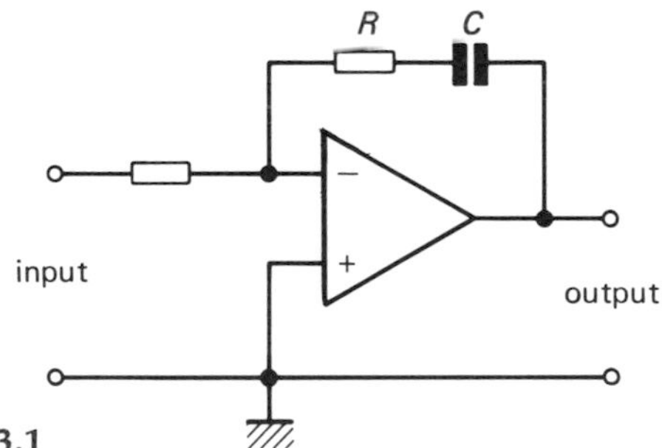

Fig. 83.1

At high audio frequencies, X_C is small, so n.f.b. is greater and there is a larger fall in A. The amplifier thus, in effect, boosts bass frequencies more than treble frequencies and by suitably choosing component values, gives the equalization required.

Tone control circuits are designed on similar principles, variable resistors being used to allow adjustments to be made.

Power amplifier: single-ended

(a) Introduction. The small amount of power produced by a preamplifier could not operate a loudspeaker. A power (or large-signal) amplifier, taking the output of a preamplifier as its input, is required. It converts d.c. power from the supply into large swings of output current and voltage, i.e. large a.c. power in the load. Basically it is a 'robust' voltage amplifier which uses a transistor (a power transistor) capable of supplying high output currents. However its matching requirement is for maximum power transfer rather than maximum voltage transfer.

The maximum power theorem (p. 18) states that a generator (e.g. a transistor amplifier) delivers maximum power to an external load (e.g. a loudspeaker) when the output impedance Z_1 of the generator equals the input impedance Z_2 of the load. The impedance of a loudspeaker is typically 8Ω, while the output impedance of a transistor amplifier is several thousand ohms. Direct connection of the speaker as the collector load would give poor results.

(b) Transformer matching. One solution to the matching problem is to connect a step-down transformer as in Fig. 83.2a with its primary winding of n_1 turns as the collector load and its secondary winding of n_2 turns supplying the speaker. It can be shown that Z_2 will 'look' equal to Z_1 and the maximum power transferred if the turns ratio n_1/n_2 is chosen so that

$$\frac{n_1}{n_2} = \sqrt{\frac{Z_1}{Z_2}}$$

For example, if $Z_1 = 8\,\text{k}\Omega$ and $Z_2 = 8\,\Omega$, $n_1/n_2 = \sqrt{8000/8} = \sqrt{1000} \approx 32/1$. A matching transformer with a step-down ratio of about 32/1 would be suitable.

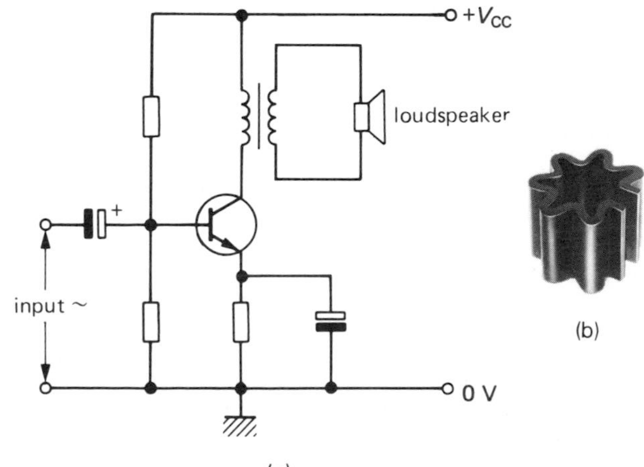

Fig. 83.2

Single-ended power amplifiers (i.e. with one output transistor) are not used much except where low powers (up to about 1 W) are required, for two reasons. First, transformers are costly, bulky, heavy and cause distortion. Second, the d.c. operating point for the transistor has to be near the middle of the load line to prevent distortion of the output (p. 106); the quiescent collector current is therefore comparatively large and represents wasted d.c. power which produces unwanted heat, possibly requiring the transistor to have a heat sink, Fig. 83.2b. Overall, the efficiency is low.

Power amplifier: push–pull

Most audio power amplifiers employ the push–pull principle. Two transistors are used and are called a *complementary pair* because one is an n-p-n type and the other a p-n-p type. The basic circuit is given in Fig. 83.3.

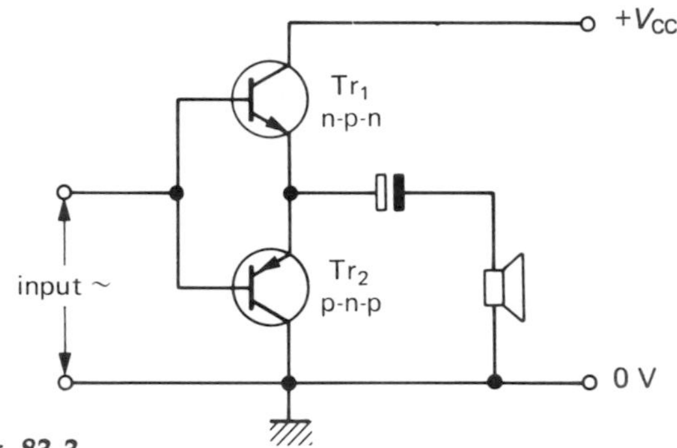

Fig. 83.3

(a) Emitter-follower matching. Tr_1 is the n-p-n transistor, its collector is positive with respect to the emitter, the opposite being the case for Tr_2, the p-n-p type. The bases of both transistors are connected to the input terminal. The load (loudspeaker) is joined to both emitters via a d.c. blocking capacitor. The circuit consists of two emitter-followers, which as we saw earlier (p. 114), have a low output impedance. In fact, it is comparable with that of a loudspeaker, so solving the impedance matching problem.

(b) Push–pull action. When an input is applied, the positive half-cycles forward bias the base-emitter junction of Tr_1, which conducts and 'pushes' a series of positive half-cycles of output current through the loudspeaker. During this time Tr_2 is cut off because its base-emitter junction is reverse biased when the base is positive with respect to the emitter.

Negative half-cycles of the input reverse bias Tr_1 (i.e. base negative with respect to emitter) and cut it off but forward bias Tr_2. Tr_2 conducts and 'pulls' a series of negative half-cycles of output current through the speaker. Both halves of the input current are thus 'processed'.

Each transistor only conducts for one half of each input cycle and is cut off during the other half. The quiescent base and collector currents are zero since the base-emitter junctions of neither Tr_1 nor Tr_2 are forward biased. Hence, no d.c. power is wasted when there is no input, which accounts for the high efficiency of a push–pull amplifier.

Tr_1 and Tr_2 should be a 'matched pair' of power transistors, i.e. their characteristics should be identical, otherwise if one has a greater gain than the other, one half of the output waveform will be amplified more and cause distortion. Matched pairs are available for different power outputs.

(c) Crossover distortion. A disadvantage of the simple circuit in Fig. 83.3 is that each transistor does not turn-on until the input is about 0.6 V. As a result there is a 'dead-zone' in the output waveform, as shown in Fig. 83.4, producing crossover distortion.

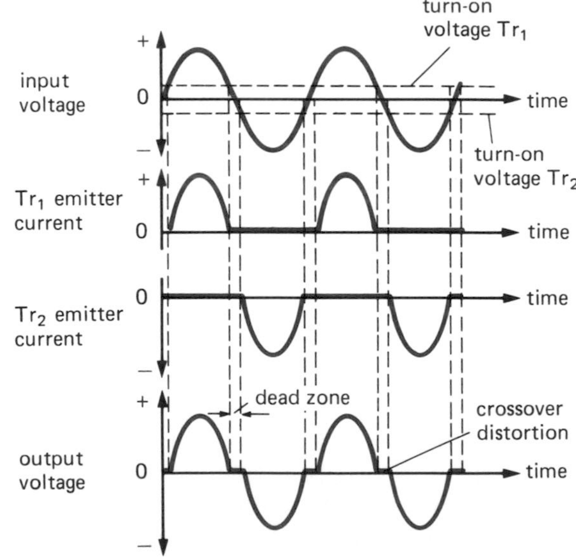

Fig. 83.4

Matters are improved by applying some forward bias to the base-emitter junctions of both transistors so that small quiescent currents flow and the smallest inputs make them conduct.

The bias required can be provided by connecting two resistors R_1 and R_2 as in Fig. 83.5 so that in the

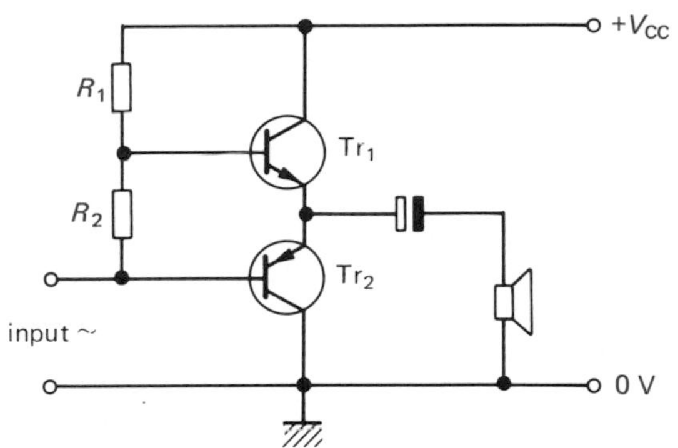

Fig. 83.5

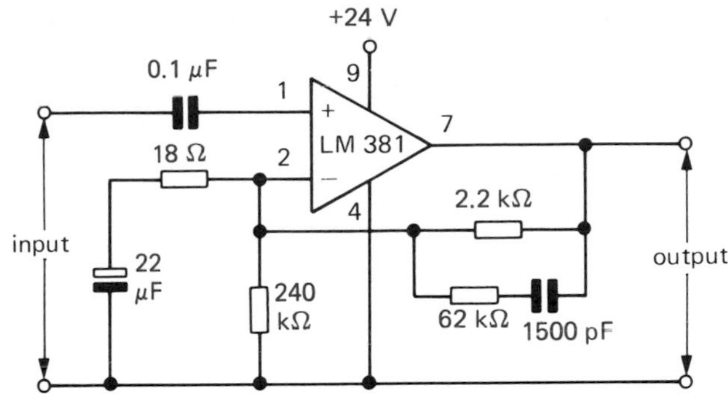

Fig. 83.6

quiescent state, the bases of Tr_1 and Tr_2 are, with respect to their emitters, 0.6 V positive and 0.6 V negative respectively. Both transistors are then forward biased.

Audio amplifier ICs

(a) Preamplifier. The LM381 is a dual preamplifier IC designed to amplify small signals from, for example, a magnetic pick-up or a tape playback head. A circuit for the latter, giving equalization, is shown in Fig. 83.6. If required the output could be applied to a tone control circuit before the power amplification stage.

(b) Power amplifier. The LM380 is a power amplifier IC with a typical output of 2 W into a 8 Ω speaker on a 20 V supply and using as a heat sink, about 6 cm square of copper strip connected to the centre pins on its d.i.l. package. The input can be taken directly from a crystal or ceramic pick-up or microphone or from the output of an LM381 preamplifier.

84 Complete hi-fi system

Music centre

A typical hi-fi (high fidelity) music centre and its block diagram are shown in Fig. 84.1a, b. It comprises

(i) a record player,
(ii) a 3-band vhf/LW/MW radio tuner (p. 182),
(iii) a tape cassette deck,
(iv) an amplifier (pre- and power) into which the outputs from (i), (ii) and (iii) can be switched separately,
(v) a loudspeaker system.

Fig. 84.1

(a)

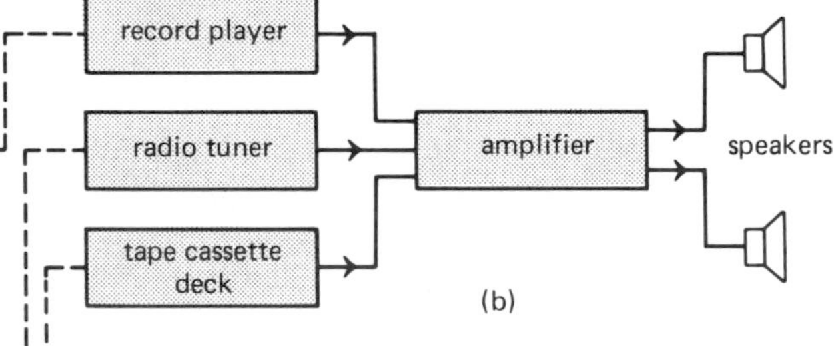

(b)

Loudspeaker systems

(a) Baffles and cabinets. When the paper cone of a loudspeaker (p. 44) moves forward it produces a compression (i.e. greater air pressure) in front and a rarefaction (i.e. lower air pressure) behind. If these meet (as they will if the distance from the front to the back of the cone is small), they partly cancel (especially the low frequencies), thereby reducing the sound.

Mounting the speaker on a flat board, called a *baffle*, Fig. 84.2a, helps by increasing the front-to-back path length. A further improvement is to totally enclose the speaker in a cabinet (an infinite baffle), Fig. 84.2b. However this can lead to unwanted resonance of the air in the cabinet and also 'loads' the speaker due to the air being compressed and resisting when the cone moves back into it. Efficient coupling between a speaker and the air in the room requires careful cabinet design.

(b) Woofers, tweeters and crossover networks. More than one loudspeaker is needed to handle the full range of audio frequencies efficiently; in practice there are often two or three in the same cabinet. A large-coned one, the *woofer*, deals with low (bass) frequencies, while a small one, the *tweeter*, handles high (treble) frequencies, Fig. 84.2c. Middle frequencies may also be catered for by a third speaker.

The correct range of frequencies is fed to each speaker by a *crossover network*. In the simple arrangement of Fig. 84.2d, L and C act as a voltage divider with high frequencies developing a voltage across L, for application to the tweeter, and low frequencies creating a voltage across C, for application to the woofer. (Remember that $X_L = 2\pi fL$ and $X_C = 1/(2\pi fC)$.) A typical crossover frequency is about 3 kHz, i.e. frequencies above this go mostly to the tweeter, those below mostly to the woofer.

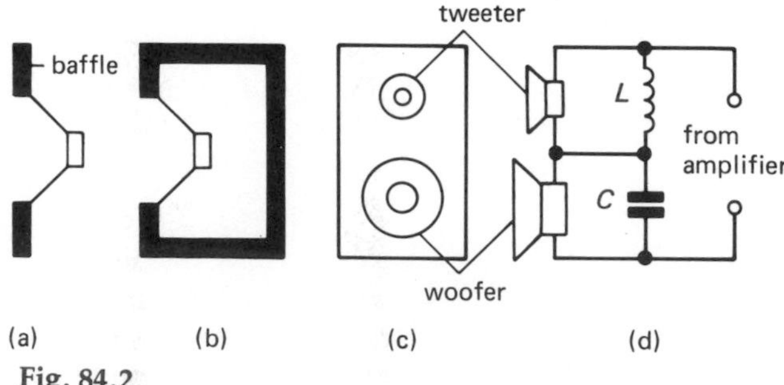

Fig. 84.2

85 Progress questions

1. Explain how *either* a record *or* a magnetic tape contains the audio signal which has been recorded on it. (*L.*)

2. Describe briefly the purpose in a preamplifier of (a) equalization, (b) tone controls.

3. Describe how the following defects of sound reproduction occur: (a) distortion, (b) rumble. (*C.*)

4. Many tape recorders include, in the recording circuit, an automatic overload control. This device prevents the recording head being overloaded and so producing distortion. The performance of an automatic overload control is shown by the graph in Fig. 85.1.

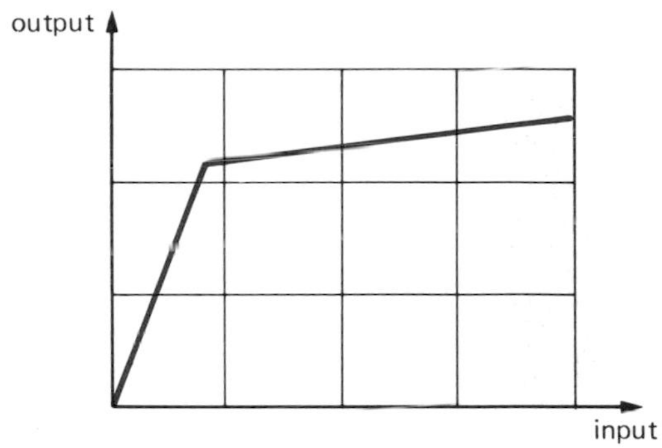

Fig. 85.1

(a) Use this graph to explain how the overloading of the recording head can be prevented.

One type of automatic recording level control uses a preamplifier of the design in Fig. 85.2. In the dark the light dependent resistor (LDR) has a resistance very much larger than 1 MΩ, but when in the light its value falls to near 100 kΩ.

(b) What will be the approximate gain of the preamplifier in the dark?

(c) When the LDR goes from the dark into the light:

(*i*) what will be the approximate combined resistance of the 1 MΩ resistor and the LDR?

Fig. 85.2

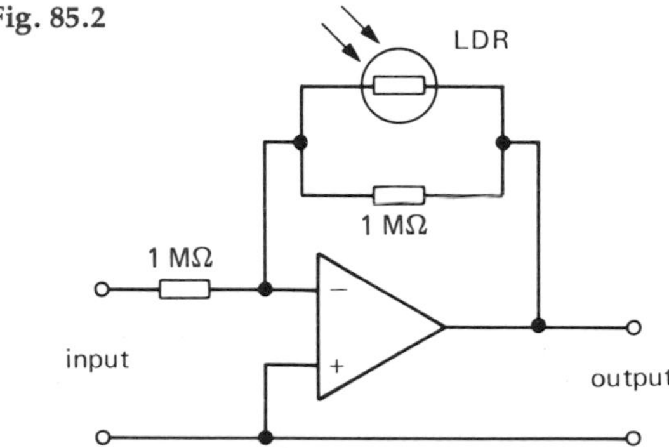

(*ii*) what will be the approximate gain of the preamplifier?

The preamplifier is connected to the rest of the automatic overload control as in Fig. 85.3.

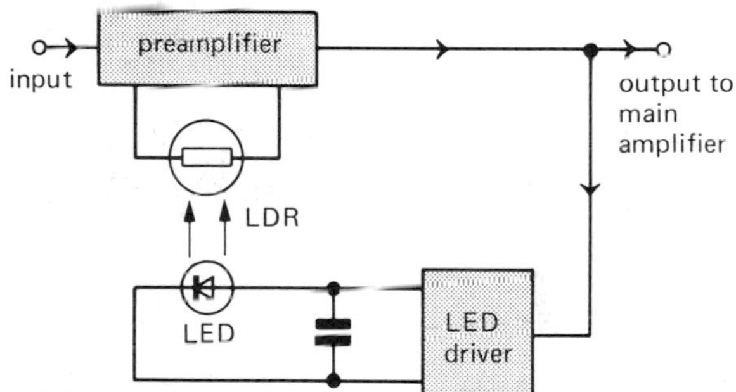

Fig. 85.3

(d) Explain how such an arrangement can produce the performance shown by the graph. In your explanation describe what you will expect to happen when the input is very low, when the input is enough to light the light emitting diode (LED) and when the input is very high and would overload the recording head.

(e) This circuit does not respond to sudden changes of input immediately. Explain why this might be.

(*A.E.B. 1982 Electronics*)

Radio and television

86 Radio waves

Electromagnetic waves

Electromagnetic (e.m.) waves are produced when electrons suffer an energy change, e.g. by being accelerated in an aerial. They are energy-carriers which are associated with a varying electric field accompanied by a varying magnetic field of the same frequency and phase, the fields being at right angles to each other and to the direction of travel of the wave. A diagrammatic representation of an e.m. wave is given in Fig. 86.1.

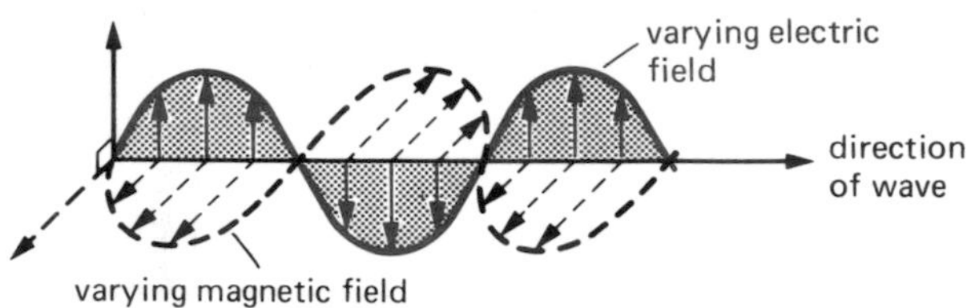

Fig. 86.1

(a) Electromagnetic spectrum. A whole family of e.m. waves exists, extending from gamma rays of very short wavelength to very long radio waves. Fig. 86.2 shows the wavelength ranges of the

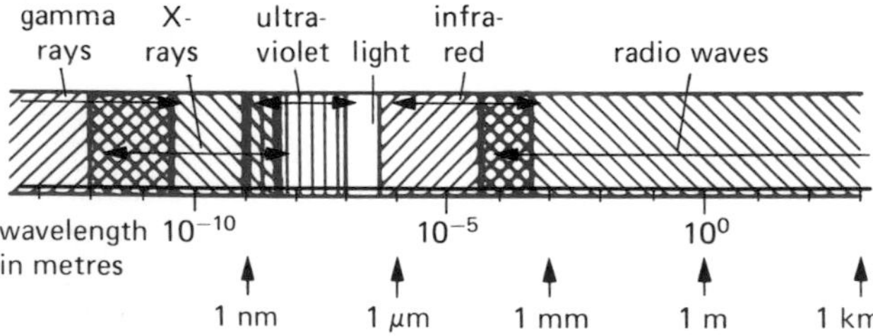

Fig. 86.2

various types but there is no rapid change of properties from one to the next. Although methods of production and detection differ from member to member, all exhibit typical wave properties such as interference and travel in space at a speed $c \approx 3.0 \times 10^8$ m/s, i.e. the speed of light.

(b) Wave equation. For any wave motion, its speed v is related to its wavelength λ and frequency f by

$$v = f\lambda$$

For example for violet light of $\lambda = 0.4\,\mu\text{m} = 4 \times 10^{-7}$ m, $f = v/\lambda = c/\lambda = 3 \times 10^8/(4 \times 10^{-7}) = 7.5 \times 10^{14}$ Hz. For a radio wave of $\lambda = 300$ m, $f = 10^6$ Hz = 1 MHz. The smaller λ is, the larger will be f, since c is fixed. Electromagnetic waves may be distinguished either by their frequency or wavelength but the former is more fundamental since unlike λ (and c), f does not change when the wave travels from one medium to another.

Radio frequency bands

Radio frequency (r.f.) waves have frequencies extending from about 30 kHz upwards and are grouped into bands, as in the table below.

Frequency band	Some uses
Low (l.f.) 30 kHz–300 kHz	Long wave radio and communication over large distances
Medium (m.f.) 300 kHz–3 MHz	Medium wave, local and distant radio
High (h.f.) 3 MHz–30 MHz	Short wave radio and communication, amateur and CB radio
Very high (v.h.f.) 30 MHz–300 MHz	FM radio, police, meteorology devices
Ultra high (u.h.f.) 300 MHz–3 GHz	TV (bands 4 and 5) aircraft landing systems
Microwaves (s.h.f.) Above 3 GHz	Radar, communication satellites, telephone and TV links

Propagation of radio waves

Radio waves from a transmitting aerial can travel in one or more of three different ways, Fig. 86.3.

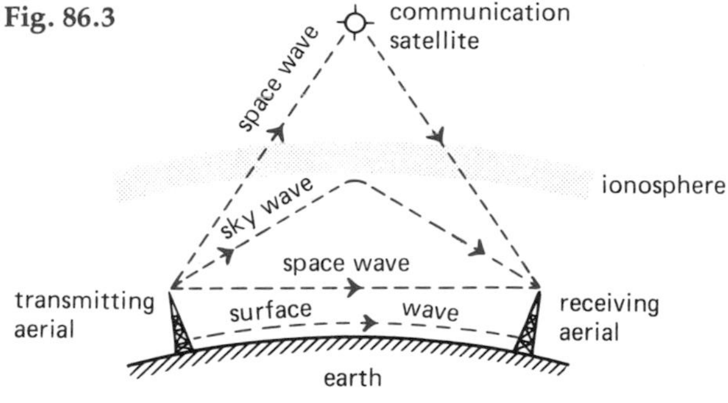

Fig. 86.3

(a) Surface or ground wave. This travels along the ground, following the curvature of the earth's surface. Its range is limited mainly by the extent to which energy is absorbed from it by the ground. Poor conductors such as sand absorb more strongly than water and the higher the frequency the greater the absorption. The range may be about 1500 km at low frequencies (long waves) but much less for v.h.f.

(b) Sky wave. It travels skywards and, if it is below a certain *critical frequency* (typically 30 MHz), is returned to earth by the *ionosphere*. This consists of layers of air molecules (the D, E and F layers), stretching from about 80 km above the earth to 500 km, which have become positively charged through the removal of electrons by the sun's ultraviolet radiation. On striking the earth the sky wave bounces back to the ionosphere where it is again gradually refracted and returned earthwards as if by 'reflection'. This continues until it is completely attenuated.

The critical frequency varies with the time of day and the seasons. Sky waves of low, medium and high frequencies can travel thousands of kilometres but at v.h.f. and above they usually pass through the ionosphere into outer space.

If both the surface wave and the sky wave from a transmitter are received at the same place, interference can occur if the two waves are out of phase. When the phase difference varies, the signal 'fades', i.e. goes weaker and stronger. If the range of the surface wave for a signal is less than the distance to the point where the sky wave first reaches the earth, there is a zone which receives no signal.

(c) Space wave. For v.h.f., u.h.f. and microwave signals, only the space wave, giving line of sight transmission, is effective. A range of up to 150 km is possible on earth if the transmitting aerial is on high ground and there are no intervening obstacles such as hills, buildings or trees.

Aerials

An aerial radiates or receives radio waves. Any conductor can act as one but proper design is necessary for maximum efficiency.

(a) Transmitting aerials. When a.c. from a transmitter flows in a transmitting aerial, radio waves of the same frequency f, as the a.c. are emitted if the length of the aerial is comparable with the wavelength λ of the waves. For example, if $f = 100\,\text{MHz} = 10^8\,\text{Hz}$, $\lambda = c/f = 3 \times 10^8/10^8 = 3\,\text{m}$; but if $f = 1\,\text{kHz}$, $\lambda = 300\,000\,\text{m}$. Therefore if aerials are not to be too long they must be supplied with r.f. currents from the transmitter.

The *dipole* aerial consists of two vertical or horizontal conducting rods or wires, each of length one-quarter of the wavelength of the wave to be emitted, and is centre fed, Fig. 86.4a. It behaves as a series LC

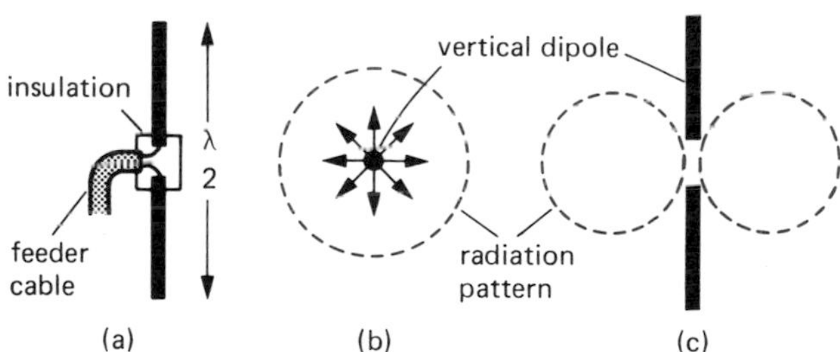

Fig. 86.4

circuit (p. 36) whose resonant frequency depends on its length, since this determines its inductance L (which all conductors have) and its capacitance C (which is distributed along it and arises from the capacitor formed by each conductor and the air between them). The radiation pattern of Fig. 86.4b shows that a vertical dipole emits equally in all horizontal directions, but not in all vertical directions, Fig. 86.4c. The electric field of the wave is emitted parallel to the aerial and, if the latter is vertical, the wave is said to be *vertically polarized*.

When required (e.g. in radio telephony but not in most sound or television broadcasting systems) concentration of much of the radiation in one direction can be achieved by placing another slightly longer conductor, a *reflector*, about $\lambda/4$ behind the dipole and a slightly shorter one, a *director*, a similar

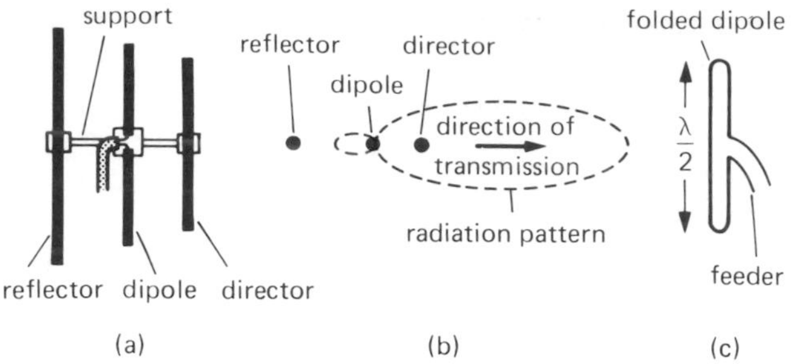

Fig. 86.5

distance in front, Fig. 86.5a. The radiation these extra elements emit arises from the voltage and current induced in them by the radiation from the dipole. It adds to that from the dipole in the wanted direction and subtracts from it in the opposite direction. Fig. 86.5b shows the radiation pattern given by the array, called a *Yagi aerial*. The extra elements reduce the input impedance of the dipole and causes a mismatch with the feeder cable unless a *folded dipole* is used, Fig. 86.5c, to restore it to its original value.

In the microwave band *dish* aerials in the shape of large metal 'saucers' are used. The radio waves fall on them from a small dipole at their focus, which is fed from the transmitter, and are reflected as a narrow, highly directed beam, Fig. 86.6.

Dish aerials using frequencies of several gigahertz (1 GHz = 1000 MHz) play an important part in the reception and retransmission of signals between different earth stations and geostationary communication satellites which orbit the earth at 36 000 km in 24 hours and appear at rest. Two-thirds of all intercontinental telephone calls, television broadcasts giving worldwide coverage of major

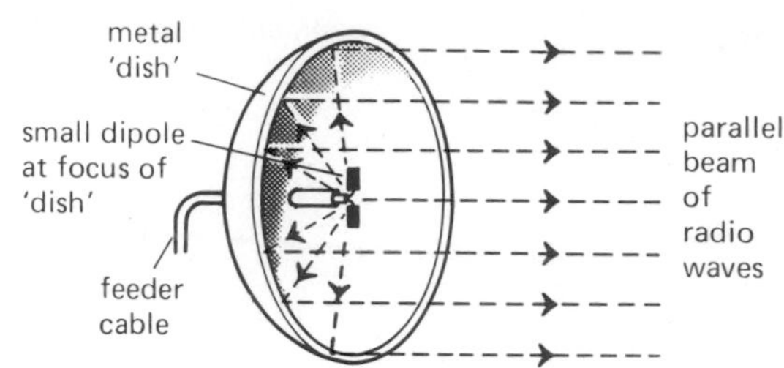

Fig. 86.6

events as well as other data, go by the satellite network.

(b) Receiving aerials. Whereas transmitting aerials may handle many kilowatts of r.f. power, receiving aerials, though basically similar, deal with only a few milliwatts due to the voltages (of a few microvolts) and currents induced in them by passing radio waves. Common types are the *dipole*, the *vertical whip* and the *ferrite rod*.

Dipole arrays are used to receive TV (where the reflector is often a slotted metal plate) and v.h.f. radio signals. They give maximum output when they are (i) lined up on the transmitter and (ii) vertical or horizontal if the transmission is vertically or horizontally polarized.

A vertical whip aerial is a tapering metal rod (often telescopic), used for car radios. In most portable medium/long wave radio receivers the aerial consists of a coil of wire wound on a ferrite rod, as in Fig. 11.1d (p. 29). Ferrites are non-conducting magnetic substances which 'draw in' nearby radio waves. The coil and rod together act as the inductor *L* in an *LC* circuit tuned by a capacitor *C*. For maximum signal, the aerial, being directional, should be lined-up so as to be end-on to the transmitter.

87 Radio systems

Transmitter

An aerial must be fed with r.f. power if it is to emit radio waves effectively (p. 179) but speech and music produce a.f. voltages and currents. The transmission of sound by radio therefore involves *modulating* r.f. so that it 'carries' the a.f. information, as we saw earlier (p. 168). Amplitude modulation (AM) is used in medium, long and short wave broadcasting.

A block diagram for an AM transmitter is shown in Fig. 87.1. In the *modulator* the amplitude of the r.f. carrier from the *r.f. oscillator* (usually crystal-

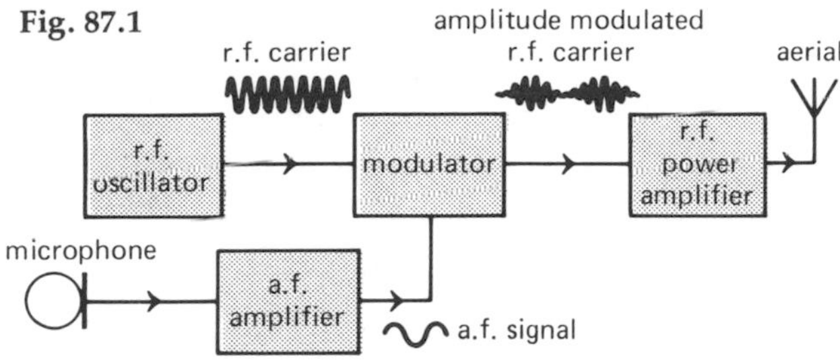

controlled) is varied at the frequency of the a.f. signal from the *microphone*, Fig. 87.2.

The *modulation depth m*, is defined by:

$$m = \frac{\text{signal peak}}{\text{carrier peak}} \times 100\%$$

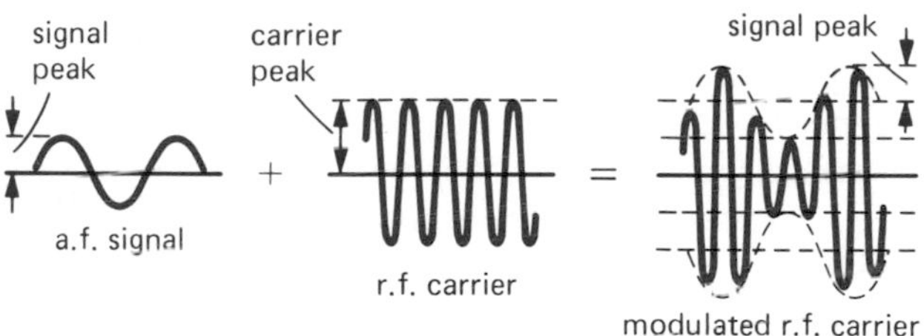

Fig. 87.2

For a signal peak of 1 V, and a carrier peak of 2 V, $m = \frac{1}{2} \times 100 = 50\%$. If m exceeds 100%, distortion occurs, but if it is too low, the quality of the sound at the receiver is poor; a value of 80% is satisfactory.

The modulated signal consists of the carrier and the upper and lower sidebands (p. 168, Fig. 81.4b). The bandwidth it requires to transmit a.fs. up to 5 kHz, is 10 kHz. In practice, in the medium waveband, which extends from about 500 kHz to 1.5 MHz, 'space' is limited if interference between stations is to be avoided and the bandwidth is restricted to 9 kHz.

Straight receiver

(a) Block diagram. The various parts of a *straight* or *tuned radio frequency* (TRF) receiver are shown in Fig. 87.3. The wanted signal from the *aerial* is selected and amplified by the *r.f. amplifier*, which is a voltage amplifier with a tuned circuit (p. 36) as its load, and should have a bandwidth (p. 106, Fig. 49.7) of 9 kHz to accept the sidebands.

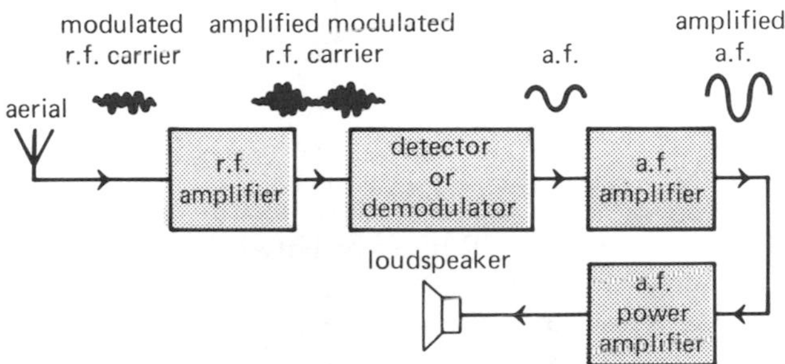

Fig. 87.3

The a.f. is next separated from the modulated r.f. by the *detector* or *demodulator* and amplified by the *a.f. pre-* and *power amplifiers* to operate the *loudspeaker*. The bandwidths of the a.f. amplifiers need not exceed 4.5 kHz, i.e. the highest a.f. allowed.

(b) Detection. The basic detector circuit is shown in Fig. 87.4a. If the AM signal, voltage V, of Fig. 87.4b is applied to it, the diode produces rectified pulses of r.f. current I, Fig. 87.4c. These charge up C whenever V exceeds the voltage across C, i.e. during part of each positive half-cycle, as well as flowing through R. During the rest of the cycle when the diode is non-conducting, C partly discharges through R. The voltage V_C across C (and R) is a varying d.c. which, apart from the slight r.f. ripple (much exaggerated in Fig. 87.4d), has the same waveform as the modulating a.f.

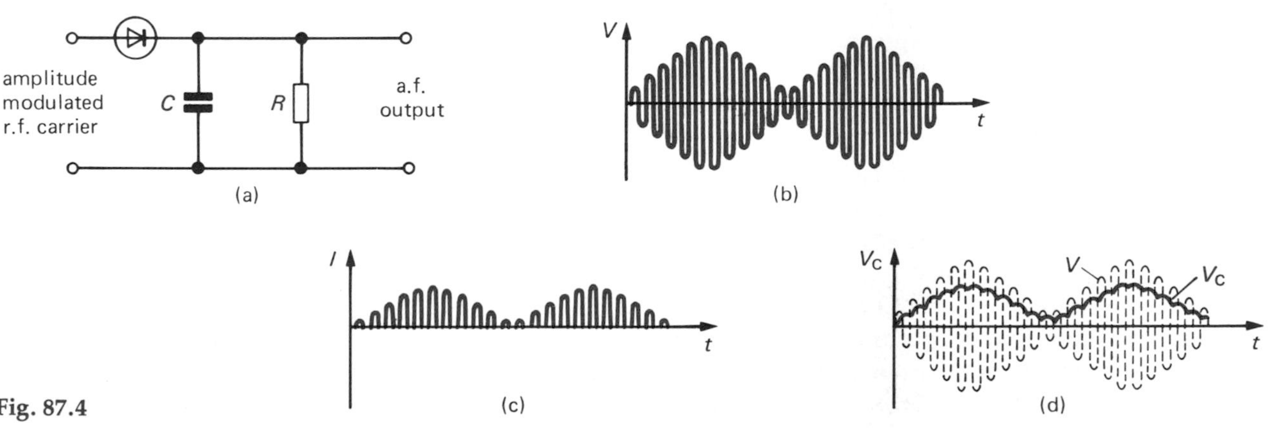

Fig. 87.4

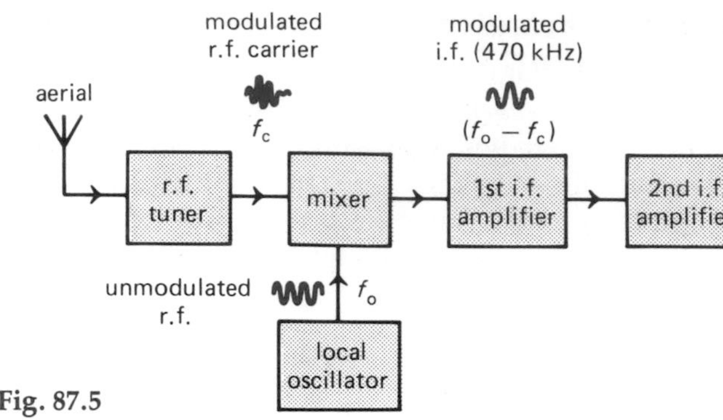

Fig. 87.5

The time constant CR must be between the time for one cycle of r.f., e.g. 10^{-6} s (1 μs), and one cycle of a.f. (average), e.g. 10^{-3} s (1 ms). If it is too large, C charges and discharges too slowly and V_C does not respond to the a.f.; too small a value allows C to discharge so rapidly that V_C follows the r.f. Typical values are $C = 0.01\,\mu$F and $R = 10\,\text{k}\Omega$ giving $CR = 10^{-4}$ s. Point-contact germanium diodes are used as explained before (p. 60).

Superheterodyne receiver

The 'superhet' is much superior to the straight receiver and is used in commercial AM radios. Fig. 87.5 shows a typical block diagram.

(a) Action. The modulated r.f. signal of frequency f_C from the wanted station is fed via an *r.f. tuner* (which may not be an amplifier but simply a tuned circuit that selects a band of r.fs) to the *mixer* along with an r.f. signal from the *local oscillator* which has a higher frequency f_O.

The output from the *mixer* contains an r.f. oscillation of frequency $f_O - f_C$, called the *intermediate frequency* (i.f.), having the a.f. modulation. Whatever the value of f_C, f_O is always greater than it by the *same* amount, i.e. the i.f. The i.f. ($= f_O - f_C$) is therefore fixed, and in an AM radio is usually 470 kHz.

To achieve this, two variable capacitors, one in each of the tuned circuits of the *r.f. tuner* and the *local oscillator*, are 'ganged' so that their resonant frequencies change in step, i.e. track. Long, medium and short wavebands are obtained by switching a different set of coils into each tuned circuit. Longer waves (smaller frequency) require greater inductance, i.e. more turns on the coil.

The *i.f. amplifiers* are in effect, r.f. amplifiers with tuned circuits having a fixed resonant frequency of 470 kHz. They amplify the modulated i.f. before it goes to the *detector* which, like the *a.f. amplifiers*, acts as in the straight receiver.

(b) Heterodyning. The frequency changing of f_C to $f_O - f_C$ is done in the *mixer* by *heterodyning*. This is similar to the production of beats in sound when two notes of slightly different frequencies f_1 and f_2 ($f_1 > f_2$), create another note, the *beat* note, of frequency $f_1 - f_2$. In radio, the 'beat' note must have a supersonic frequency (or it will interfere with the a.f.), hence the name *super* (sonic) *heterodyne* receiver.

(c) Advantages. By changing all incoming carrier frequencies to one lower, fixed frequency (the i.f.) at which most of the gain and selectivity occurs, we obtain greater:

(i) *stability* because positive feedback is less at lower frequencies and so there is less risk of the receiver going into oscillation,
(ii) *sensitivity* because more i.f. amplifiers can be used (two are usually enough) as a result of **(i),** so giving greater gain,
(iii) *selectivity* because more i.f. tuned circuits are possible without creating the tracking problems that arise when several variable capacitors are ganged.

Frequency modulated radio

Frequency modulation (FM) is used in v.h.f. radio.

(a) Transmitter. An FM transmitter is similar to the AM one in Fig. 87.1 but the *frequency* of the r.f. carrier is changed by the a.f. signal (see p. 168, Fig. 81.3). The change or *deviation* is proportional to the amplitude of the a.f. at any instant.

Each a.f. modulating frequency produces a large number of side frequencies (not two as in AM) but their amplitudes decrease the more they differ from the carrier. In theory therefore, the bandwidth of an FM system should be very wide but in practice the 'outside' side frequencies can be omitted without noticeable distortion. The BBC uses a 250 kHz bandwidth which is readily accommodated in the v.h.f. radio broadcasting band (band II covering frequencies 88 to 97.6 MHz) and allows the full range of a.fs. needed for 'high quality' sound.

(b) Receiver. FM receivers use the superhet principle. To ensure adequate amplification of the v.h.f. carrier, the first stage is an r.f. amplifier with a fairly wide bandwidth. The mixer and local oscillator change all carrier frequencies to an i.f. of 10.7 MHz. These three circuits often form the *FM tuner*.

The detector, called a *ratio detector*, is more complex and quite different from an AM detector. First, it limits any changes in amplitude of the carrier due to unwanted 'noise' being picked up. This makes 'quiet' reception another good feature of FM and produces a signal that varies in frequency only. Second, it contains tuned circuits which create an a.f. voltage that is proportional to the deviation.

88 Black-and-white television

Television camera

A black-and-white (monochrome) TV camera changes light into electrical signals.

(a) Action. The *plumbicon* tube, shown simplified in Fig. 88.1, consists of an electron gun, which emits a narrow beam of electrons, and a target of photoconductive material (lead monoxide) on which a lens system focuses an optical image.

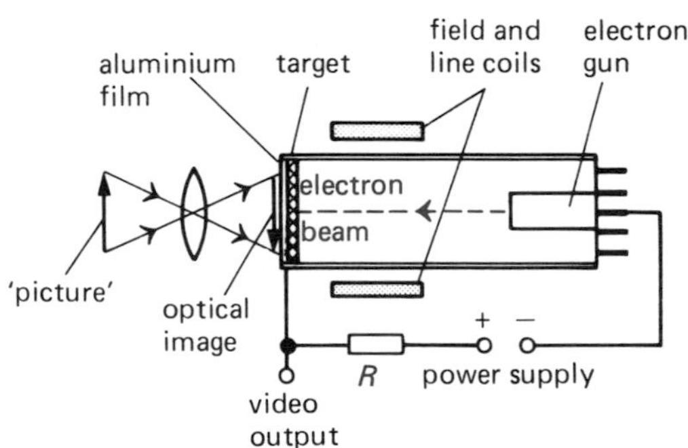

Fig. 88.1

The target behaves as if the resistance between any point on its back surface and a transparent aluminium film at the front depends on the brightness of the image at the point. The brighter it is, the lower is the resistance. The electron beam is made to scan across the target and the resulting beam current (i.e. the electron flow in the circuit consisting of the beam, the target, the load resistor R, the power supply and the gun) varies with the resistance at the spot where it hits the target. The beam current thus follows the brightness of the image and R turns its variations into identical variations of voltage shown in Fig. 88.2, for subsequent transmission as the video signal.

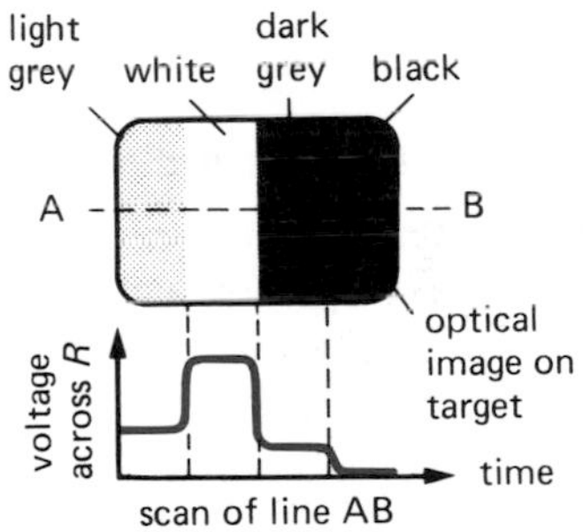

Fig. 88.2

(b) Scanning. The scanning of the target by the electron beam is similar to the way we read a page of print, i.e. from left to right and top to bottom. In effect the picture is changed into a set of parallel lines, called a *raster*.

Two systems are needed to deflect the beam horizontally and vertically. The one which moves it steadily from left to right and makes it 'flyback' rapidly, ready for the next line, is the *line scan*. The other, the *field scan*, operates simultaneously and draws the beam at a much slower rate down to the bottom of the target and then restores it suddenly to the top. Magnetic deflection is used in which relaxation oscillators (p. 117) act as *time bases* and generate currents with sawtooth waveforms, Fig. 88.3a, at the line and field frequencies. These are passed through two pairs of coils mounted round the camera tube.

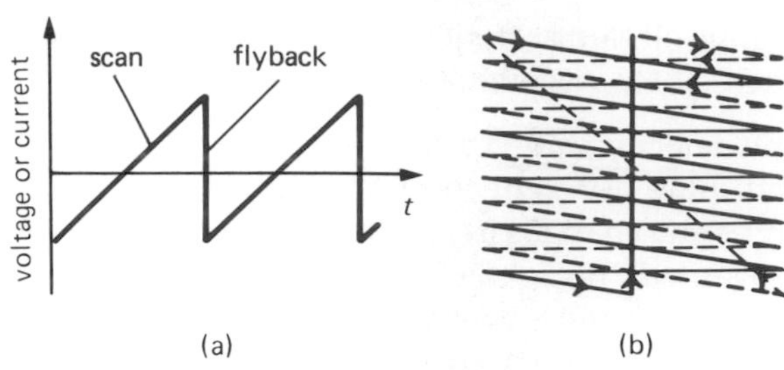

Fig. 88.3

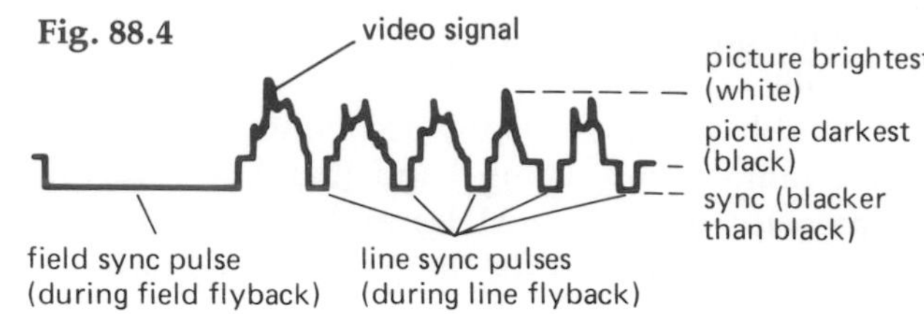

Fig. 88.4

For the video signal to produce an acceptable picture at the receiver, the raster must have at least 500 scanning lines (or it will seem 'grainy') and the total scan should occur at least 40 times a second (or the impression of continuity between successive scans due to persistence of vision of the eye, will cause 'flicker'). The European TV system has 625 lines and a scan rate of 50 Hz.

It can be shown that for such a system, the video signal would need a bandwidth of about 11 MHz, owing to the large amount of information that has to be gathered in a short time. This high value would make extreme demands on circuit design and for a broadcasting system would require too much radio wave 'space'.

However, it can be halved by using *interlaced* scanning in which the beam scans alternate lines, producing half a picture (312.5 lines) every 1/50 s, and then returns to scan the intervening lines, Fig. 88.3b. The complete 625 lines or *frame* is formed in 1/25 s, a time well inside that allowed by persistence of vision to prevent 'flicker'.

Transmission

(a) Synchronization. To ensure that the scanning of a particular line and field starts at the same time in the TV receiver as in the camera, synchronizing pulses are also sent. These are added to the video signal during the flyback times when the beam is blanked out. The field pulses are longer and less frequent than the line pulses (50 Hz compared with $312.5 \times 50 = 15\,625$ Hz). Fig. 88.4 shows a simplified video waveform with line and field sync pulses.

(b) Bandwidth. In broadcast television the video signal is transmitted by amplitude modulation of a carrier in the u.h.f. band. Since the video signal has a bandwidth of 5.5 MHz, the bandwidth required for the transmission would be at least 11 MHz owing to the two sidebands on either side of the carrier.

In practice a satisfactory picture is received if only one sideband and a part (a vestige) of the other is transmitted. This is called *vestigial sideband* transmission. The part used contains the lowest modulating frequencies closest to the carrier frequency, for it is the video information they carry that is most essential. The video signal is therefore given 5.5 MHz for one sideband and 1.25 MHz for the other, Fig. 88.5.

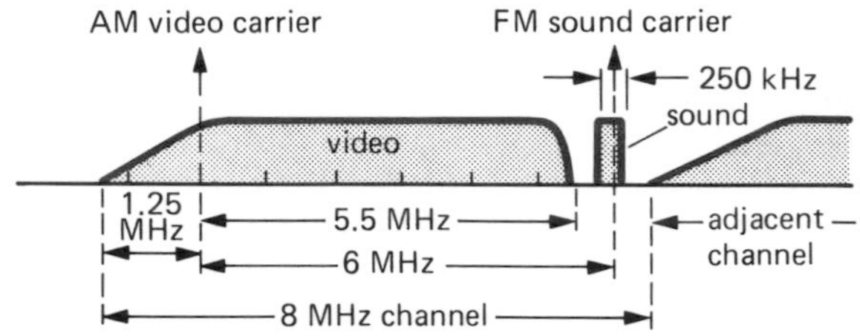

Fig. 88.5

The accompanying audio signal is frequency modulated on another carrier, spaced 6 MHz away from the video carrier. The audio carrier bandwidth is about 250 kHz and is adequate for good quality FM sound. The complete video and sound signal lies within an 8 MHz wide channel.

Television receiver

In a black-and-white TV receiver the incoming video signal controls the number of electrons travelling from the electron gun of a cathode ray tube (p. 48) to its screen. The greater the number, the brighter the picture produced by interlaced scanning as in the camera.

The block diagram in Fig. 88.6 for a broadcast receiver shows that the early stages are similar to those in a radio superhet, but bandwidths and frequencies are higher (e.g. the i.f. is 39.5 MHz). The later stages have to demodulate the video and audio signals as well as separate them from each other and from the line and sync pulses.

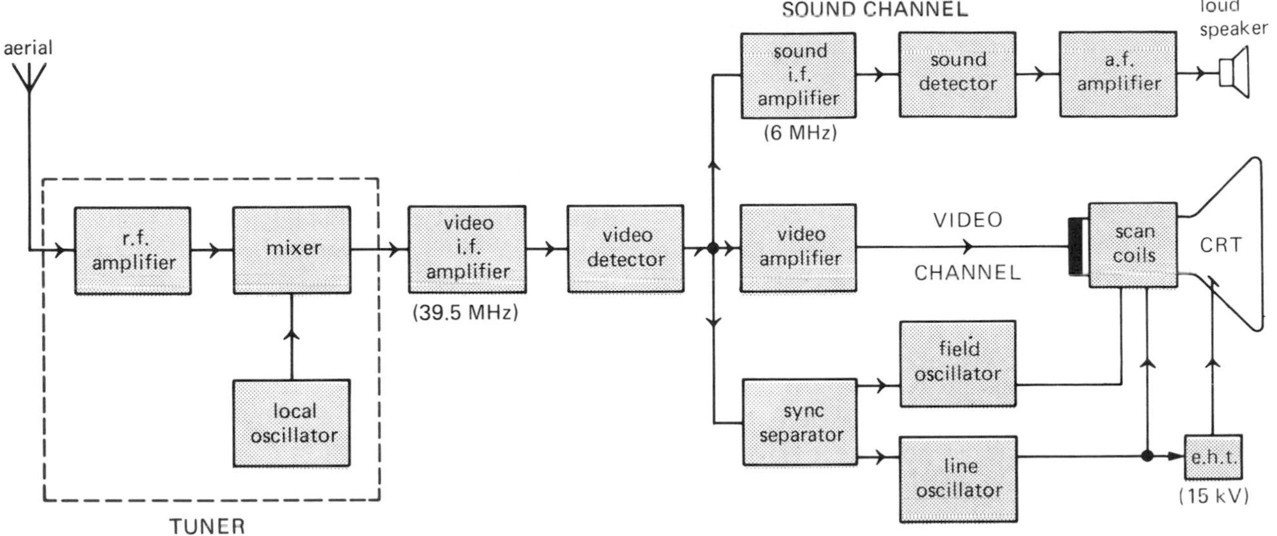

Fig. 88.6

Separation occurs at the output of the *video detector*. There, the now demodulated AM video signal is amplified by the *video amplifier* and applied to the modulator (grid) of the *CRT* to control the electron beam. The still-modulated FM sound signal, having been heterodyned (p. 182) in the *video detector* with the video signal to produce a sound i.f. of 6 MHz (i.e. the frequency difference between the sound and video carriers, shown in Fig. 88.5), is fed into the sound channel where, after amplification and FM detection, it drives the *loudspeaker*.

The mixed sync pulses are processed in the time base channel by the *sync separator* which produces two sets of different pulses at its two outputs. One set is derived from the line pulses and triggers the *line oscillator*. The other set is obtained from the field pulses and synchronizes the *field oscillator*. The oscillators produce the deflecting sawtooth waveforms for the *scan coils*. The *line oscillator* also generates the extra high voltage or tension (e.h.t.) of about 15 kV required by the final anode of the *CRT*.

Some facts and figures

(a) Television channels and frequencies (MHz)

Band	Channel	Vision	Sound	Receiving aerial group
IV	21	471.25	477.25	A (21, 22, 34)
	22	479.25	485.25	
	34	575.25	581.25	W = wideband
V	39	615.25	621.25	E (39, 68)
	68	847.25	853.25	

(b) Transmitting (T) and relay (R) stations

Station	Channel				Polarization	Power (kW)
	BBC1	BBC2	ITV	Ch 4		
Crystal Palace (T)	26	33	23	30	Horizontal	1000
Woolwich (R)	57	63	60	67	Vertical	0.63
Winter Hill (T)	55	62	59	65	Horizontal	500
Windermere (R)	51	44	41	47	Vertical	0.5

89 Colour television

Transmission

Colour TV uses the fact that any other colour of light can be obtained by mixing the three primary colours (for light) of red, green and blue in the correct proportions. For example, all three together give white light; red and green give yellow light.

The principles of transmission (and reception) are similar to those for black-and-white TV but the circuits are more complex. A practical requirement is that the colour signal must produce a black-and-white picture on a monochrome receiver; this is called *compatibility*. Therefore in a broadcast system the bandwidth must not exceed 8 MHz despite the extra information to be carried.

A monochrome picture has only brightness or *luminance* variations ranging from black through grey to white. In a colour picture there are also variations of colour or *chrominance*. The signals for both are combined without affecting each other or requiring extra bandwidth.

In a colour TV camera, three plumbicon tubes are required, each viewing the picture through a different primary colour filter. The 'red', 'green' and 'blue' signals so obtained provide the chrominance information, which is modulated by encoding circuits on a carrier. If added together correctly, they give the luminance as well, Fig. 89.1.

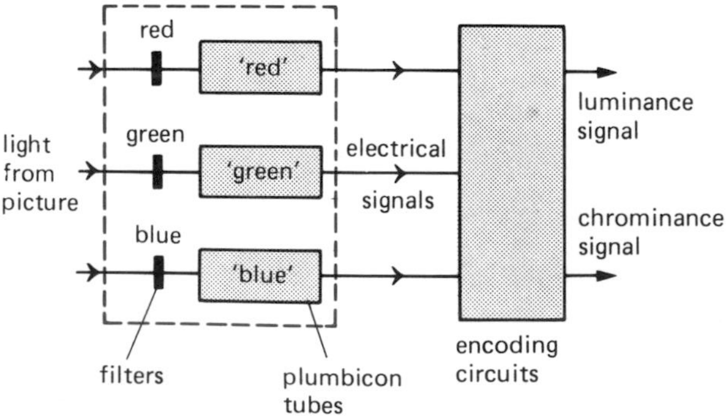

Fig. 89.1

Receiver

In a colour TV receiver, decoding circuits are needed to convert the luminance and chrominance signals back into 'red', 'green' and 'blue' signals and a special CRT is required for the display.

One common type of display is the *shadow mask tube* which has three electron guns, each producing an electron beam controlled by one of the primary colour signals. The principle of its operation is shown in Fig. 89.2. The inside of the screen is coated with many thousands of tiny dots of red, green and blue phosphors, arranged in triangles containing a dot of each colour. Between the guns and the screen is the shadow mask consisting of a metal sheet with about half a million holes (for a 26 inch diagonal tube).

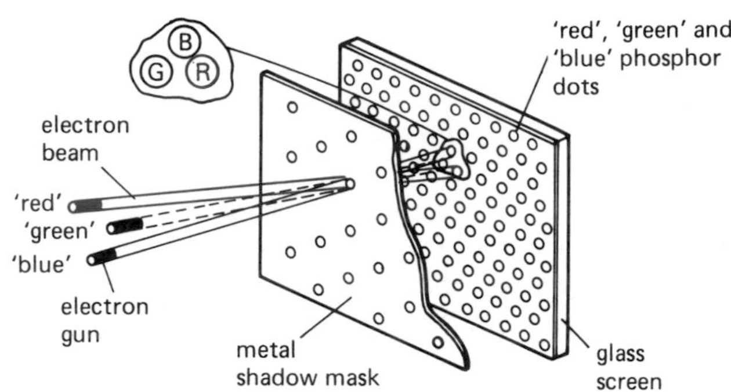

Fig. 89.2

As the three electron beams scan the screen under the action of the same deflection coils, the shadow mask ensures that each beam strikes only dots of one phosphor, e.g. electrons from the 'red' gun strike only red dots. When a particular triangle of dots is struck, it may be that the red and green electron beams are intense but not so the blue. In this case the triangle would emit red and green light strongly and so appear yellowish. The triangles of dots are excited in turn and since the dots are so small and the scanning so fast, we see a continuous colour picture.

The holes in the mask occupy only 15% of the total mask area; 85% of the electrons emitted by the three guns are stopped by the mask. The beam current in a colour tube therefore has to be much greater than in a monochrome tube for a similar picture brightness. Also, the final anode voltage in the tube is higher, about 25 kV.

Teletext

This is a system which, aided by digital techniques, displays on the screen of a modified domestic TV receiver, up-to-the-minute facts and figures on news, weather, sport, travel, entertainment and many other topics. It is transmitted along with ordinary broadcast TV signals, being called CEEFAX ('see facts') by the BBC and ORACLE by ITV.

During scanning, at the end of each field (i.e. 312.5 lines), the electron beam has to return to the top of the screen. Some TV lines have to be left blank to allow time for this and it is on two of these previously blank lines in each field (i.e. four per frame of 625 lines) that teletext signals are transmitted in digital form.

One line can carry enough digital signals for a row of up to 40 characters in a teletext page. Each page can have up to 24 rows and takes about $\frac{1}{4}$ second (i.e. $12 \times 1/50$ s) to transmit. The pages are sent one after the other until, after about 25 seconds, a complete magazine of 100 pages has been transmitted before the whole process starts again.

The teletext decoder in the TV receiver picks out the page you asked for (by pressing numbered switches on the remote control keypad) and stores it in a memory. It then translates the digital signals into the sharp, brightly coloured words, figures and symbols that are displayed a page at a time on the screen.

90 Cable and satellite TV

Cable television (CATV)

In cable television, pictures are sent to the homes of viewers via a cable, as distinct from over-the-air transmissions picked up by roof-top aerials.

In the *star* distribution system, Fig. 90.1, every home is linked directly to a local cable station from which many channels could be available to choose from. Each home has its own junction box and control which would enable the subscriber to send information back to the cable station. This is called an *interactive system* and opens the way for the television set to provide additional facilities such as:

(i) shopping by ordering directly from home after the shop shelves have been viewed on the TV screen,

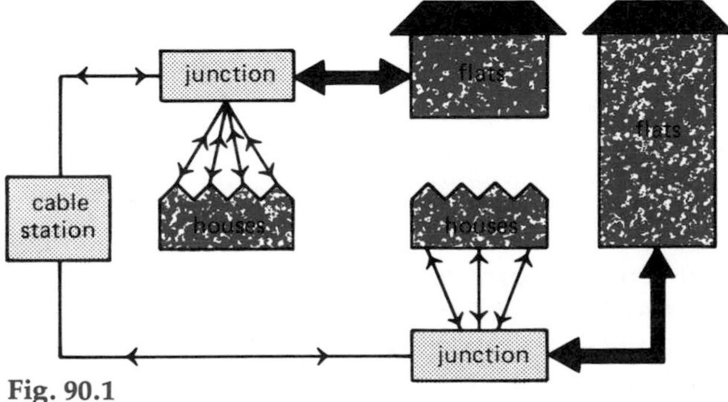

Fig. 90.1

(ii) home banking in which bank accounts could be debited automatically at the push of a button,
(iii) two-way teaching,
(iv) booking holidays via the 'box' at home, after browsing through travel brochures seen on it,
(v) advertising articles for sale electronically,
(vi) taking public opinion poll samples very quickly and even replacing the polling booth by the TV set,
(vii) providing a burglar or fire alarm system if linked to detection sensors in the home and the police or fire station,
(viii) reading of domestic meters (e.g. electricity) remotely, and
(ix) viewing TV programmes beamed live via satellites from across the world.

These developments form the basis of what is known as the *information society*.

Cable systems using modern coaxial cables (p. 197) or optical fibres (p. 197) can carry simultaneously, several TV channels, ordinary telephone links and other telecommunication services because of their wide bandwidth (p. 169). The cables are laid in existing telephone line ducts (pipes) under the street. For economic reasons, CATV is likely to develop only in major population centres.

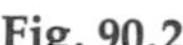

Fig. 90.2

Fig. 90.3

Satellite television

British television viewers saw their first live broadcast from America on 11 June 1962. The microwave signals were beamed from a large steerable dish aerial on the east coast of the U.S.A. to the satellite *Telstar* which amplified and retransmitted them back down to a 26 m diameter dish at Goonhilly Downs earth station in Cornwall, Fig. 90.2. *Telstar* orbited the earth in $2\frac{1}{2}$ hours at a height varying from 320 to 480 km (200 to 300 miles) but signals were received for only about 20 minutes when it could be 'seen' by the aerials on both sides of the Atlantic.

(a) Geostationary (synchronous) satellites. Today, the idea proposed by the space science writer Arthur C. Clarke in 1945 has been realized because rockets are powerful enough to place communication satellites in *geostationary* orbit 36 000 km (22 500 miles) above the equator where they circle the earth in 24 hours and appear at rest.

Early Bird, launched in 1965 over the Atlantic, was the first geostationary satellite to give round-the-clock use. At present, throughout the world there are over 200 earth stations like the one at Goonhilly (in the U.K. there is a second one at Madley, near Hereford). The system is managed by the 105-nation International Telecommunications Satellite Organization (INTELSAT). Satellites are launched for different countries either by the American *Space Shuttle* or by the European rocket *Ariane.* INTELSAT V, shown in Fig. 90.3 with its power-generating solar panels, was launched into geostationary orbit in 1980 with an expected life of 7 years. It can handle two TV channels plus 12 000 telephone circuits using microwave frequencies of 4, 6, 11 and 14 GHz.

(b) Direct broadcasting by satellite (DBS). This development will enable homes in any part of the U.K. to become low-cost 'earth stations' and receive TV programmes from the national geostationary satellite if they have a small roof-top dish aerial (0.9 m in diameter), pointing towards the satellite. In this way, one satellite, having had its programme beamed from an earth station, will give countrywide coverage. The dish will need to be steerable to obtain reception from the satellites of other countries although some overlap is likely. Alternatively the programmes might be received via cable from a cable station with dish aerials.

To receive satellite signals on one of today's domestic TV receivers extra circuits are needed as well as a dish aerial. First a frequency converter in the base of the aerial has to convert the 12 GHz signals from the satellite down to around 1 GHz before they go indoors to the receiver. Here a second converter must reduce the frequency to u.h.f. Existing terrestrial TV transmissions are AM but satellite transmitters use FM to carry the programmes (because of the limited power available). The conversion circuits in the receiver must therefore change the signal from FM to AM. Finally, as the sound signals accompanying the TV signals are digitally encoded, conversion to analogue form is also required.

91 Progress questions

1. Draw a diagram illustrating the various regions of the electromagnetic spectrum in order of increasing wavelength.

To which regions of the spectrum do the following radiations belong?

(i) Radiation of 632.8 nm wavelength from a helium-neon laser.

(ii) 3 cm waves from a klystron transmitter.

(iii) Radiation of 300 MHz.

(iv) Radiation of 0.1 nm wavelength produced in the electron bombardment of a metal surface.

(C. part qn.)

2. (a) What do radio waves consist of?

(b) Describe briefly the ways in which radio waves travel.

(c) Models are often controlled by radio signals of approximately 27 MHz which are received by an aerial of one-eighth of the wavelength. If the speed of radio waves is 3×10^8 m/s, calculate the length of the aerial.

3. (i) Explain why modulation is necessary in radio broadcasting.

(ii) Draw a simple sketch to show the nature of an amplitude-modulated signal. Why must a definite bandwidth be allowed for such a signal?

(iii) Explain how the signal is demodulated at the receiver. *(A.E.B.)*

4. (a) If a radio set was tuned to receive a radio signal of constant amplitude and frequency, state and explain what sound, if any, would be heard from the loudspeaker.

(b) What is done to a radio wave in order to *amplitude-modulate* it with a sound wave?

(c) What is meant by *frequency modulation*?

(O.L.E. part qn.)

5. A circuit for a very simple diode radio receiver is shown in Fig. 91.1.

Which components enable the receiver to (i) pick up a weak signal, (ii) select the signal required, (iii) produce a.f. signals from the modulated r.f. signals?

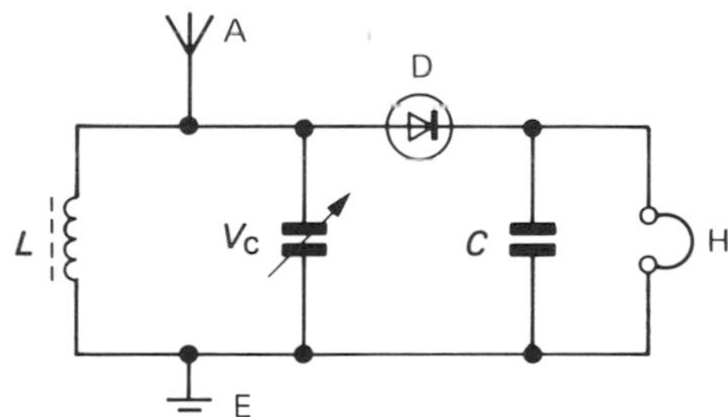

Fig. 91.1

6. (a) State *three* advantages of the superheterodyne radio receiver.

(b) Draw a block diagram of a superhet radio receiver.

(c) If a superhet has an intermediate frequency of 10 MHz, what is the frequency of the local oscillator to receive a signal of 30 MHz?

7. State the advantages and disadvantages of AM and FM radio systems.

8. (a) Explain the terms (i) raster, (ii) line timebase, (iii) field timebase, in relation to a television picture.

(b) Draw a block diagram of a black-and-white *television system*, and give a brief account of the basic elements included in your system.

(c) A television picture with 625 lines uses interlaced scanning in which half the lines are scanned every 1/50th second. What is (i) the line frequency, (ii) the field frequency? (C.)

Telephone system

92 Simple telephone circuits

Speech circuits

(a) Basic circuit. This consists of a microphone (p. 43) connected by wires (the 'line') to an earpiece (p. 43) and a battery of a few volts which drives a current (d.c.) round the circuit, Fig. 92.1. When someone speaks into the microphone, its resistance varies in response to the sound and causes corresponding *changes* in the current. These changes reproduce the sound in the earpiece.

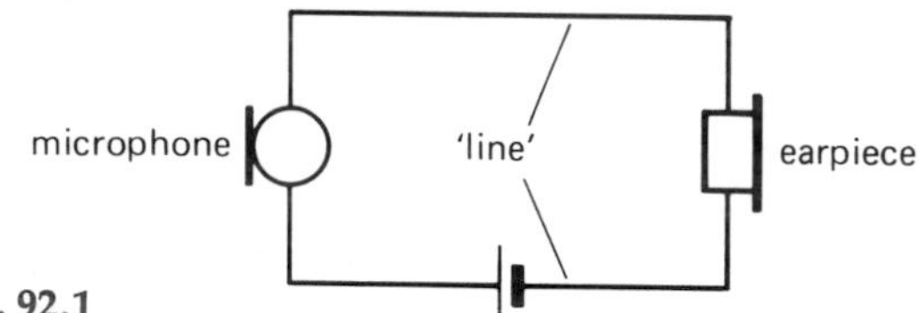

Fig. 92.1

If the line is long, the *total* resistance of the circuit is high and it changes only slightly when the microphone resistance varies. The current changes are therefore small and the sound in the earpiece is very faint.

(b) Two-way circuit. Speech between two telephones many miles apart is possible with the circuit of Fig. 92.2, in which the battery current does not flow in the line but only in the microphone and the primary of a transformer. When the resistance of the microphone in this low resistance local circuit varies, it causes large current changes (varying d.c.). These induce in the secondary winding of the transformer, a current varying in the same way (but a.c.), which flows in the line and through the earpiece at the distant end, Fig. 92.3. Communication can occur in both directions. The microphone and earpiece at each end is combined in a single unit, the handset.

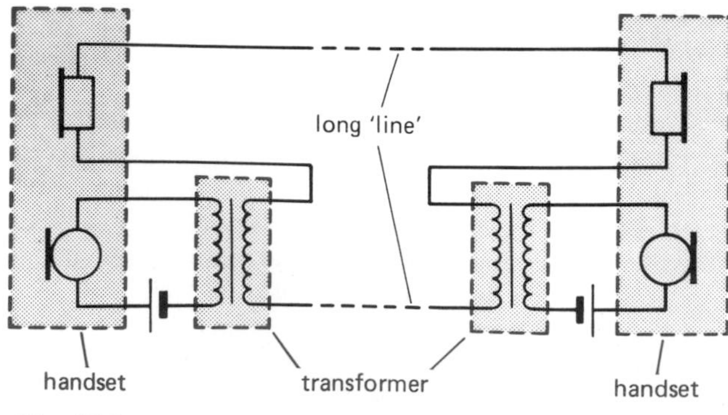

Fig. 92.2

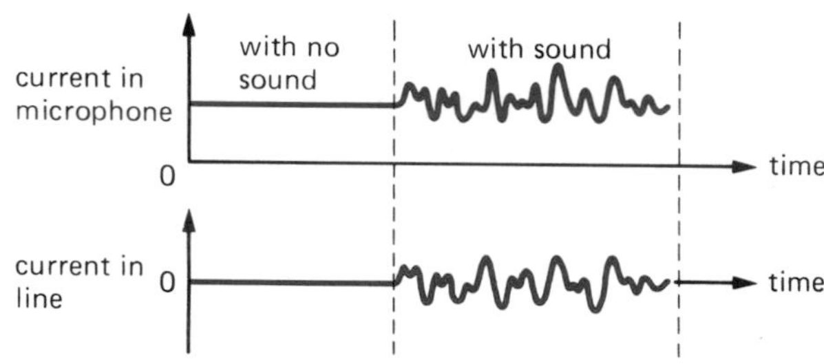

Fig. 92.3

Today most telephones are connected to an exchange with a central battery (of higher voltage, e.g. 50 V) which replaces the local battery at each telephone. The system requires some rearrangement of the circuit and contains a capacitor and a specially designed transformer.

Bell circuit

As well as being able to speak to the person at the other end of the line, it is necessary to call them first.

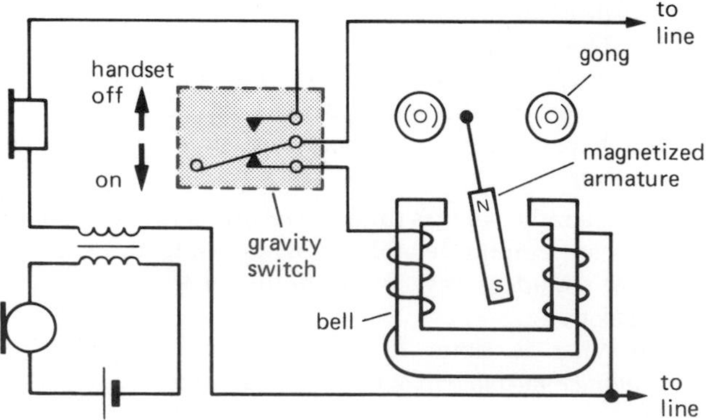

Fig. 92.4

This is generally done by an electric bell and a two-way *gravity switch* operated by the handset. The switch connects the line to the bell when the telephone is not in use (handset on its rest) and to the speech circuit when it is in use (handset off), Fig. 92.4.

The bell is not a d.c. doorbell type. It works on a.c. (supplied by the exchange), which makes the hammer on the armature strike each gong (of different size) in turn, every cycle, to produce the distinctive tremolo sound. Lifting the handset also disconnects the a.c. supply to the bell.

93 Telephone dial and keypad

Telephone dial

Lifting a telephone handset connects the speech circuit to the line (as we have just seen) and allows d.c. to flow from the central battery at the exchange through the line and the telephone. When someone dials, this current is interrupted, for example, five times if the number dialled is 5. Dialling 0 produces ten interruptions. Equipment at the exchange sets up the call in response to the number dialled.

In more detail, the action is as follows. When the caller puts a finger in one of the holes in the dial and turns it clockwise, there are no current interruptions but a spring is wound up, Fig. 93.1. The interruptions occur on releasing the dial when the spring returns it at *constant* speed to its normal position under the control of a device called a 'governor'. They are due to the opening and closing of electrical contacts by a notched wheel or cam which revolves with the dial and engages with a jointed arm. Fig. 93.2a to h shows what happens.

In (a) the dial is at rest. In (b) the cam engages the arm and moves it away from the contacts when the caller turns the dial clockwise. In (c) the arm is fully

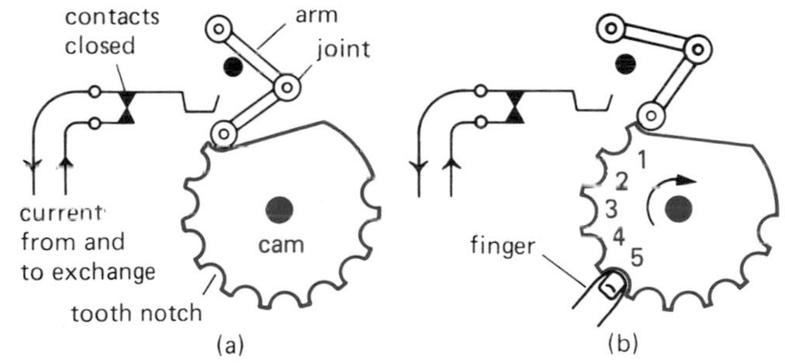

Fig. 93.2

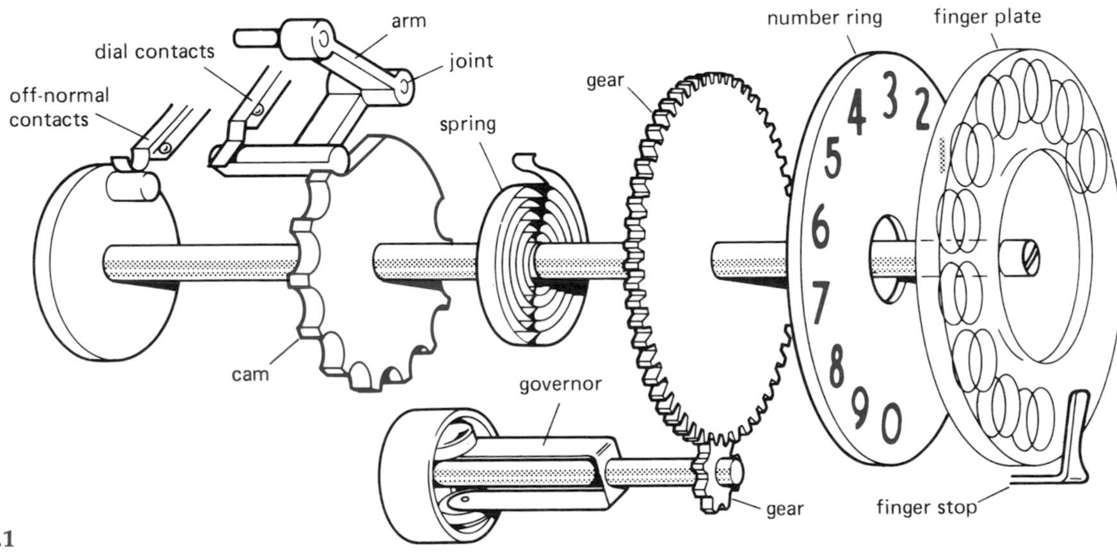

Fig. 93.1

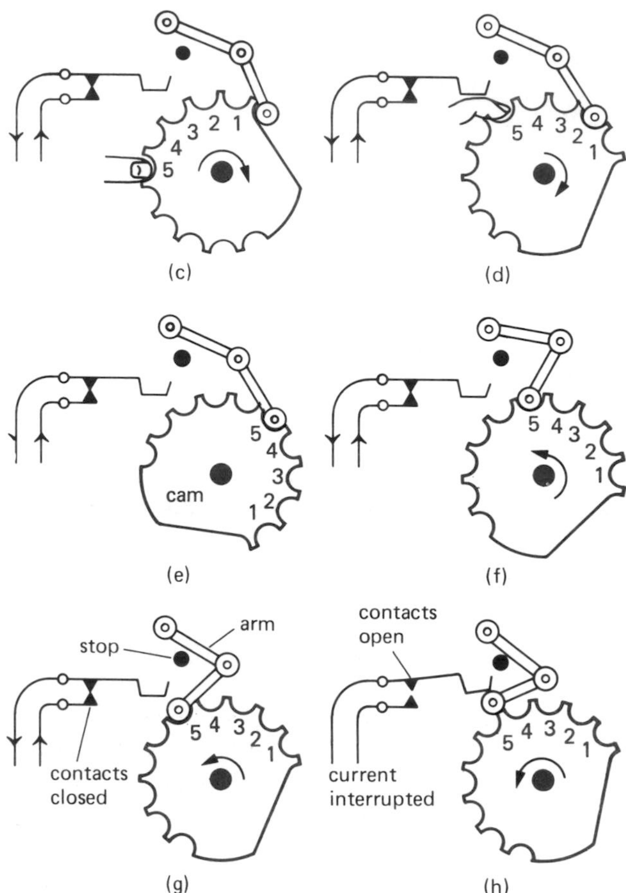

Fig. 93.2 cont.

extended and in (d) it falls into each of the notches in the cam as it passes. If 5 is dialled, the arm stops in the fifth notch, as shown in (e). Releasing the dial allows the arm to be carried back, still in the fifth notch, as in (f). On reaching its rest position, the arm is halted by a stop, see (g). However the cam continues turning anticlockwise and as each tooth between two notches passes, it lifts the end of the arm so opening the dial contacts and interrupting the current, as (h) shows. The cam eventually returns to its position in (a).

Fig. 93.3

To stop the caller hearing the loud clicks when he dials (due to the current interruptions), the dial has an extra set of contacts, called 'off-normal' contacts. These close if the dial is used, i.e. is not in its normal position, and the current from the exchange goes through them, bypassing the speech circuit.

The current interruptions produced by the dial (called 'loop-disconnect pulses') are converted into current pulses by relays at the telephone exchange and operate the equipment (e.g. selectors) which set up the call.

Telephone keypad

The keypad on a press button telephone, Fig. 93.3, is the electronic equivalent of the dial. When a caller keys a number it is stored in an IC in code or as tones of different frequencies (called 'multi-frequency' signalling) depending on the type of keypad. A second IC then arranges for the correct number of 'loop-disconnect pulses' to be sent to the exchange, e.g. two if 2 is keyed.

94 Telephone exchanges

Introduction

(a) The exchange network. Communication between just two telephones can be achieved quite satisfactorily by simple circuits like those in Fig. 92.1 and 2. In a telephone system involving many subscribers, connection of each one directly to every other subscriber would be extremely complex and costly. The number of connections can be greatly reduced if the subscribers in one area are each connected to a *local exchange*, located centrally. Any subscriber in that area can then be connected to any other subscriber in the area or to subscribers in other areas via other exchanges. Power supplies, other essential equipment and staff can also be conveniently based at the exchange.

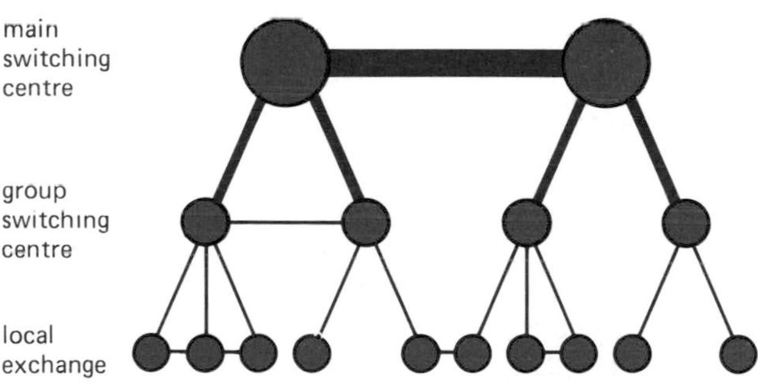

Fig. 94.1

In the same way, and for similar reasons, local exchanges are connected to *group switching centres* which in turn are linked to *main* or *trunk switching centres*, Fig. 94.1. In the U.K. there are several thousand local exchanges, a few hundred group centres and about ten main centres, the last in large cities. Subscribers can also be connected via one or more switching centres to the *international switching centre* for calls abroad.

(b) Switching. To reach its destination a telephone call must be directed or *routed*. This involves *switching* at the exchange to ensure the correct lines are connected.

For an automatic exchange, the destination is indicated by the number dialled or in the case of a press button telephone, by the keys pushed on the keypad. The switching would be done usually by electromechanical *selectors* (see below) or in some exchanges by *reed switches* (p. 51) or increasingly in the U.K. as the telephone system goes 'digital', by *electronic logic gates* (p. 132).

For a manual exchange, the call required would be given orally to the operator who, in early switchboards would plug a connecting cord into one of the many sockets to which the lines were connected, Fig. 94.2. In a modern cordless switchboard, Fig. 94.3, the connection would be made by pushing buttons and switches.

Fig. 94.2

Fig. 94.3

Fig. 94.4

Strowger automatic exchange

The first practical automatic exchange was due to Almon B. Strowger, an undertaker from Kansas City, U.S.A. and was patented in 1889. His system was adopted by the Post Office in the U.K. in 1926 and uses selectors for switching. Fig. 94.4 shows racks of selectors, of which there are two main types, in a Strowger exchange.

(a) Uniselector. This is the simplest type and consists of a shaft carrying a contact arm which can be rotated so as to make electrical connection with any one of several contacts in an arc around it, Fig. 94.5.

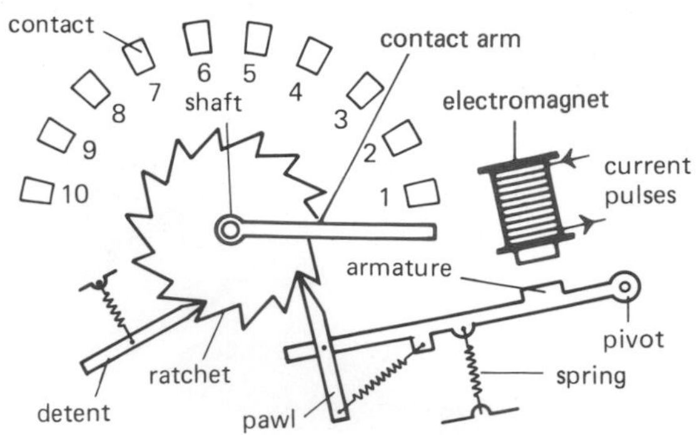

Fig. 94.5

When the number 1 is dialled, or keyed, one electrical pulse passes through the electromagnet which attracts the armature. The pawl engages in one of the notches of the ratchet and moves it round one step so that the contact arm is on the first contact. At the end of the pulse, when the armature is released, the pawl moves back over the ratchet tooth and falls into the next notch where it is kept by the detent.

If number 2 is dialled or keyed, two pulses move the contact arm through two steps to make contact with the second contact, and so on. Dialling 0 gives ten rapid pulses. The uniselector is so-called because its step-by-step rotary motion is in *one* plane.

(b) Two-motion selector. It has ten arcs of contacts set one above the other and the contact arm can be moved vertically (by a 'vertical' electromagnet, pawl and ratchet) to any of the arcs and then rotated horizontally (by a 'horizontal' electromagnet, pawl and ratchet) to any contact on that arc, Fig. 94.6.

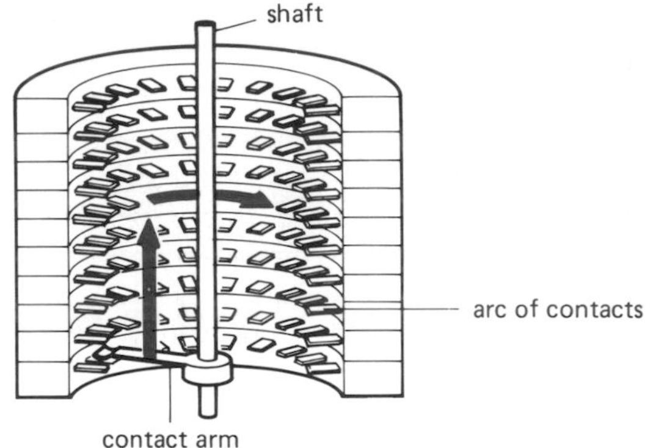

Fig. 94.6

(c) Making a call. At a four-digit (up to 9999) exchange, every subscriber has the exclusive use of a uniselector whose contacts are connected to the contact arms of a *group* of two-motion selectors, called *first group selectors*, which deal with the first digit in the number called. These in turn are connected via the *second group* (of two-motion) *selectors*, which respond to the second digit, to the *final selectors* that handle the last two digits and, being connected to the lines of the various subscribers, make the connection to the required number, Fig. 94.7. Switching is thus a step-by-step process.

Suppose the number 2345 is called. The action, in more detail, is as follows. When the caller lifts his handset, the contact arm of his uniselector starts to rotate and stops on the contact connected to the first

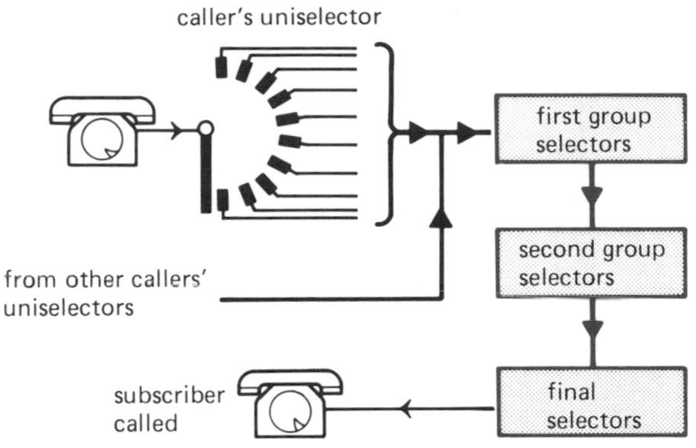

Fig. 94.7

free first group selector. The uniselector thus 'hunts' for a free first selector and when it has found one, the caller hears a 50 Hz *dialling tone* which is the signal to start dialling.

When he dials the first digit, 2 in this case, the first selector's contact arm steps up to the second row of contacts and rotates on this row till it finds a contact connected to a free second group selector. The second digit dialled, i.e. 3, makes the second selector step up to the third row and hunt for a free final selector in the group handling calls to numbers starting with 23. It finds this before the last two digits, i.e. 4 and 5, are dialled; these cause the final selector to step to row 4 and rotate, without hunting, to the fifth contact so making the connection to number 2345.

The exchange then sends a 17 Hz *ringing current* to the called number and a two-pulse 400 Hz *ringing tone* (repeated burr-burr) to the caller to let him know the number is being rung.

(d) Advantages. By using two-motion selectors in groups, the Strowger system makes them available to any subscriber when required and so economizes on the number of selectors used. Even so, there must be enough to cope with peak periods or callers will have difficulty getting through. Careful planning of the number of selectors in each group is necessary to give a good service but yet avoid waste.

It also offers flexibility since a call between two numbers can take any one of many routes through the exchange. Even when a caller makes successive calls to the same number, the routes may be different, which is a great advantage when a selector is faulty.

(e) Supervisory signals. In addition to dialling and ringing tones, other signals must be sent from the exchange to let a caller know if a circuit is available before he dials.

The *engaged tone*, consisting of an intermittent 400 Hz note with a mark-to-space ratio of 1, is heard if the number is engaged.

The *equipment busy tone* is also a 400 Hz note but it is in the form of pairs of pulses every 1.5 s, one pulse being twice as long as the other.

The *number unobtainable tone* is a continuous 400 Hz note.

Subscriber Trunk Dialling (STD)

(a) National numbers. STD, first introduced in 1958 in Bristol, enables any telephone subscriber in the U.K. to call any other subscriber in the country (and in over one hundred other countries) by dialling or keying the appropriate national (or international) number.

A national number has two parts. The first part is the STD code or national prefix and every exchange in the country has one. All start with 0, for example for London it is 01, for Liverpool 051. The second part is the ordinary local number, e.g. 2345.

(b) Making an STD call. The equipment used is shown in Fig. 94.8. When the caller lifts his handset, his uniselector at the local exchange hunts and finds a free first group selector in the normal way. The contacts of level 0 (the tenth level) of the first selector are connected, not to second selectors as at other levels, but to the STD equipment.

When the caller dials 0 as the first digit of a national number, the first selector steps up to the tenth level and hunts for a free STD unit. This records the remaining digits of the national number called, identifies the route to be taken and dials into the trunk exchange which is connected by trunk lines to other towns and cities. The trunk exchange works in the same way as other exchanges and is often in the same building as the local exchange but it can only be accessed by callers via the STD equipment.

Electronic exchanges

Strowger exchanges contain moving parts that wear out. They are costly to maintain and are liable to faults such as crossed lines, buzzes, crackles and

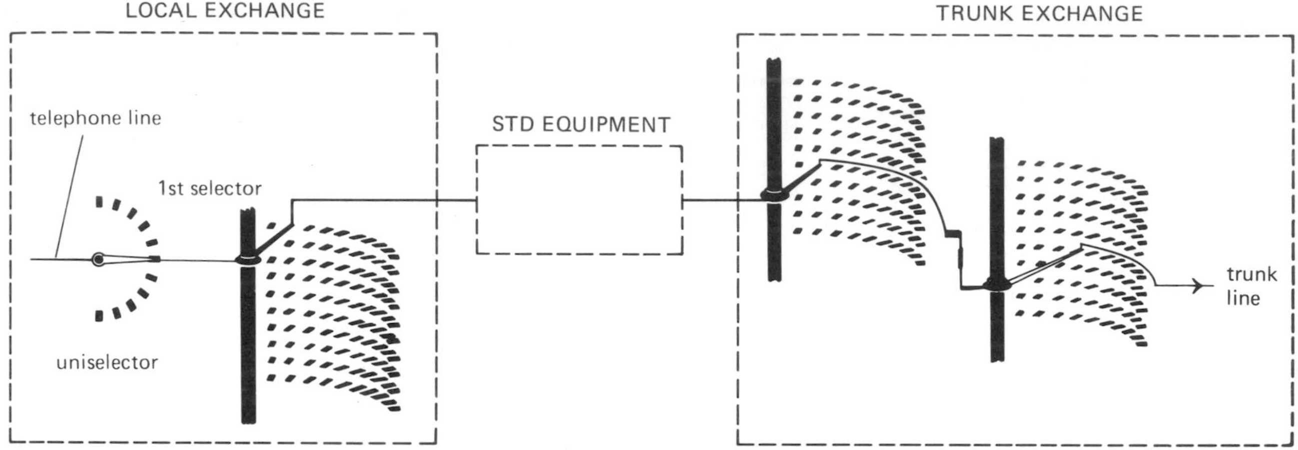

Fig. 94.8

wrong numbers. Expensive conversion equipment is needed to handle computer data. Also, the selectors are the 'brain' and all those used for a call are 'tied up' during the call. They are being replaced by electronic exchanges.

(a) Reed electronic. This half-way stage to a full electronic exchange operates on analogue signals and the 'brain' is an electronic circuit which has the advantage of becoming free as soon as it has set up a call. The switches, which stay engaged during the call, are electromechanical reed switches (p. 51 and Fig. 94.9) activated by currents in the coils surrounding them. Since high operating speed is a feature of electronic circuits, their use in a common

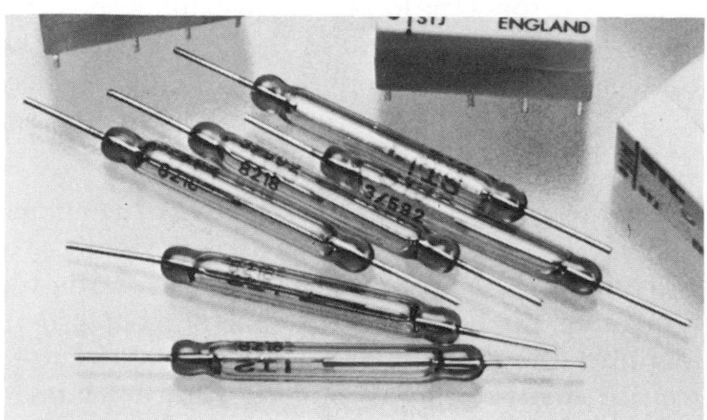

Fig. 94.9

control system allows calls to be handled much more rapidly.

The *crossbar* principle is employed in which, instead of step-by-step switching, incoming lines are connected to rows of horizontal wires and outgoing ones are fed from columns of vertical wires with reed switches where columns cross rows.

(b) System X. Developed by British Telecom in co-operation with industry, System X is a family of electronic exchanges ranging in size from small local exchanges to large international exchanges that is more compact and cheaper than existing exchanges. It uses pulse code modulated (PCM: p. 167) *digital signals*, *electronic switching* and *computers* to control the routing of calls.

The first System X exchange (Fig. 94.10) was opened in the City of London in 1980 and by the mid 1990s about half of all telephones in the U.K. will be connected to this type of exchange.

Fig. 94.10

In addition to clearer speech, less noise, faster connection, fewer wrong numbers, System X will offer new facilities to subscribers, including the following.

(i) *Code calling.* Frequently used numbers can be stored and called quickly using a short code.
(ii) *Call diversion.* Incoming calls can be transferred to another number when required.
(iii) *Three-way calling.* While holding one call, another can be made, then a three-way conversation held.
(iv) *Reminder call.* The exchange can be programmed by the subscriber to ring back at a certain time.
(v) *Charge advice.* At the end of a call the exchange calls back giving the cost.

Eventually, by the early years of the 21st century, System X will provide a multi-purpose, *Integrated Digital Network* (IDN) handling in digital form, all communications whether they be speech, text, photographs, drawings, music, television or computer data.

Private exchanges

Many large organizations have their own exchanges, called Private Branch Exchanges (PBX) or Private Automatic Branch Exchanges (PABX), if they are automatic.

95 Telephone links

Copper cables

The earliest telephone links were copper wires carried overhead on wooden poles. Today most cables are underground and a typical one may contain hundreds of pairs of wires, each covered with plastic insulation and all bunched together inside a thicker plastic covering called the cable sheath, Fig. 95.1. The sheath provides protection and excludes moisture. The cables are laid in earthenware or plastic pipes, called 'ducts', and are joined either in 'joint boxes' just under footpaths or in 'manholes' (as large as small rooms) often under the roadway.

Fig. 95.1

On local links, the cables carry a.f. currents; on trunk lines amplifiers, called *repeaters*, are required to boost the signals which may be multiplexed (p. 169). That is, the a.f. speech currents are modulated on different high frequency carriers, enabling many telephone calls to be carried by one circuit.

Multiplexing requires cables to transmit r.f. currents with minimum loss, i.e. to have a wide bandwidth. Low-loss coaxial cables, Fig. 95.2, can handle frequencies in the u.h.f. range. The inner conductor is solid copper and the outer one is copper braid which also acts as a screen against stray signals and 'noise'.

Fig. 95.2

Optical fibres

(a) Principle. The suggestion that information could be carried on light and sent over long distances in thin fibres of very pure (optical) glass, was first made in 1966. Eleven years later, the world's first fibre optic telephone link was working in Britain and it is expected that eventually trunk circuits and submarine cables to other countries, as well as local lines from homes and offices will change over from copper to glass cables.

Fig. 95.3

Light, like radio waves, is electromagnetic radiation but because of its much higher frequency (typically 10^{14} Hz = 10^5 GHz), it has a considerably greater information-carrying capacity, i.e. it has a very wide bandwidth (p. 168). When modulated and guided by glass fibre cables installed in (existing) cable ducts, it escapes the severe attenuation (weakening) it would suffer from rain and fog if sent through the air. It is also free from 'noise' due to electrical interference and distances of at least 30 km can be worked without regenerators. (Coaxial cables require repeaters to be much closer.)

Compared with copper cables, optical fibre cables are lighter, smaller and easier to handle; the one round the lady technician's neck in Fig. 95.3 has the same capacity as the drum of coaxial cable behind her.

(b) Lasers. The 'light' used is infrared radiation in the region just beyond the red end of the visible spectrum; for the earliest fibres the wavelength was 850 nm (0.85 μm). Later fibres employ 1300 or 1500 nm since the longer the wavelength, the less is the attenuation of the radiation by the glass, which is why infrared is preferred to 'visible' light.

The infrared is generated by a tiny semiconductor *laser* made from gallium, aluminium and arsenic. A laser (standing for *l*ight *a*mplification by the *s*timulated *e*mission of *r*adiation) produces a very narrow *coherent* beam of electromagnetic radiation of one particular frequency. Coherent light, in contrast to light from other sources (e.g. a lamp) consists of waves vibrating in phase with each other, rather like the radiation, at much lower frequencies, from a radio transmitter.

(c) Modulation. The infrared is *pulse code modulated* (p. 167) by the speech or other data to be transmitted, i.e. digital signals are sent in the form of pulses of radiation, being on for a '1' and off for a '0'. An optical fibre communications system is shown in Fig. 95.4.

Fig. 95.4

(d) Optics. The optical fibres (which are about 0.1 mm in diameter) have a glass core of higher refractive index than the glass cladding around it. As a result, the infrared beam is trapped in the core by *total internal reflection* at the core-cladding boundary (just as light is in the prisms of binoculars when it strikes the back surface of the prism where

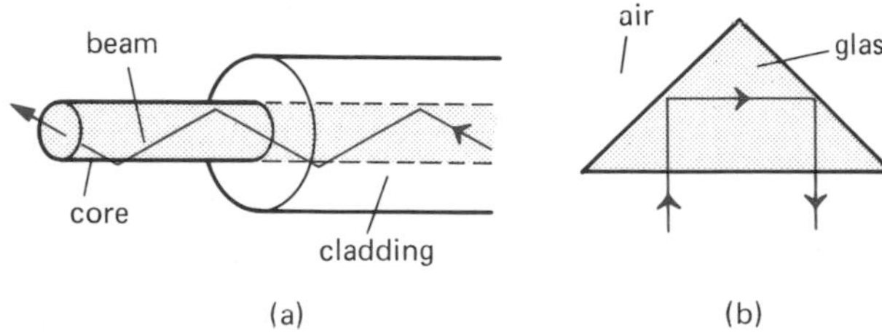

Fig. 95.5

the refractive index is high in the glass but low in the air), Fig. 95.5a,b. The glass in optical fibres is so pure that a 2 km length absorbs less 'light' than a sheet of window glass.

(e) Capacity. The information carrying capacity of an optical fibre system is typically 140 M bit/s at present. (1 M bit/s = 1 million binary bits of information, i.e. 1's or 0's, per second.) A 140 M bit/s system can carry 1920 telephone channels, or the same number of A4 pages of electronic mail per second (p. 202), or 4 colour TV channels or a mixture of these. 140 M bit/s is a very high rate of information transfer, being equivalent to delivering 8 average length books every second!

Microwave links

Microwaves are radio waves with frequencies of the order of GHz (and wavelengths less than 10 cm or so); their information-carrying capacity is therefore quite high but not as high as light. They are easily focused into a narrow beam by a dish aerial (p. 180), 2 or 3 metres in diameter. This makes better use of the power from the transmitter and reduces the risk of interference from other transmitters on the same frequency.

About half of the trunk telephone calls in the U.K. go by microwave links which cross the country in line-of-sight hops between dish aerials mounted on tall towers, like that in Fig. 95.6, containing equipment for boosting the signals before they are retransmitted on their next hop. The nerve centre of the system is the British Telecom Tower in London which is connected to sixteen Network Switching Centres (NSCs) throughout the country.

Television programmes are also carried by the microwave system between studios and transmitters, the NSCs rerouting the programmes as required, Fig. 95.7. Microwave transmissions across the English Channel link the U.K. and European TV networks.

Fig. 95.6

Fig. 95.7

Satellites

Two-thirds of all intercontinental telephone calls now pass via the satellite network (p. 188). A typical satellite call goes from the caller to a local exchange, to an international exchange and then by either cable or terrestrial microwave link to an earth station. There it is beamed up to the satellite by microwaves, Fig. 95.8, and retransmitted down to another country.

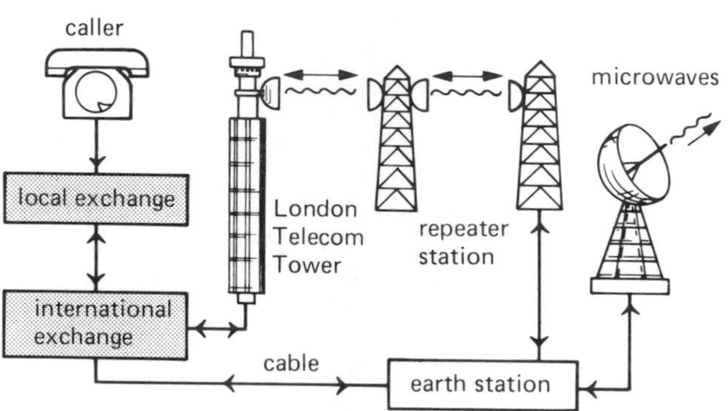

Fig. 95.8

Another important development (SatStream) offers businesses a small-dish satellite service in which they can communicate within Britain and with Europe directly, through rooftop aerials placed at or near their offices in cities, Fig. 95.9. Using digital techniques the system offers flexibility and diversity in that many different services such as speech,

Fig. 95.9

telex (p. 201), facsimile (p. 202) or computer data can be sent over the same transmission path. An A4 page of type or a colour photograph can be transmitted in 1 second, the contents of the *Concise Oxford Dictionary* in 32 seconds and of the entire *Encyclopaedia Britannica* in 30 minutes.

INMARSAT (International Maritime Satellite Organization) uses satellites to provide direct communication with ships at sea.

96 Other telephone services

Viewdata

Like teletext (p. 187), viewdata uses digital techniques to display on the screen of a modified domestic TV receiver, up-to-the-minute information on a wide range of topics. It uses the telephone network and is called *Prestel* by its operators, British Telecom. The information (text and diagrams) is presented a screenful (page) at a time, as in a newspaper or magazine. From the index page(s) the viewer selects the page required on a remote control keypad, Fig. 96.1.

Viewdata signals are sent serially, i.e. one after the other, from a central computer which is the database. Units called *modems* (*mo*dulator-*dem*odulator) are needed at each end of the telephone line to convert analogue to digital signals and vice versa for transmission and reception. The modem encodes digital data by representing a '1' as a low audio tone and a '0' by a high tone.

Viewdata information is more detailed and comprehensive than that from teletext because of the large number of 'pages' ($\frac{1}{4}$ million and increasing) of

Fig. 96.1

data the computer can store. It is an *interactive* system (p. 187), allowing the subscriber to communicate with the database. Instant and easy access is given to almost unlimited information and so it provides the 'information society' with facilities such as 24-hour banking, teleshopping and paying bills, like those listed before (p. 187)—all at the touch of a button from an armchair at home. While teletext is 'free' and available to all simultaneously, viewdata has to be paid for and the number of subscribers that can be connected at any one time is limited.

Confravision

Using this British Telecom service, several groups of people in different parts of the country can see and speak to each other at the same time without travelling, i.e. they can have a 'conference'. It is achieved by linking nine studios in main cities to Network Switching Centres (p. 199).

Radiopaging and radiophoning

Radiopaging allows people who are out and about to be contacted. By dialling a ten-digit number on any phone, a low-power radio transmitter in one of forty zones covering the U.K. is triggered. The signal it emits causes the small radio receiver carried by the person whose code has been dialled, to bleep. Contact is then made with the caller from the nearest telephone.

Radiophoning enables telephone calls to be made and received from a car or boat.

Telex

This is of interest mainly to firms, especially those with overseas connections. The message to be transmitted is typed by an operator at the sender's end on the typewriter-style keyboard of a teleprinter which translates letters, figures, signs and punctuation marks into code. In the modern telex terminal in Fig. 96.2, the VDU (visual display unit) permits editing and a printer gives a permanent copy of the message if it is required.

The receiving teleprinter decodes the message and types it automatically on to a sheet of paper without anyone needing to be present.

Connection between telex subscribers is by a network of *tel*eprinter *ex*changes.

Fig. 96.2

Datel

The Datel service enables business organizations to send computerized information, for example about sales, to other offices, often overseas. It is much used by banks to transfer data about money transactions. Modems are required since transmission is over the ordinary telephone network.

Facsimile

Facsimile (meaning 'exact copy') allows a document to be 'sent' over the telephone system and is a kind of electronic mail. When the document is inserted in the sender's facsimile machine, the 'optical eye' scans it and produces a string of electrical pulses which, on reaching the receiver's machine, makes it print out the same document which is then available for discussion by both parties if required. An A4 page of text can be transmitted and printed in 1 minute.

97 Progress questions

1. Draw the circuit diagram of the equipment needed at *one* end of a simple telephone circuit which enables an electric bell to be rung when a call is made and then allows a two-way conversation.

2. (i) Describe in detail the operation of the telephone dial.
(ii) Explain how the dial is used to generate electrical signals and how these signals are used to control the routing of a trunk telephone call. *(A.E.B.)*

3. Write brief notes on the use of the following in telecommunications: (a) coaxial cables, (b) optical fibres, (c) microwaves and (d) satellites.

4. In 1978 Post Office Telecommunications (now called British Telecom) began work on replacing the existing telephone network by a new system, called System X. It was decided that this new system would use digital transmission, electronic switching and computers to control the routing of calls. Some of the advantages listed for System X are shown in Table 1.

Explain in each of the four cases, what causes the problems in the existing network and why they can be overcome in System X. *(A.E.B. 1982 Electronics part qn.)*

Table 1

Problems in existing network	Characteristics of System X
(a) Transmission loss, producing a weaker signal, varies with call routing	(a) Transmission loss is independent of circuit length
(b) Signals are affected by noise and distortion	(b) Low noise and distortion
(c) Mechanical switches are prone to wear	(c) Electronic switches are reliable
(d) Routing of calls is not easily arranged	(d) Routing of calls can be changed in response to loading

5. Some telecommunication services are: viewdata, datel, telex, facsimile, radiophone, confravision, radiopaging.
(a) Describe how *two* of these services are useful to business organizations.
(b) Describe how a *third* service could have a considerable effect on a person's everyday life.

Computers and microprocessors

98 Digital computers

About computers

Computers are playing an increasing role in our everyday lives at home, at work and in our leisure. This is due to the speed with which they can undertake almost any task involving the processing of information (data). Some of their many uses are shown in Fig. 98.1. Whereas machines replaced muscles in the first industrial revolution, in the second, now upon us, computers are replacing brain power.

A digital computer performs arithmetical and logical operations on data, in the form of digits (0 and 1), that has been converted into a sequence of electrical pulses. It is built from a large number of switching-type circuits which can be programmed to provide different pathways for the pulses when different logic gates are opened and closed. The computer is thereby able to perform in a variety of ways.

The power of a computer depends on its speed of working and the amount of data it can handle at the same time. There are three very broad classes.

(i) *Mainframe* computers are usually the most powerful and contain several large units, housed in an air-conditioned room and operated by a team of people. They are used by big organizations to prepare bills and payrolls and for large calculations, e.g. in weather forecasting.

(ii) *Minicomputers* are made from smaller units, only require one or two operators and are used, for example, by government departments, hospitals and businesses and to control manufacturing operations such as steel-making (i.e. in automation). They were developed for the space industry.

(iii) *Microcomputers* are the personal computers found today in homes, schools and many small offices. They have been made possible by the development of microprocessors and consist of just one or two connected units.

Central processing unit (CPU)

All computers are basically similar, whatever their size and power. They contain the units, called *hardware*, shown in the block diagram of Fig. 98.2.

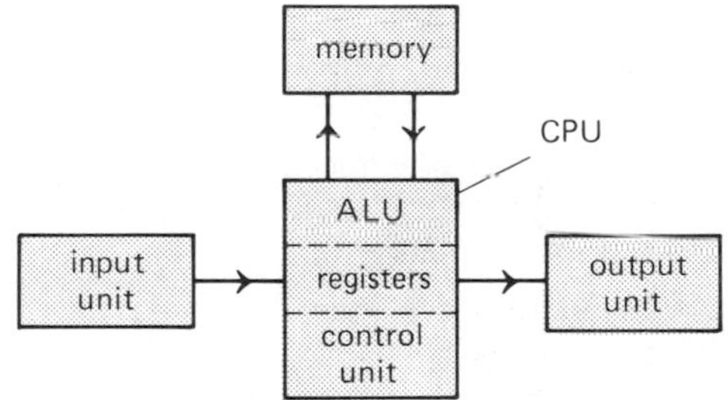

Fig. 98.2

The CPU is the 'brain' of the computer. It accepts digital signals from the input unit and uses them to work out the answer that is subsequently displayed on the output unit. What it can do depends on the set of instructions (typically fifty or more), called the *instruction set*, which it is designed to interpret and obey.

A CPU consists of an *arithmetic and logic unit* (ALU; p. 146), a number of *shift registers* (p. 160) and a *control unit*. The ALU performs arithmetic calculations and logical operations. The registers are temporary stores; one of them, the *accumulator*, contains the data actually being processed. Two important components of the control unit are the *clock* and the *program counter*. The clock is usually a crystal-controlled oscillator which generates timing pulses at a frequency of several megahertz to syn-

Computers in the office.

A teaching aid in which the tutor 'questions' the student using a computer-generated picture. The student 'replies' by touching the screen with a light-sensitive pen which sends signals back to the tutor.

An arc-welding computer-controlled robot on a car production line.

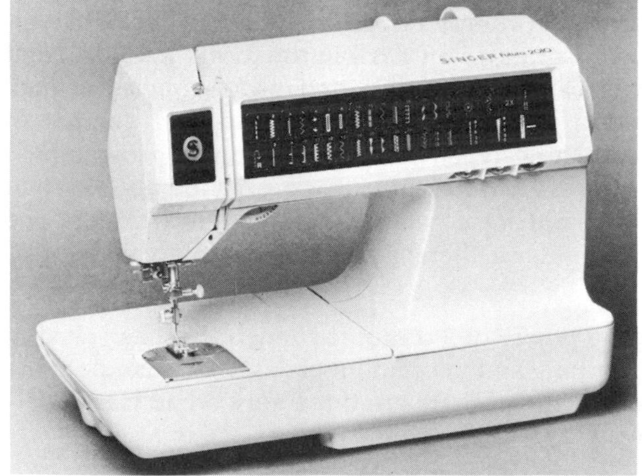

A computer-based sewing machine.

Fig. 98.1

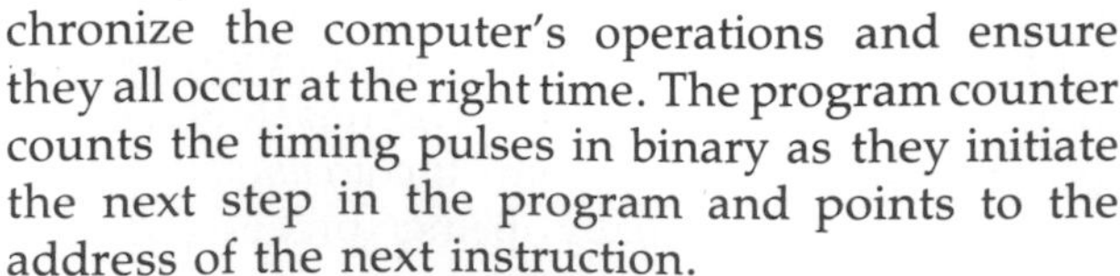

chronize the computer's operations and ensure they all occur at the right time. The program counter counts the timing pulses in binary as they initiate the next step in the program and points to the address of the next instruction.

The operation of the CPU and some of its other components will be described more fully when the microprocessor is considered (p. 210).

Memory

A computer needs a memory to store the program of *instructions* it has to execute. If the program is fixed (as in a computer-controlled washing machine or electronic game), the memory has only to be 'read' and a ROM (p. 160), programmed by the manufacturer, is used since its contents are retained when the power is switched off. A PROM or

EPROM (p. 161) would also be suitable if the user wanted to construct the program himself and keep it permanently in the memory.

If the program has to be changed, the 'read' and 'write' facilities of a RAM (p. 160) are required to allow instructions to be written in, read out and altered at will. A RAM is also needed to store the *data* for processing because it too may change. Sometimes in small computers a RAM is used for both the program of instructions and the data and so all is lost at switch off. As we saw earlier (p. 160) each location in a memory has its own address which allows us to get *directly* to any program instruction or item of data.

Single ROM and RAM ICs with storage capacities up to 64 K bits ($2^{16} = 65\,536$) are available and, as the development of semiconductor memories continues, this value will rise.

99 Computer peripherals

The input and output units of a computer and extra memory devices are called *peripherals*.

External memories

In addition to internal ROM and RAM memories, a back-up store (a database) is required when large amounts of data has to be stored, for example, for a payroll. External memories generally use some form of magnetic storage on tapes and on 'floppy' or 'rigid' plastic discs coated with magnetic material.

Magnetic tape is the simplest and cheapest backing store, the data being stored serially as 1s and 0s by magnetizing small areas of the tape in either of two directions, Fig. 99.1a. *'Floppy' discs*, Fig. 99.1b, are common for small computers; they have limited storage (less than 4 million bits) and resemble small audio records. They are inserted one at a time into a read and write device with a magnetic head. *'Rigid' discs* are used in larger systems and several are stacked permanently on a spindle. They have capacities up to 8000 million bits. Compared with semiconductor memories, access to such devices is slow, especially with tape, but storage is 'permanent', i.e. they are non-volatile.

Bubble memories offer storage of high capacity (greater than 1 million bits) and high bit-packing density with much faster access times than magnetic tapes or discs. Their action is complex but they are made of crystals in which the bits are stored as minute magnetized regions, called 'bubbles', which can be moved around by a magnetic field. They are mounted in a d.i.l. package like an IC and at present

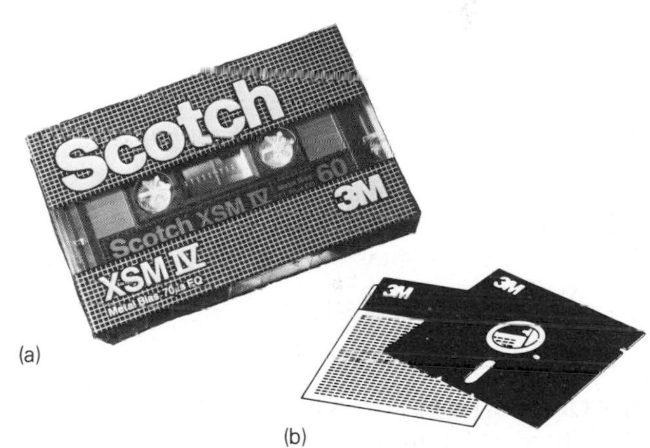

Fig. 99.1

are used mostly for computers in military aircraft and ships where discs would be affected by dust and gunfire.

Input and output devices

These enable the computer to communicate with the outside world. Their form depends on the role of the computer.

(a) Visual display unit (VDU). In data processing the VDU is commonly used as both the input and output unit. It consists of a keyboard on which data can be entered manually as the input and a cathode ray tube to display the output (and input). The keyboard is similar to that on a typewriter and contains the usual range of alphanumeric characters (A to Z and 0 to 9) as well as some 'command' keys for giving instructions to the computer. One is shown in Fig. 98.1 (Computers in the Office).

(b) Teletype. This also has a keyboard as the input unit but the output (and input) is displayed on a printer. It is slower and noisier than a VDU, but a permanent record ('hard copy') is obtained.

Printers are of two main types.

(i) *Dot-matrix* printers form letters and characters from tiny dots, at a rate of 300 per second.

(ii) *Daisy-wheel* printers print one character after the other using a daisy-wheel mechanism (the characters are arranged on it like the petals of a daisy); they are slower but the printing is of a high quality.

(c) Laser-scanner. Bar codes like the one on the back cover of this book and on tins and packets are read at supermarket check-outs by a laser-scanner. The narrow beam of light from the laser is reflected back from the pattern of bands of various widths, and converted into a series of electrical pulses. These identify the item and enable a computer connected to the scanner to produce a bill for the purchase. They are also used by libraries to record the issue and return of books on loan.

Laser-scanner systems are also used with special typewriter faces which can be recognized by *optical character readers* (OCRs). Banks use magnetic ink character recognition (MICR) for cheques.

(d) Other sensors. In industry, science, medicine and other areas, inputs may come from transducers producing analogue voltages that represent continuously varying quantities such as temperature, fluid flow rate or movement. Outputs may be used to control electric motors, industrial processes or robots (those in a car factory have a sensor in their arm telling the computer its position).

Interfacing

Unfortunately data in its original form cannot usually be fed directly from an input device into the CPU. It has to be presented in digital form. Similarly digital signals from the CPU may not be acceptable to an output device. Very often *interfacing* electronic circuits are required between the CPU and its peripherals.

For example, to interface a teletype to a computer, an encoder (p. 145) is needed to produce a different pattern of binary bits, according to the ASCII code (p. 166), for every key operated. Sensors creating analogue voltages require analogue-to-digital converter (A/D) ICs, while some output devices must have a digital-to-analogue converter (D/A) as the interfacing circuit.

Another consideration is whether a peripheral is a *serial* or *parallel* device. Teletypes and VDUs are serial types which produce and must receive a string of bits following one after the other. A seven-segment LED display requires parallel interfacing so that the bits for a particular output are supplied together, i.e. in parallel. Computers handle 'parallel' data as 4-, 8- or 16-bit words and so interfacing a peripheral may involve series to parallel conversion or the reverse by a shift register.

Buses

The CPU is connected to other parts by three sets of parallel wires which are called *buses* because they 'transport' information. The *data bus* carries data for processing and is a two-way system with 4, 8 or 16 lines, each line carrying one bit at a time. The one-way *address bus* conveys addresses and has anything from 4 to 32 lines depending on the number of memory addresses (8 lines give $2^8 = 256$ addresses). The *control bus* deals with timing signals and could have 3 to 10 lines. In diagrams each bus is represented by one broad line.

To stop interference between parts that are not sending signals to a bus and one that is, all parts have in their output, a circuit called a *tri-state gate.* Each gate is enabled or disabled by the control unit. When enabled, the output is 'high' or 'low' and pulses can be sent to the bus or received from it. When disabled, the output has such a high impedance that it is in effect disconnected. Then it can neither send nor receive signals and does not upset those passing along the bus between other parts.

100 Analogue/digital converters

Digital-to-analogue (D/A) conversion

There are many occasions when digital signals have to be converted to analogue ones. For example, a digital computer is often required to produce a graphical display on the screen of a VDU. This involves using a D/A converter to change the two-level digital output voltage from the computer, into a continuously varying analogue voltage for the input to the CRT, so that it can deflect the electron beam and make it 'draw pictures' at high speed.

The principle of the *binary weighted resistor* D/A converter is shown in Fig. 100.1 for a four-bit input. It is so called because the values of the resistors, R, $2R$, $4R$ and $8R$ in this case, increase according to the binary scale, the l.s.b. of the digital input having the largest value resistor. The circuit uses an op amp as a *summing amplifier* (p. 123) with a feedback resistor R_f. S_1, S_2, S_3 and S_4 are digitally controlled electronic SPDT switches.

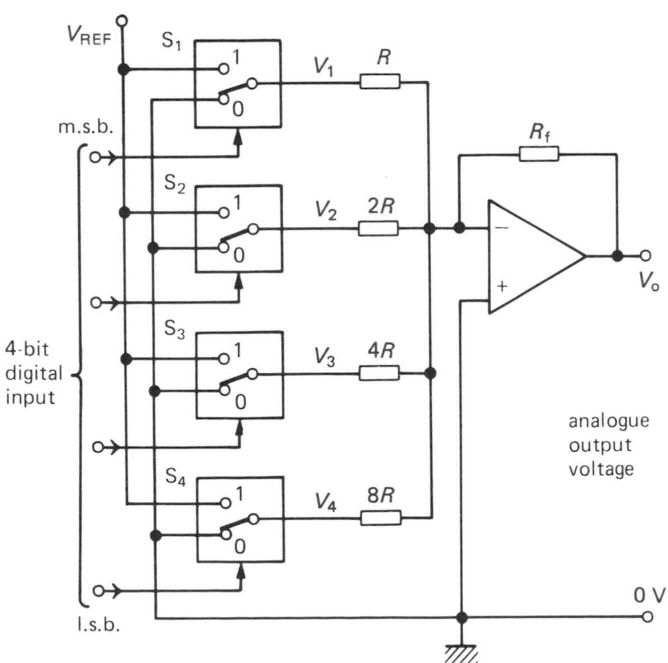

Fig. 100.1

Each switch connects the resistor in series with it to a fixed reference voltage V_{REF} when the input bit controlling it is a 1 and to ground (0 V) when it is a 0. The input voltages V_1, V_2, V_3 and V_4 applied to the op amp by the four-bit input (via the resistors) therefore have one of two values, either V_{REF} or 0 V.

Using the summing amplifier formula (p. 123), the analogue output voltage V_o from the op amp is:

$$V_o = -\left(\frac{R_f}{R} \cdot V_1 + \frac{R_f}{2R} \cdot V_2 + \frac{R_f}{4R} \cdot V_3 + \frac{R_f}{8R} \cdot V_4\right)$$

If $R_f = 1\,\text{k}\Omega = R$, then

$$V_o = -(V_1 + \tfrac{1}{2}V_2 + \tfrac{1}{4}V_3 + \tfrac{1}{8}V_4)$$

With a four-bit input of 0001 (decimal 1), S_4 connects $8R$ to V_{REF}, making $V_4 = V_{REF}$. S_1, S_2 and S_3 connect $2R$, $4R$ and $8R$ respectively to 0 V, making $V_1 = V_2 = V_3 = 0$. And so, if $V_{REF} = -8$ V, we get

$$V_o = -\left(0 + 0 + 0 + \frac{(-8)}{8}\right) = -\frac{(-8)}{8} = +1\text{ V}$$

With a four-bit input of 0110 (decimal 6), S_2 and S_3 connect $2R$ and $4R$ to V_{REF}, making $V_2 = V_3 = V_{REF} = -8$ V; S_1 and S_4 connect R and $8R$ to 0 V, making $V_1 = V_4 = 0$ V. Therefore

$$V_o = -\left(0 + \frac{(-8)}{2} + \frac{(-8)}{4} + 0\right)$$
$$= -(-4 - 2) = +6\text{ V}$$

From these two examples we see that the analogue output voltage V_o is directly proportional to the digital input. V_o has a 'stepped' waveform with a 'shape' that depends on the binary input pattern, as shown in Fig. 100.2, but it does vary.

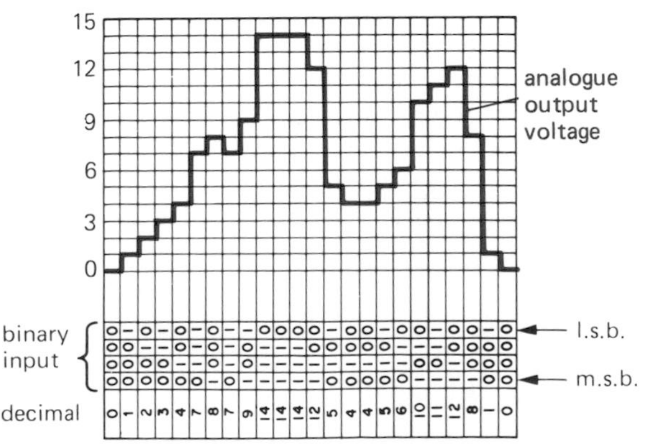

Fig. 100.2

Analogue-to-digital (A/D) conversion

In many modern measuring instruments (e.g. a digital voltmeter), the reading is frequently displayed digitally, but the input is in analogue form. An A/D converter is then needed.

The block diagram for a four-bit *counter* type A/D conversion circuit is shown in Fig. 100.3a, with waveforms to help explain the action. An op amp is again used but in this case as a *voltage comparator* (p. 124). The analogue input voltage V_2 (shown here as a steady d.c. voltage) is applied to the non-inverting (+) input; the inverting (−) input is supplied by a ramp generator with a repeating sawtooth waveform voltage V_1, Fig. 100.3b.

The output from the comparator is applied to one input of an AND gate and is 'high' (a 1) until V_1 equals (or exceeds) V_2, when it goes 'low' (a 0), as in Fig. 100.3c. The other input of the AND gate is fed by a steady train of pulses from a pulse generator, as shown in Fig. 100.3d. When both these inputs are 'high', the gate 'opens' and gives a 'high' output, i.e. a pulse.

From Fig. 100.3e, you can see that the number of pulses so obtained from the AND gate depends on the 'length' of the comparator output pulse, i.e. on the time taken by V_1 to reach V_2. This time is proportional to the analogue voltage if the ramp is linear. The output pulses from the AND gate are recorded by a binary counter and, as shown in Fig. 100.3f, are the digital equivalent of the analogue input voltage V_2.

In practice the ramp generator is a D/A converter which takes its digital input from the binary counter, shown by the dashed lines in Fig. 100.3a. As the counter advances through its normal binary sequence, a staircase waveform with equal steps (i.e. a ramp) is built up at the output of the D/A converter, like that shown by the first four steps in Fig. 100.2.

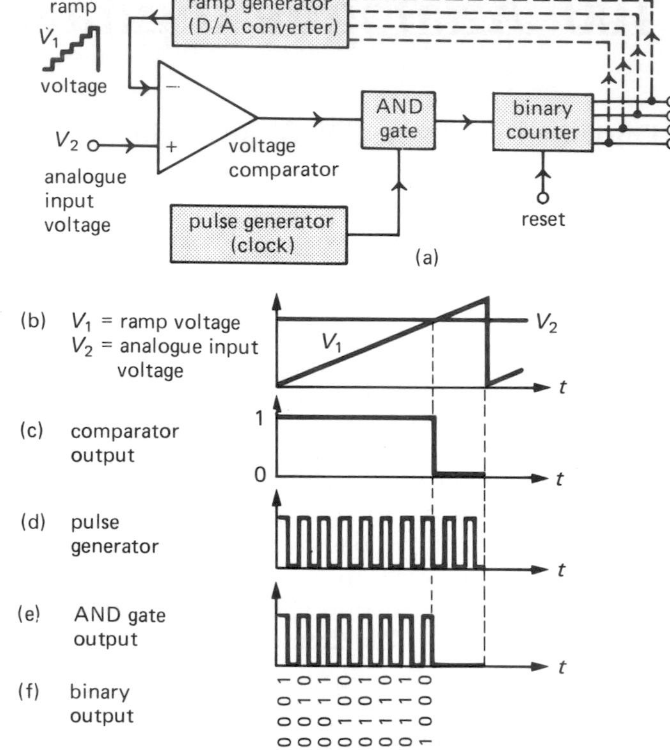

Fig. 100.3

Similar conversion techniques are used in digital voltmeters (p. 97).

101 Programs and flowcharts

Program languages

A computer must be programmed so that it knows what to do. A *program*, referred to as *software* (or *firmware* if it is stored in a ROM) consists of a series of instructions (from the CPU's instruction set), each followed by an address for data and involves the computer in a 'fetch-and-execute' process.

Programs can be written in *machine code*, i.e. the 0s and 1s of the binary system in which the computer must work. But it is a tedious, time-consuming process, liable to error. Program 'languages' have therefore been developed to make the job easier.

(a) Low-level or assembly languages. These are close to the binary system. Programs are written in mnemonics (i.e. memory aids) and by referring to the instruction set for the CPU we might find, for example, that the mnemonic for 'load data from memory into accumulator' is LDA. It would also give the binary code for this operation, say 1001 1110. To enable the CPU to understand the instruction LDA, i.e. to convert the input into machine code, a special program called an *assembler* is stored in a ROM.

(b) High-level languages. They are more like everyday English and are easier to understand and work with since they use terms such as LOAD, ADD, FETCH, PRINT, STOP. However they need more memory space and computer time because each statement converts into several machine code instructions, not just one as is usual in a low-level program. A *compiler* or an *interpreter* (the equivalents of an assembler and, like it, consisting of a program) does the translation into machine code. BASIC (Beginner's All-purpose Symbolic Instruction Code) is a popular high-level language. COBOL is designed for business use, while FORTRAN is suitable for scientific and mathematical work.

Flowcharts: binary multiplication

All programs, whether high- or low-level, must give absolutely clear instructions to the computer. Writing a program is easier if a *flowchart* is drawn up initially to show the different steps required for the task.

(a) Binary multiplication. Suppose the task is the binary multiplication of 6 and 5, i.e. 110 and 101. First consider how it is done by long-multiplication. The method is the same for decimal multiplication, i.e. the multiplicand (6) is multiplied separately by each digit of the multiplier (5) in turn and the partial products properly shifted with respect to one another before they are added.

The operation is simpler in binary because we have to deal with just two digits (1 and 0) and so at each stage there are only two possible answers, i.e. 110 × 1 = 110 or 110 × 0 = 0. It is performed as follows:

multiplicand	110 = 6	
multiplier	101 = 5	
partial product	110	← (110 × l.s.b. of 101, i.e. 110 × 1)
partial product	000	← (110 × 2nd bit of 101, i.e. 110 × 0)
partial product	110	← (110 × m.s.b. of 101, i.e. 110 × 1)
final product	11110 = 30	(sum of partial products)

The process is one of 'shift-and-add' in which the partial products are either zero or a shifted version of the multiplicand. When a digital computer is used for multiplication, it is more convenient to add each partial product as it 'accumulates', rather than adding them all at the end. The result is the same but the sequence is:

	110
	101
partial product	110
partial product	000
partial sum	0110
partial product	110
final product	11110

In practice, one shift register (X) is needed for the multiplicand, another (Y) for the multiplier and a third, the accumulator, stores the partial products and sums, and contains the final product when the multiplication is complete. This event is detected by a binary counter which records the number of 'shift-and-add' cycles to enable the whole operation to be stopped when finished.

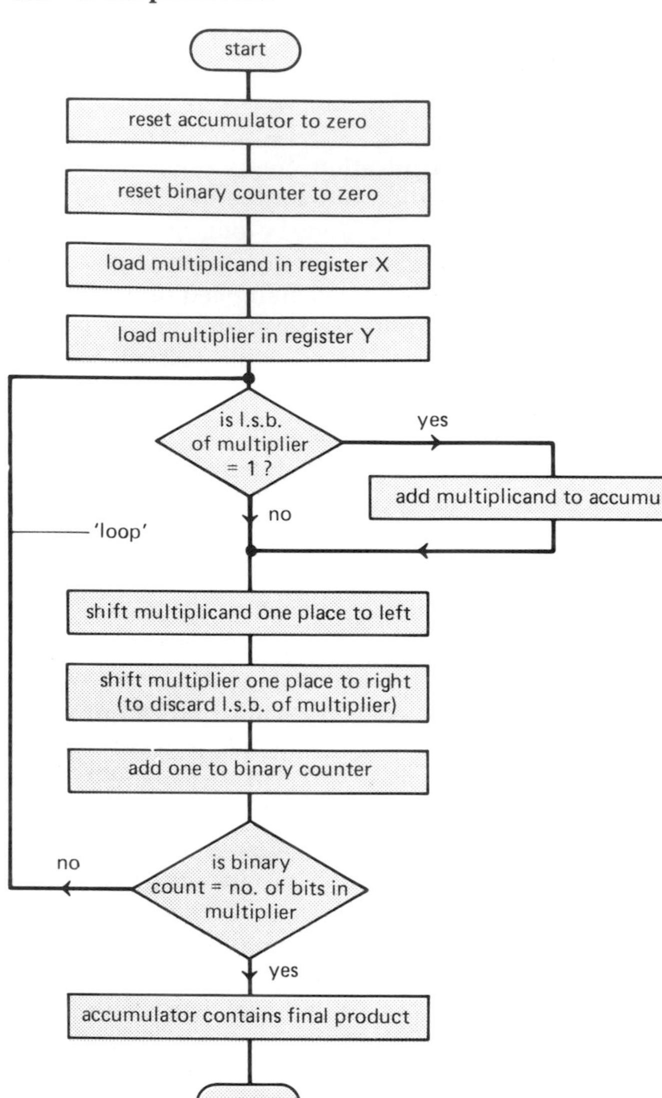

Fig. 101.1

(b) Flowchart. A flowchart for binary multiplication is given in Fig. 101.1. The 'shift-and-add' cycle is repeated (as shown by the 'loop') until every bit in the multiplier has multiplied the multiplicand. Note that 'terminal' symbols (start, stop) are oval, 'operation' ones are rectangular, while those involving a 'decision' are diamond-shaped.

102 Microprocessors

Introduction

A microprocessor (MPU or μP) is a miniature version of the CPU of a digital computer, i.e. ALU, registers and control unit. It is an LSI chip containing thousands of transistors, Fig. 102.1, developed in the early 1970s when, as ICs became more complex and specialized, the need was felt for a general purpose device, suitable for a wide range of jobs.

Fig. 102.1

Its versatility is due to the fact that it is *program-controlled*. Simply by changing the program it can be used as the 'brain' not only of a microcomputer but of a calculator, a cash register, a washing machine, a juke box or a petrol pump. Alternatively, it will control traffic lights or an industrial robot. The market for such a flexible device is much greater than for a 'dedicated' chip doing just one job.

Microprocessors can now only be designed by using computers which probably contain microprocessors. Note that an MPU is not a computer itself—to be a microcomputer it needs memories and input/output units. However the trend is to incorporate as many as possible of the peripheral support chips into the MPU package.

There are many MPUs on the market, with instruction sets for various tasks and which operate on words of different lengths (usually four-, eight- or sixteen-bit). Some cost just a few pounds and are often the cheapest part of a system. They are often housed in 40 pin 0.6 inch wide d.i.l. packages and operate from 5 V and/or 12 V power supplies. The first useful MPU was the 8080; two popular types are the Z80A (used in the Sinclair ZX *Spectrum*) and the 6502 (used in the BBC *Acorn*).

Architecture and action of an MPU

The simplified block diagram in Fig. 102.2 of a typical microprocessor chip can be used to outline how it works. Assume it is programmed with the necessary instructions and data to add two numbers and that it has been 'reset', either manually by a switch or automatically when the power is applied, by a signal to its reset input. The *program counter* then reads zero, all *registers* contain 0s and the *clock* has stopped.

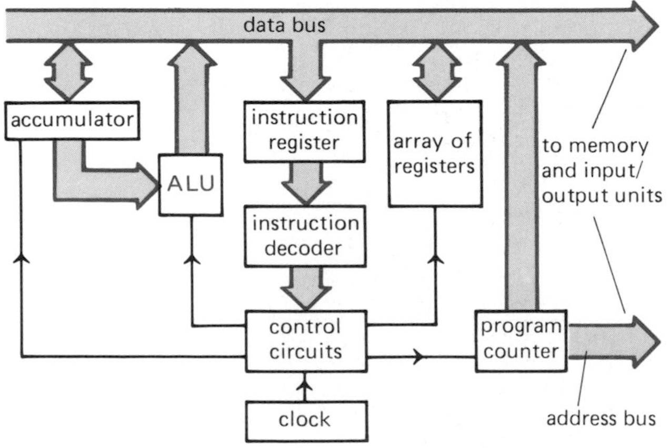

Fig. 102.2

(a) Action. The *program counter* is connected by the *address bus* to the *memory*, e.g. a ROM or PROM, in which the instructions are stored. The binary output (count) from the *counter* is the address input to the *memory* and is initially all 0s (e.g. 0000 0000 for an MPU with an eight-bit *address bus*). The instruction in this first address is thus 'read' out of the *memory* into the MPU via the *data bus*.

The instruction is held by the *instruction register* (until another is received) whose outputs are changed by the *instruction decoder* (p. 145) into a signal that goes to the *control circuits*. By opening and closing logic gates these set up the routes that enable the MPU to perform the operation required by the first instruction. Suppose it is LOAD.

If the *clock* is now started and advances the *program counter* by one count, the program advances by one step (line) and the data stored at the second address (e.g. 0000 0001) is 'read' out of the *memory* and loaded ('copied' is more exact) via the *data bus* into the *accumulator*. (The data will have been entered previously by the programmer and transferred into a RAM).

To obtain addition of the data (number) in the *accumulator* and the data (number) stored at another address in the RAM, the program must give the necessary instructions on succeeding clock pulses to enable the first data to be shifted from the *accumulator* to one of the internal *registers* in the MPU and for the second data to be copied into the *accumulator*.

If the ADD instruction is then given, the *instruction decoder* arranges for the *ALU* to perform the addition (i.e. act as a full-adder) and to store the result in the *accumulator* for subsequent transfer to the *output unit*.

(b) Register stack and subroutines. Sometimes instead of changing the program counter by one, it is useful to 'jump' from the step-by-step sequence of the main program to what is known as a *subroutine*. For example, multiplication involves a fairly long process of 'shift-and-add' (p. 209). To save writing this out every time it is used in a program, it can be written just once as a subroutine, stored and recalled when needed, Fig. 102.3a.

The store used is called a *register stack* because the data is 'stacked' on top of each other in order and then recovered from the top in reverse order, i.e. last-in, first-out, Fig. 102.3b. Only one address is needed, that of the top and if the stack is an external RAM, a *stack pointer* is used to give the first location in the RAM chosen by the programmer as stack.

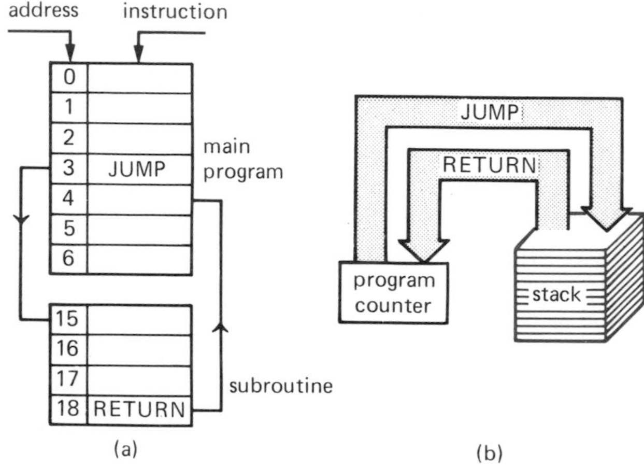

Fig. 102.3

(c) Flags. A flag is a flip-flop which is set to the 'high' (1) state to show that a particular operation has been completed. For instance, in some MPUs the *carry flag* is set to 1 if there is a carry bit in the accumulator and the *zero flag* to 1 if an instruction puts a 0 in it.

Automatic control by an MPU

Many systems and manufacturing processes can be controlled automatically by suitable transducers supplying (via appropriate interface circuits, p. 206) inputs to an MPU whose output takes corrective action when the quantity being monitored goes outside the permitted range.

The principle involved can be shown for the temperature control of a boiler system, Fig. 102.4. Suppose the output from the temperature sensor causes

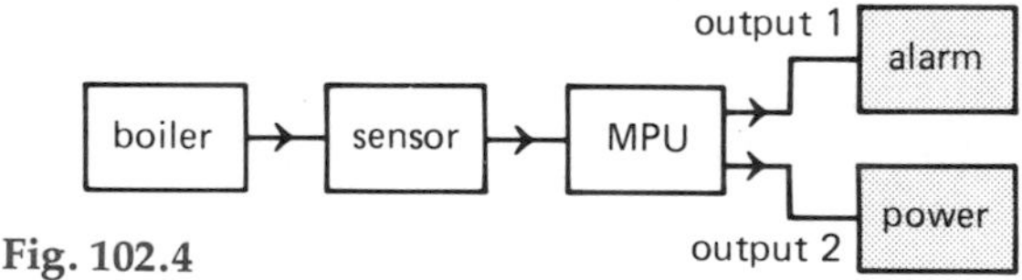

Fig. 102.4

the input to the MPU to become a 1 when the boiler temperature is too high and a 0 when the temperature is normal.

The control program has to ensure that when the input to the MPU is

(i) 1, output 1 is 1, turns *on* the *alarm* AND output 2 is 0, turns *off* the *power*,

(ii) 0, output 1 is 0, turns *off* the *alarm*, AND output 2 is 1, turns *on* the *power*.

This simple problem can be more easily solved in other ways (e.g. using a thermostat and relays) but it demonstrates 'automation by feedback'.

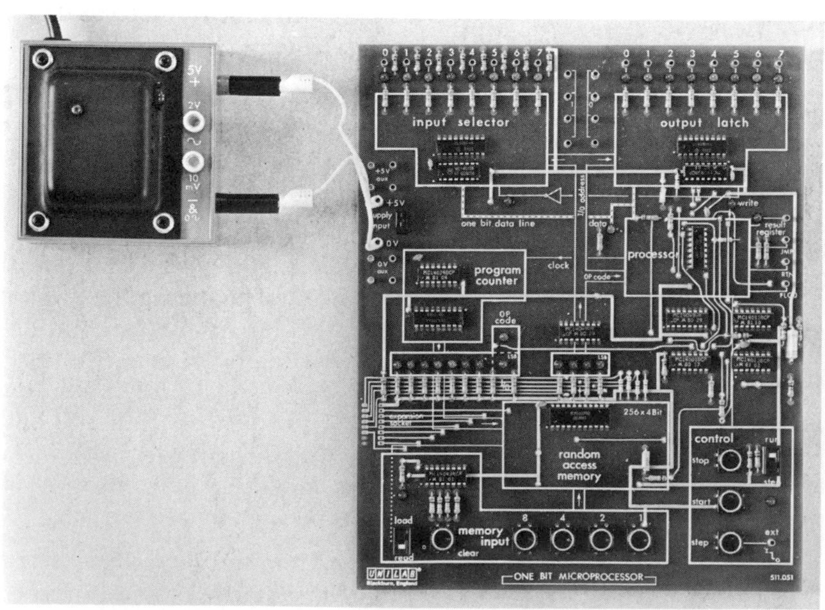

Fig. 102.5

One-bit microprocessor

The microprocessor board in Fig. 102.5 is designed to show how a simple MPU works and to demonstrate some of its uses. Only one bit of data is handled at a time, i.e. the data bus is a single wire, and there are just sixteen instructions in its instruction set (none are arithmetical). Instructions and data addresses are both stored (by the user) in a 256 × 4 bit RAM. It can be programmed to act as any type of combinational logic circuit (and the progress of the program followed in slow motion by observing LEDs that reveal logic levels at different points) or to show automatic control applications, like that described above.[1]

[1] A program is given in *Inside the Micro*, by David L. Thompson, published by Unilab Ltd, Blackburn BB1 9TA, the makers of the board.

103 Progress questions

1. (a) Draw the block diagram for a simple digital computer showing its *four* main parts. Say what each part does.
(b) Distinguish between mainframe, mini- and micro-computers.

2. Explain in a sentence the function of each of the following in a CPU:
(a) clock, (b) program counter, (c) instruction register, (d) instruction decoder, (e) ALU, (f) accumulator, (g) data bus, (h) address bus.

3. (a) State how RAMs, ROMs and PROMs are used in computer systems.
(b) Briefly describe three kinds of back-up memory.

4. Distinguish between the following:
(a) hardware and software,
(b) high-level and low-level languages, and
(c) microprocessor and microcomputer.

5. What is a flowchart? Construct one to get a microcomputer to (i) switch off an electric kettle when it boils and another for it to (ii) control automatically the temperature of the boiler system of Fig. 102.4.

6. What is meant by interfacing? Give an example of where and why it is necessary in a computer system.

7. A Bar Code, like the one on the back of this book, is found on products sold in supermarkets.
(a) Explain how the data from the Bar Code is input to the computer.
(b) Say how this method of data input is beneficial:
(i) to the customer, (ii) to the supermarket.
(*A.E.B. 1982 Control Tech.*)

8. A school possesses a microcomputer that has an integral RAM and an output to a seven-segment LED display.
A company selling computer systems suggests that the school should improve its microcomputer by adding:
(i) a cassette tape recorder backing store,
(ii) an alphanumeric VDU,
(iii) EPROM for routine programs.

Explain carefully how these improvements will affect
(a) the amount of data that can be stored,
(b) the speed at which programs can be dealt with,
(c) the number and complexity of programs that can be used and the ease with which those required frequently can be recalled,
(d) the output that can be obtained from the system.

Explain how such an improved system could be used for:
(e) one task in the running of a school, other than teaching,
(f) one task that would be of benefit at home.
(*A.E.B. 1982 Electronics*)

Practical electronics

Practical work in electronics usually has two broad aims:

(i) to aid the *understanding of basic circuits and theory*,
(ii) to enable *projects to be tackled* involving the design and construction of simple electronic systems.

The 'hardware' available takes many forms.

Fig. A2

Electronic kits

Two kits are briefly described as typical examples.

(a) Locktronics[1], Fig. A1. Circuits are built by plugging components (discrete and ICs) and connecting links into sockets on a baseboard, to give a layout that is easy to follow. It is particularly suitable for experiments on basic electronics, being easily assembled, and flexible.

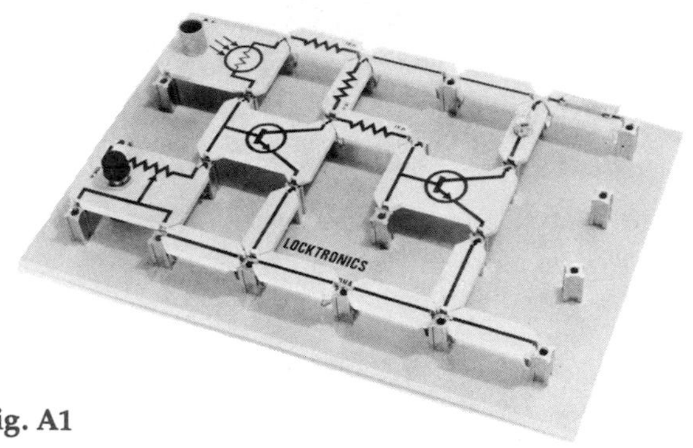

Fig. A1

(b) Alpha system[2], Fig. A2. It consists of a number of separate modules, each with a certain function and a clearly-marked circuit. Their properties can be studied individually or systems can be constructed by joining two or more together, mechanically and electrically, using 'alphalinks'.

Prototype boards

These are solderless circuit boards ('breadboards') on which connections are made by pushing wires and component leads into holes. They are useful for building temporary circuits and have a bracket for mounting switches etc. Several can be interlocked. There are two types.

(a) S-DeC, Fig. A3. It is designed for discrete components only (not ICs) and has 70 contact points arranged in two sections, each having 7 parallel rows of 5 connected contacts.

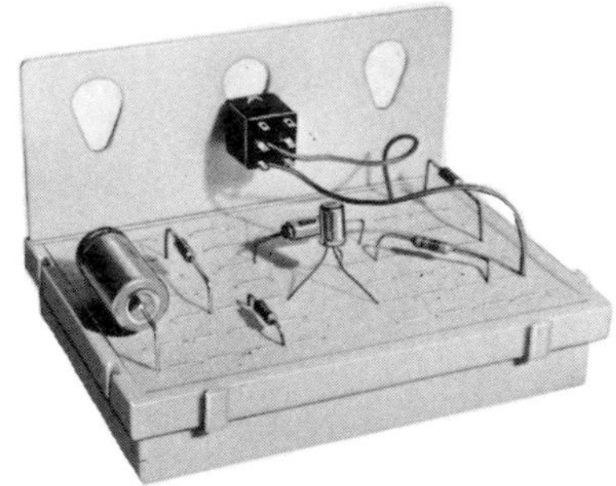

Fig. A3

(b) Professional, Fig. A4. This accepts discrete components and ICs and the large number of contact points (550 on a 6 inch long board) form a 0.1 inch grid, arranged as in S-DeC but also having

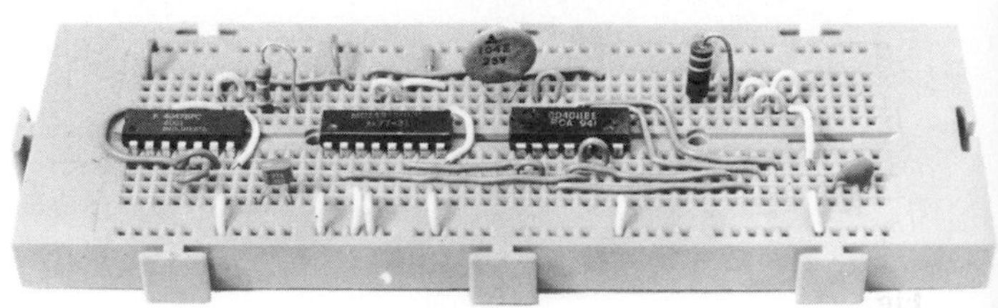

Fig. A4

[1] A. M. Lock, Oldham, OL9 6BR.
[2] Unilab Ltd., Blackburn, BB1 9TA.

continuous contact rows at the top and bottom for use as power supply rails. With the 'holes' being so close together, care is needed to ensure the correct connection is being made.

Soldered boards

'Permanent' circuits require components to be soldered together on some type of insulating (resin-bonded) board pierced with regularly-spaced holes (typically 0.1 inch apart).

(a) Matrix board, Fig. A5a. Single- or double-sided press-fit terminal pins, Fig. A5b, are pushed through the appropriate holes and components soldered to them.

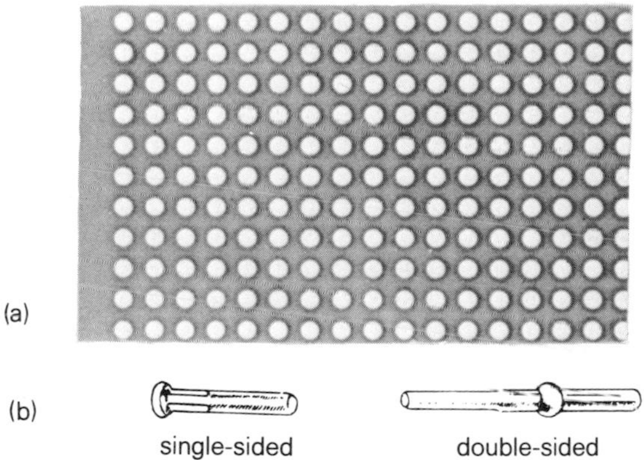

Fig. A5

(b) Stripboard, Fig. A6a. This has pierced copper strips bonded to the board on one side to form the connections between the components. Leads from the latter are inserted from the other side of the board and soldered in place on the coppered side (see Fig. A7). When the copper strip has to be broken (to prevent unwanted connections), a stripboard cutter like that in Fig. A6b is inserted into the hole where the break is required and rotated clockwise a few times. Alternatively an ordinary twist drill held between the fingers and rotated, will do.

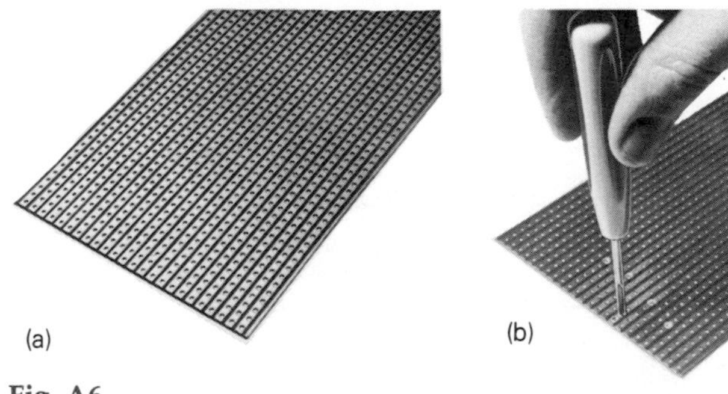

Fig. A6

Soldering

A badly soldered joint can cause trouble which may be hard to trace. It has a dull surface and forms a 'blob', a good joint is shiny with only a little solder in the joint. A 15 W iron with a 1.5 mm or 2 mm tip is suitable for most purposes but with ICs a 1 mm tip is better.

The steps in soldering a resistor to stripboard are shown in Fig. A7. The crocodile clip acts as a 'heat sink' (to prevent heat damage to the resistor and is essential when soldering diodes and transistors). ICs are best mounted on holders soldered to the board.

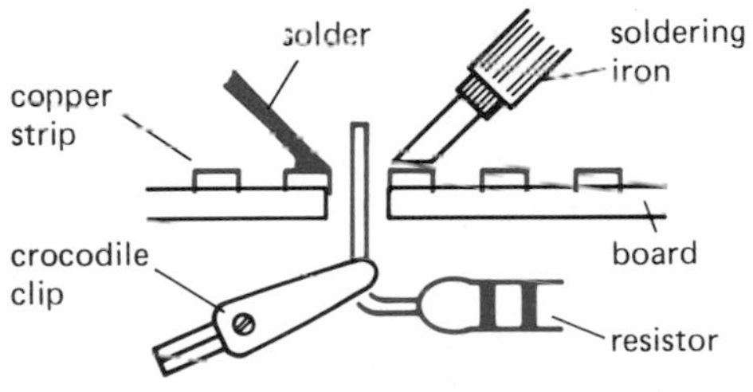

① Tip of iron held on copper strip to heat it

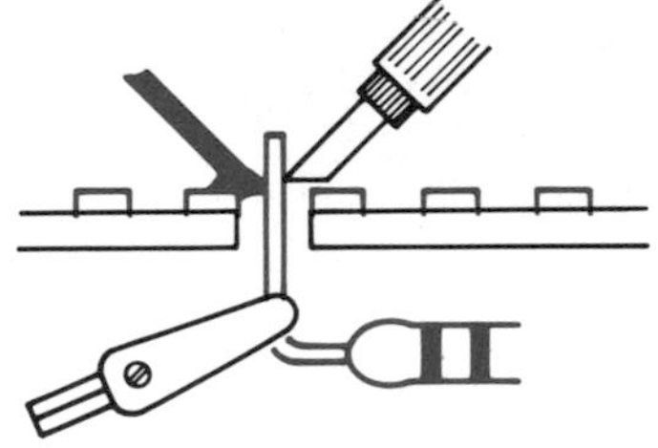

② Tip of iron moved to touch copper strip and resistor till molten solder fills joint

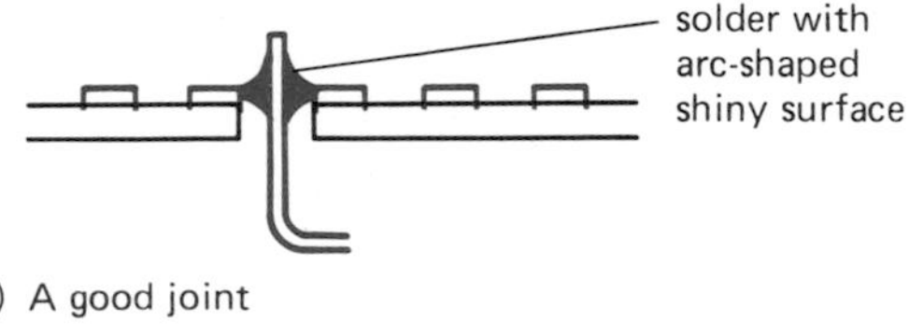

③ A good joint

Fig. A7

Parts to be soldered should be clean (rubbing with emery cloth will ensure this) and grease-free and the joint should not be moved before the solder solidifies.

Answers

1 Electric current
1. **(i)** 5 C **(ii)** 50 C **(iii)** 1500 C
2. **(a)** 5 A **(b)** 0.5 A **(c)** 2 A
3. **(a)** **(i)** 1000 mA **(ii)** 500 mA **(iii)** 20 mA
   (b) **(i)** 2000 μA **(ii)** 400 μA **(iii)** 5 μA
4. $A_2 = A_3 = A_4 = A_5 = 0.5\,A : A_6 = 1\,A$

2 E.M.F., P.D. and voltage
1. **(a)** 12 J **(b)** 36 J
3. **(a)** 1.5 V **(b)** 1.2 V **(c)** 0.3 V
4. **(a)** 4.5 V **(b)** 1.5 V
5. 18 V, 2 V, 8 V
6. **(i)** A = +6 V, B = +3 V, C = 0 V
   (ii) A = 0 V, B = −3 V, C = −6 V
   (iii) A = +3 V, B = 0 V, C = −3 V

3 Resistance and Ohm's law
1. **(a)** 1.8 kΩ **(b)** 5 V **(c)** 2 MΩ
2. **(a)** 4000 Ω **(b)** 2400 Ω **(c)** 6400 Ω
3. **(a)** 75 Ω **(b)** 2 V
4. $I = 3\,A, I_1 = 2\,A, I_2 = 1\,A$
5. **(a)** A, 5 V; B, 5 V **(b)** A, 5 V; B, 0 V

4 Meters and measurement
1. **(i)** Shunt 0.05 Ω **(ii)** Multiplier 9950 Ω
2. **(i)** 1000 Ω **(ii)** 5000 Ω. The 5 V range because its resistance is higher
3. **(i)** 0.5 mA **(ii)** 50 μA
4. **(i)** 4 V **(ii)** 3 V

5 Potential divider
1. **(a)** 1 V **(b)** 2 V **(c)** 2 V **(d)** 6 V
2. 6 V
3. 4 V

6 Electric power
1. **(a)** 1/10 A **(b)** 2/10 A **(c)** 1/60 A
2. **(a)** 3 kΩ **(b)** $27/10^3$ W = 27 mW
3. **(a)** 4.0 V, 2.5 Ω **(b)** 2.5 Ω, 1.6 W

7 Alternating current
1. **(a)** 2 ms **(b)** 500 Hz **(c)** 3 V **(d)** 2.1 V
2. **(a)** 12 V **(b)** 17 V **(c)** 24 W
3. **(a)** 3 : 1 **(b)** 1 : 4
4. 10 V steady d.c. + 5 V peak a.c.

8 Progress questions
2. A1 = 0.6 A; A2 = 0.4 A; A3 = 0.2 A; A4 = 0.6 A
3. **(b)** 3 mA, 6 V **(c)** 12 mA, 12 V **(d)** 2 mA, 4 V **(e)** 4 mA, 4 V
4. **(i)** Resistance of metal wire increases, therefore current decreases
   (ii) Resistance of semiconductor decreases, therefore current increases
5. 22.5 Ω; 22.2 cm
6. Multiplier of 4995 Ω in series with meter
7. **(i)** Fig. 8.4a **(ii)** Fig. 8.4b
8. $\alpha = (R_\theta - R_0)/R_0\theta = (3150 - 320)/(320 \times 1700) = 5.2 \times 10^{-3}$ per °C
9. **(a)** 9 V **(b)** 3 V **(c)** 0 V
10. **(a)** **(i)** 6 V, 4 Ω **(ii)** 0.25 A **(iii)** 1.25 W
    (b) **(i)** 4 Ω **(ii)** 2.25 W
11. **(a)** 4 Ω **(b)** 0.125 A (1/8) **(c)** 1.75 V
12. With voltmeter (8/13) × 9 = 5.54 V
    Without voltmeter 6 V
13. **(c)** 2*E*, *E* **(d)** **(i)** 2.4 V **(ii)** 1.6 V
14. 7 V; 50 Hz

9 Resistors
1. $R_1 = 1\,k\Omega \pm 10\%; R_2 = 47\,k\Omega \pm 5\%; R_3 = 560\,k\Omega \pm 20\%$
2. **(a)** brown green brown silver **(b)** brown black black gold
   (c) orange white red silver **(d)** brown black orange silver
   (e) orange orange yellow **(f)** brown black green silver
3. **(a)** 2.2 kΩ ±20% **(b)** 270 kΩ ±5% **(c)** 1 MΩ ±10% **(d)** 15 Ω ±10%
4. **(a)** 100RJ **(b)** 4K7M **(c)** 100KK **(d)** 56KM
5. **(a)** 1.2 kΩ **(b)** 4.7 kΩ **(c)** 68 kΩ **(d)** 330 kΩ

10 Capacitors
1. **(a)** 1 C **(b)** 2 μF **(c)** 5 V
2. **(a)** 1000 μC = 10^{-3} C **(b)** 0.25 J
3. **(a)** 6.9 μF **(b)** 50 μF
4. **(a)** 159 Ω **(b)** 0.159 Ω
5. **(b)** 7.5 mA

11 Inductors
1. Resistance
2. R same: X_C decreases: X_L increases
3. **100 mH**
4. **(a)** 9.4 kΩ **(b)** 6.3 kΩ
5. **(a)** 3 Ω **(b)** 90 Ω **(c)** 90 Ω **(d)** 0.29 H (290 mH)

12 CR and LR circuits
1. 12 V
2. **(a)** 1 s **(b)** 5 s
3. **(a)** **(i)** 4 V **(ii)** $5\frac{1}{3}$ V **(iii)** 6 V
   (b) **(i)** 2 V **(ii)** 2/3 V **(iii)** **0 V**
4. **(a)** **(i)** 3 V **(ii)** 1 V **(iii)** 0 V
   (b) **(i)** −3 V **(ii)** −1 V **(iii)** 0 V
5. **(a)** 10 mA **(b)** 1500 μC **(c)** 0.15 s
6. **(a)** −6 V **(b)** +6 V **(c)** −12 V

13 LCR circuits
1. 10 Ω
2. 159 kHz

14 Transformers
1. **(a)** 20 : 1 **(b)** 1600 **(c)** $V_p \times I_p = V_s \times I_s, \therefore 240 \times I_p = 12 \times 2$, i.e. $I_p = 24/240 = 0.1$ A

16 Progress questions
1. 22.4 mA
2. **(a)** brown, black, red
   (b)
   (i) **(ii)**

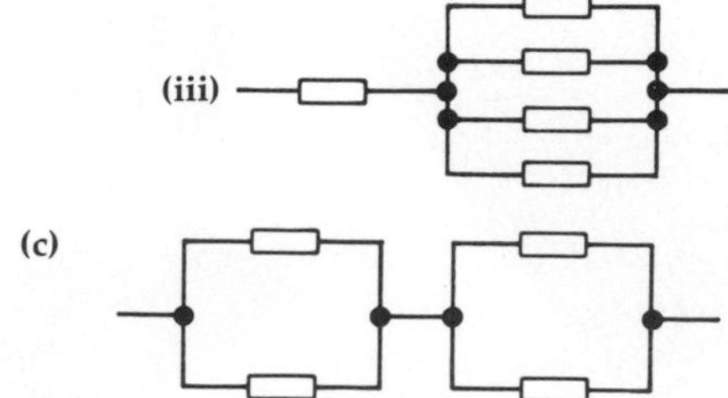

3. 2 μC
4. **(a)** 200 μC = 2×10^{-4} C **(b)** 10^{-2} J
5. **(a)** 4 μF **(b)** 200 μC **(c)** 40 V

6. **(a)** 17.5 mA **(b)** 20 Ω **(c)** 0.25 A
7. 4.5 mA; 100 Hz
8. **(i)** 1 mA **(ii)** 10 μC **(iii)** 10^{-2} s
9. **(b)** 9 V **(c)** 3 V
12. 1.27 μF
13. 15.9 MHz to 7.1 MHz
14. **(i) (a)** 6.3 μA **(b)** 100 μA
    (ii) (a) leads by about 90° since $X_C(10^6/2\pi\ \Omega)$ is large compared with R ($10^4\ \Omega$) and circuit as a whole is capacitive; **(b)** current and voltage are almost in phase since X_c ($1/2\pi\ \Omega$) is small compared with R ($10^4\ \Omega$) and circuit as a whole is resistive
15. 240/20 = 12/1
16. 80 V, 1A assuming no energy loss

25 Progress questions

3. **(b)** 45°C
4. **(ii)** 50 mA **(iii)** 100 mW

27 Junction diode

1. **(a)** L_1 bright, L_2 off;
   (b) L_1 bright, L_2 off because the diode is forward biased and offers an easy path for the current to bypass L_2;
   (c) L_1 and L_2 both dim because the reverse biased diode forces current through L_2 and there is 4.5 V across each lamp—they need 6 V (or more) to be bright
2. **(a)** 5 Ω **(b)** 1 W

28 Other diodes

1. From 3 V to 6 V the brightness gradually increases and is then 'tied' at that brightness by the Zener diode from 6 V to 9 V
2. 0.5 A (500 mA)
3. 680 Ω

29 Progress questions

2. **(a)** 1 pole 4 way **(b) (i)** L_1 and L_4 **(ii)** L_1, L_2 and L_4 **(iii)** L_3 **(iv)** L_4
3. **(b)** 0.04 A (40 mA) **(i)** current increases, p.d. = 6 V **(ii)** current zero, p.d. = 4 V
4. **(iii)** 5 V
5. **(a)** C **(c)** I = (9-2)/680 = 7/680 A ≈ 0.01 A ≈ 10 mA

30 Transistor as a current amplifier

1. **(a)** ASRBE **(b)** ALCE
2. **(b)** and **(d)**
3. **(a)** 50 **(b)** 102 mA
4. 10 mA
5. 80 × 100 = 8000

31 Transistor as a switch

1. **(a)** 0.5 mA **(b)** 40 mA **(c)** no, because at saturation, p.d. across transistor is 0 V and 5 V across 100 Ω, therefore saturation collector current = 5 V/100 Ω = 0.05 A = 50 mA
2. **(a)** 0 V **(b)** 6 V
3. **(a) (i)** 0 V **(ii)** 6 V
   (b) (i) 6 V **(ii)** 0 V
   (c) (i) large **(ii)** small
   (d) (i) 0 to +0.6 V **(ii)** ≈ +0.6 V
   (e) (i) To prevent excessive base currents destroying the transistor if R_1 is made zero. **(ii)** To act as a 'load' which causes the collector-emitter (output) voltage to change from 6 V to 0 V when the transistor switches fully on

32 More about transistors

1. **(i)** 100 **(ii)** 2 kΩ **(iii)** 50 kΩ **(iv)** 5 V
2. It is the maximum p.d. between the collector and emitter when the transistor is saturated by a collector current of 150 mA and a base current of 15 mA. Important in switching circuits where V_{CE} should be as small as possible (ideally 0 V) to reduce power loss

35 Progress questions

1. **(a)** Lamp gradually reaches full brightness (if R_2 not too large).
   (b) (i) h_{FE} = 1 A/10 mA = 1000 mA/10 mA = 100
   (ii) 1 A + 10 mA = 1.01 A
2. I_B = 10 V/500 kΩ = 1/50 mA ∴ I_C = 100 × 1/50 = 2 mA
   V_{CE} = 10 V − 2 mA × 4.5 kΩ = 10 V − 9 V = 1 V
3. Max I_C = 9 V/1 kΩ = 9 mA = 9/10³ A
   (a) (max $I_C/2)^2$ × 1 kΩ W = $(9/(2 \times 10^3))^2 \times 10^3$ W = $(81 \times 10^3)/(4 \times 10^6)$ = 81/4 × 10^{-3} W = 81/4 mW ≈ 20 mW
   (b) total power supplied − power in resistor ≈ (9 V × 4.5 mA − 20 mW) ≈ 20 mW
4. **(a)** V_{out}/V — 10 — 0 1 2 3 — t/ms **(b)** 10 kΩ 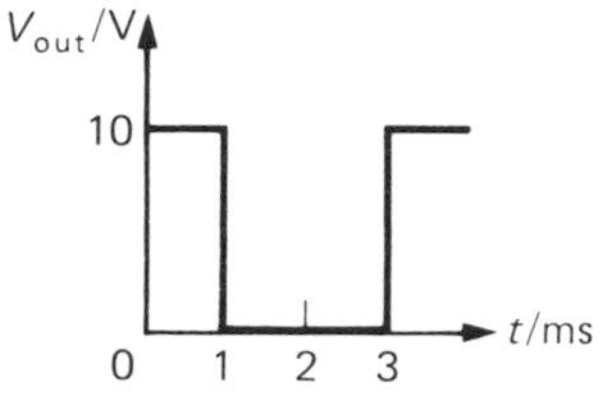
5. **(a)** 0.4 mA (400 μA) **(b)** 30 kΩ **(c)** see p. 68 **(d)** see Fig. 31.6: replace bell by lamp and make R a variable resistor
6. See p. 67
8. **(i)** 40 **(ii)** 50
9. **(a)** When S is closed the transistor (and L) is on since it can receive base current via S and R_B. When S is opened, initially the p.d. V_C across C is zero and the p.d. V_R across R, is 6 V—the latter keeps the transistor on. As C charges up through R, V_C rises and V_R falls (see chapter 12). Eventually V_R is too small to make the base-emitter p.d. V_{BE} equal to 0.6 V (for a silicon transistor) and the transistor switches off—the delay being determined by the time constant CR. Finally V_C = 6 V and V_R = 0.
   (b) See chapter 31, *Alarm circuits* (c).

36 Electricity in the home

1. No heat is produced in an inductor apart from that due to the resistance of its windings
2. Switch off at socket, remove plug, find fault and put it right
3. **(a)** 3 A **(b)** 13 A **(c)** 13 A
4. 22.5

37 Dangers of electricity: safety precautions

1. In **(a)** the fuse is correctly connected in the live wire, in **(b)** it is incorrectly connected in the neutral wire and the right-hand end is still connected to the live side of the supply
2. The fault current to earth blows the fuse in the live wire in **(a)** but by-passes the fuse in **(b)** as it flows through the earth and not the neutral wire. The fault current is large (due to the low resistance of the earth circuit)
3. The output voltage between either end of the secondary and earth is half the voltage across the whole secondary

39 Sources of E.M.F.

2. 2 A

40 Rectifier circuits

2. During each half-cycle of V_i the current supplied has to pass through two conducting diodes (either D_2 and D_4 or D_1 and D_3) across each of which there is a voltage drop (of about 1 V per diode for silicon)
   When the load current is zero (i.e. R infinite) $V_o = V_i$ since then no current flows through the diodes and so there is no voltage drop across them
3. Half-wave circuit: 50 Hz
   Full-wave circuit: 100 Hz

41 Smoothing circuits

1. **(a)** 50 Hz **(b)** 100 Hz
2. The peak value of the a.c. input, i.e. $\sqrt{2} \times$ r.m.s. value
3. **(i)** 10 V peak **(ii)** 20 V peak

42 Stabilizing circuits

2. **(a)** **(i)** 2 V **(ii)** 3 V **(iii)** 3 V
   (b) **(i)** 0 **(ii)** 10 mA **(iii)** 30 mA
3. **(a)** **(i)** 100 mA **(ii)** 10 mA **(iii)** 90 mA
   (b) **(i)** 100 mA **(ii)** 30 mA **(iii)** 70 mA
   (c) 30 Ω

44 Progress questions

1. $12.5 \times 100/625 = 2\%$
4. **(b)** $(66 \times 10^6\,\text{Js}^{-1} \times 60\,\text{s}) \times 100/(400\,\text{kg} \times 33 \times 10^6\,\text{J kg}^{-1}) = (66 \times 10^6 \times 60 \times 100)\,\text{J}/(400 \times 33 \times 10^6)\,\text{J} = 30\%$
6. **(a)** **(i)** 5 V (less p.d. across one diode)
   (ii) 10 V approx.
   (b) For 50 Hz supply, period $T = 1/50 = 0.02$ s. The time constant CR should be large compared with T for good smoothing. Take $CR = 0.1$ s. Hence $C = 0.1/100\,\text{F} = 0.1 \times 10^6/10^2\,\mu\text{F} = 1000\,\mu\text{F}$.
   If R on open circuit p.d. between P and Q = 5 V peak
7. **(i)** 8.8 mA **(ii)** Current through load = 5.6 mA and diode current falls to $(8.8 - 5.6) = 3.2$ mA
   (iii) If load too small, current through diode falls below value for it to work on breakdown part of its characteristic.
8. **(a)** 60 mA **(i)** Current in Z rises to 80 mA, **(ii)** current through load = 40 mA and current through Z falls from 60 mA to 20 mA. Minimum resistance across AB = 4 V/0.06 A = 66.7 Ω since the current through Z = 0
   (b) See p. 91
9. **(c)** 20 : 1 **(e)** 1675 Ω **(f)** 11.5 V

45 Multimeters

2. $2/3 \times 10\,\text{V} = 6.7\,\text{V} : 5\,\text{V}$

46 Oscilloscopes

1. **(a)** $2 \times 10/2 = 10$ V **(b)** $0.7 \times 10 = 7$ V
2. 50 Hz

48 Progress questions

1. 4.5 kΩ
4. Faulty: D (short-circuits R_2). Not faulty: R_1
6. Sine waveform of varying d.c. with amplitude 0.2 V and frequency 2.5 kHz (period 400 μs)
8. **(i)** 300 Hz **(ii)** 100 Hz
10.

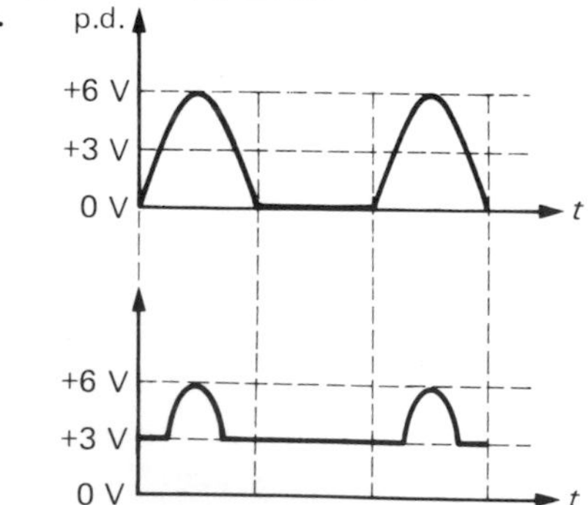

49 Transistor voltage amplifiers I

2. **(a)** 5 V **(b)** 0.02 mA (20 μA) **(c)** 420 kΩ

50 Transistor voltage amplifiers II

1. **(a)** $V_{CC} = 10$ V, $R_L = 2$ kΩ
   (b) $V_{CE} = 5$ V, $I_C = 2.5$ mA, $I_B = 20\,\mu$A
   (c) $(V_{CE} - V_{BE})/I_B = (5 - 0.6)\text{V}/20\,\mu\text{A} = 4.4\,\text{V}/0.02\,\text{mA} = 220\,\text{k}\Omega$
2. **(a)** 9 V/1.8 kΩ = 5 mA
   (b) $I_C = 2.5$ mA, $I_B = 30\,\mu$A, $V_{CE} = 4.5$ V
   (c) Power = $I_C \times V_{CE} = 2.5 \times 4.5 = 11.3$ mW
   (d) **(i)** 2.5 to 6.5 V = 4.5 V ± 2.0 V **(ii)** ±2.0 V peak
   (e) ±20 mV
   (f) 2.0 V/20 mV = 100
   (g) $R_B = \dfrac{V_{CE} - V_{BE}}{I_B} = \dfrac{(4.5 - 0.6)\,\text{V}}{30\,\mu\text{A}} = 130\,\text{k}\Omega$

53 Amplifiers and feedback

3. 40
4. **(a)** 1/100 **(b)** 1/10

54 Amplifiers and matching

2. **(a)** 80 mV **(b)** 4 V **(c)** 2 V **(d)** 1 W.

55 Impedance matching circuits

2. 5/2005 = 1/401 : step-down turns ratio of $\sqrt{2000/5} = \sqrt{400/1} = 20/1$

56 Transistor oscillators

1. 2.25 MHz, 7.12 MHz
2. 500 pF, 125 pF

57 Progress questions

1. **(a)** **(i)** npn **(ii)** A (+) and D (−) **(iii)** E and F
2. **(a)** 2 kΩ **(b)** 4.5 mW **(c)** 4.5 mW : decreased.
3. **(a)** 0.4 V **(b)** 4.6 V **(c)** 0 V
6. 9 (if the open-loop gain is much greater than the closed-loop gain)
8. 150 mV; 5.6 mW
9. **(a)** 2.5 V **(b)** 5/8 mW **(c)** 2 kΩ
11. **(i)** 3 V **(ii)** 2.4 V **(iii)** 4.8 mA
12. 0.01 μF

58 Operational amplifier

2. $\pm 15/10^5\,\text{V} = \pm 150\,\mu\text{V}$

59 Op amp voltage amplifiers

1. **(a)** −2 V **(b)** −9 V (i.e. maximum of supply negative)
2. **(a)** +3 V **(b)** +9 V (i.e. maximum of supply positive)

60 Op amp summing amplifier

1. **(a)** −10 V **(b)** +6 V

62 Op amp integrator

1. **(a)** fall; −6 V/s **(b)** nearly −18 V, after about 3 s

65 Progress questions

1. **(a)** **(i)** −5 V **(ii)** −9 V **(b)** Input resistor $R_i = 1$ kΩ
3. −10 V
8.

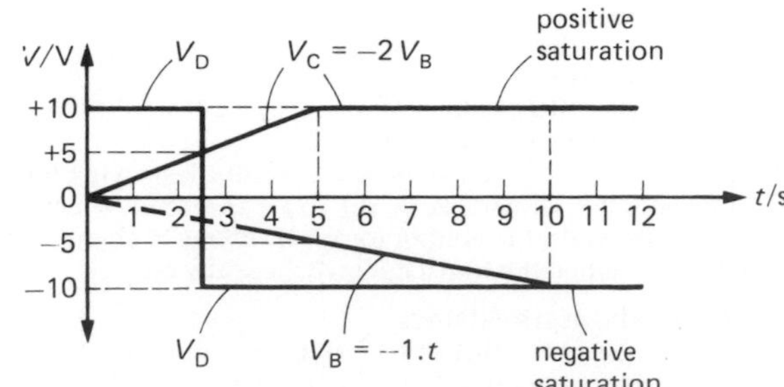

9. **(a)** $V_1 > V_2$ **(b)** **(i)** resistance large **(ii)** $V_2 > V_1$ **(iii)** goes positive **(iv)** switches on **(v)** switches on **(c)** interchange R_1 and LDR

66 Logic gates

1. (a) AND (b) OR
2.

A	B	C	D	F
0	0	0	0	0
1	0	0	0	0
0	1	0	0	0
1	1	0	1	1
0	0	1	0	1
1	0	1	0	1
0	1	1	0	1
1	1	1	1	1

3. (a) (b)

4. AND

67 Logic families

1. (a) *a* and *c*
   (b) *c*, because *a* requires the output of the TTL gate to act as a 10 mA source (i.e. provide the 10 mA required to light the LED); *c* uses the TTL gate to *sink* the 10 mA which lights the LED and this it can easily do.
   (c) 16/1.6 = 400/40 = 10
   (d) *a* and *c*

68 Binary adders

1. (a) 100; 1101; 10101; 100110; 1000000
   (b) 7; 25; 42; 50
   (c) 1001; 0001 0111; 0010 1000; 0011 0111 0000; 0110 0100 0101

69 Logic circuit design I

1. (c)

A	B	P	Q
0	0	1	0
0	1	1	1
1	0	0	0
1	1	1	0

2.

A	B	C	D	E	F	G
0	0	1	1	0	0	1
0	1	1	0	0	1	0
1	0	0	1	1	0	0
1	1	0	0	0	0	1

3. (a) AND gate (b) OR gate

70 Logic circuit design II

1. 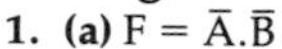
(a) $F = \bar{A}.\bar{B}$

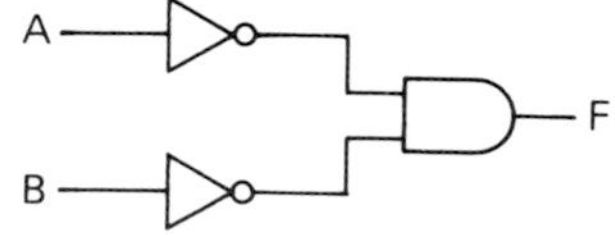

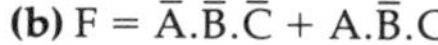
(b) $F = \bar{A}.\bar{B}.\bar{C} + A.\bar{B}.C$

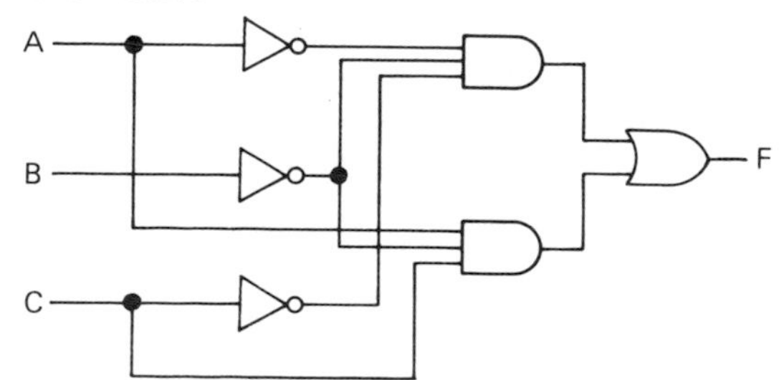

2. (a)

Number	Binary input			Output
	A	B	C	F
0	0	0	0	0
1	0	0	1	1
2	0	1	0	0
3	0	1	1	1
4	1	0	0	0
5	1	0	1	0
6	1	1	0	0
7	1	1	1	0

$$F = \bar{A}.\bar{B}.C + \bar{A}.B.C$$
$$= \bar{A}.C\,(\bar{B} + B) = \bar{A}.C$$

(b)

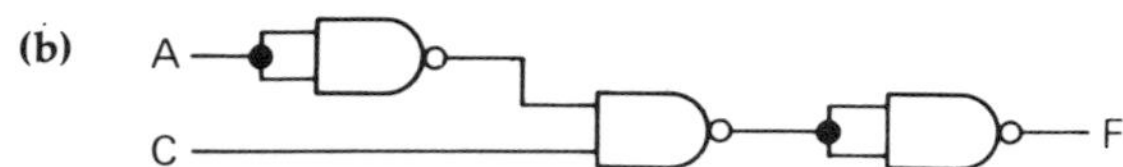

71 Logic circuit design III

1. $F = A.\bar{B}$

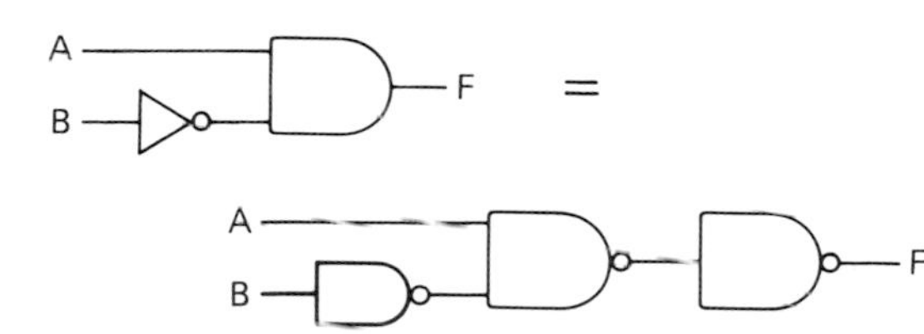

2. $F = A.\bar{B}.\bar{C} + A.B.\bar{C} + \bar{A}.B.C. + A.B.C$
$= A.\bar{C}\,(\bar{B} + B) + B.C\,(\bar{A} + A)$
$= A.\bar{C} + B.C \quad$ since $\bar{B} + B = \bar{A} + A = 1$

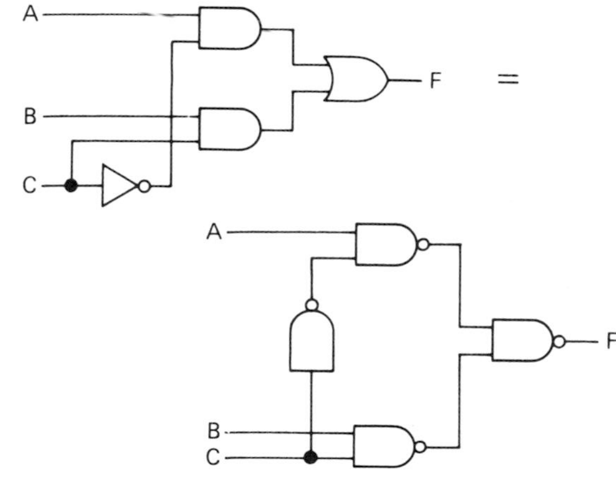

72 Progress questions

2. AND
3. (a) NOT (b) NOR
4. (c) NAND

6.

A	B	C	D	E
0	0	0	0	0
0	1	0	1	1
1	0	1	0	1
1	1	0	0	0

7.

A	B	C	D	E
0	0	1	0	1
0	1	0	0	0
1	0	0	0	0
1	1	0	1	1

8. **(a)**

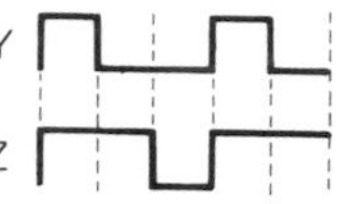

(b) 101101

9. **(b)** F = A.B̄.C + Ā.B.C + Ā.B̄.C̄

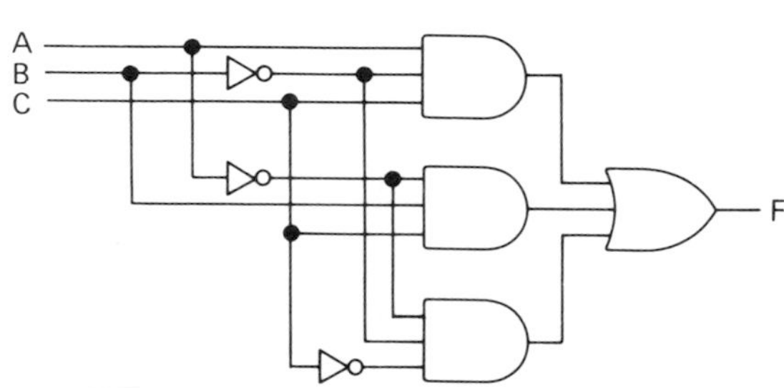

10. **(i)** F = A.B̄.C.D̄.Ē + Ā.B.C.D̄.Ē. + Ā.B̄.C̄.D.E
That is, F is 1 when
A is 1 AND B is 0 AND C is 1 AND D is 0 AND E is 0
OR
A is 0 AND B is 1 AND C is 1 AND D is 0 AND E is 0
OR
A is 0 AND B is 0 AND C is 0 AND D is 1 AND E is 1

(ii)

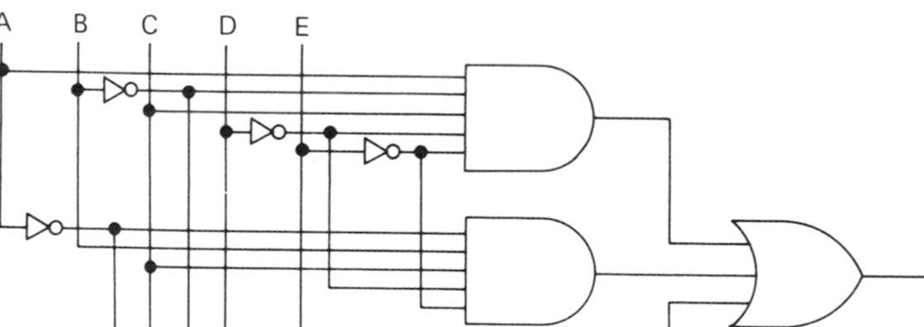

73 Bistable multivibrators I

2. *(a)* 1 **(b)** 0

3. **(b)**

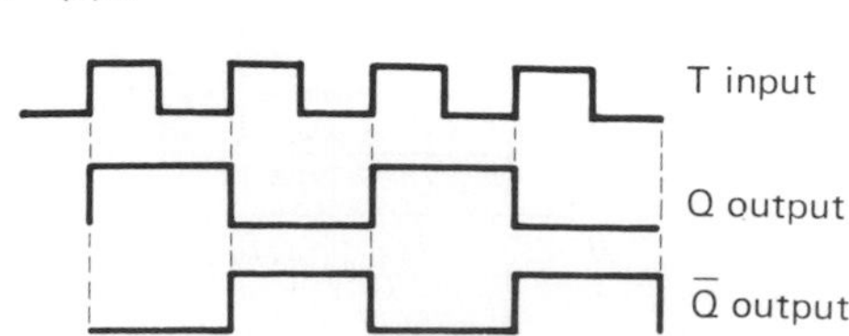

74 Bistable multivibrators II

1.

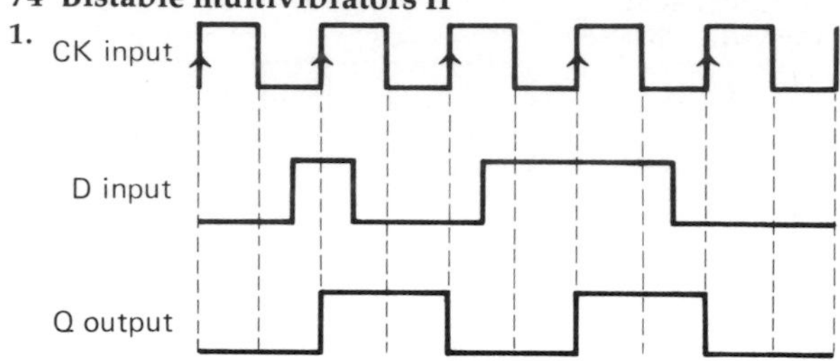

2. 1 Hz

3.

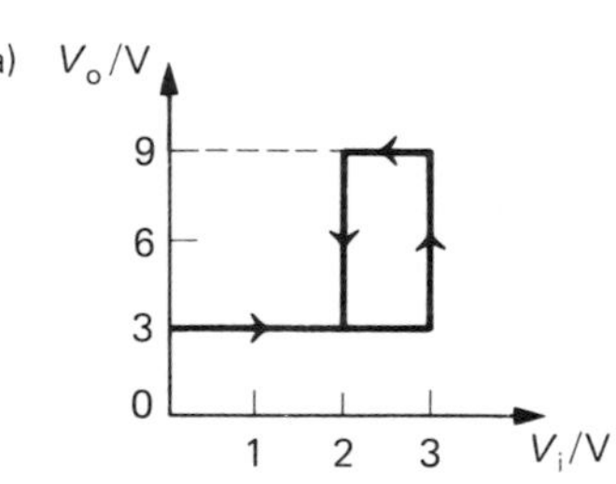

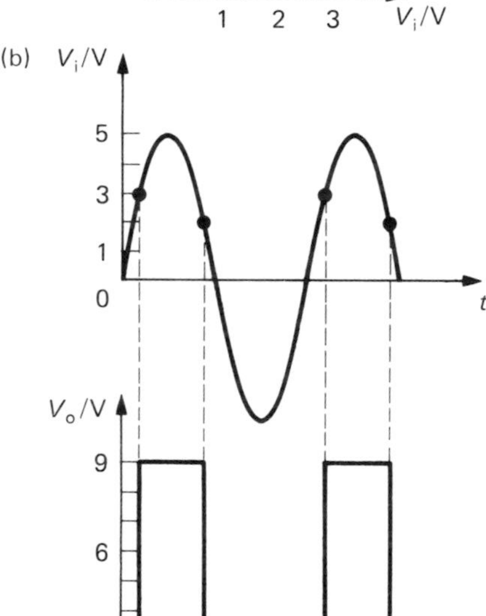

4.

D	CK	S	R	Q
0	1	1	0	0
0	0	1	1	0
1	1	0	1	1
1	0	1	1	1

Q follows D when CK is a 1, therefore it is suitable

75 Astables and monostables

1. **(a)** Astable multivibrator : for both lamps to be on and off for equal times, the time constants C_1R_1 and C_2R_2 should be equal
(b) Make, say, $C_1R_1 = 2\,C_2R_2$ by doubling the value of either C_1 or R_1
(c) Two relays with protective diodes for the transistors (see Fig. 31.6) and a 12 V power supply capable of supplying 4 A at least

2. If Tr_1 comes on first (e.g. because it has a greater h_{FE} than Tr_2), the left plate of C_1 will be near 0 V while the right plate is connected via R_1 to V_{CC}. C_1 therefore starts to charge up from V_{CC} through R_1 and Tr_1 (which is on). When the right plate reaches +0.6 V or so, Tr_2 is turned on and its collector voltage falls rapidly to near 0 V: in so doing it turns Tr_1 off

76 Binary counters

1. **(a)** **(i)** $2^1 = 2$ **(ii)** $2^2 = 4$ **(iii)** $2^3 = 8$ **(iv)** $2^4 = 16$ **(v)** $2^5 = 32$
   (b) 1, 3, 7, 15, 31
2. **(b)** **(i)** 1 **(ii)** 3 **(iii)** 3 **(iv)** 4 **(v)** 5
   (c) $f/10$

77 Registers and memories

1. **(b)** 8
2. **(c)** **(i)** 512 **(ii)** 64

79 Progress questions

2. **(a)** **(i)**

Pulse number	LED A	LED B	LED C
1	1	0	0
2	0	1	0
3	1	1	0
4	0	0	1
5	1	0	1
6	0	1	1
7	1	1	1

(b) **(i)** Tr A **(ii)** S_A **(iii)** rises to about 0.6 V for a silicon transistor **(iv)** lights **(v)** close S_B

3. **(b)** z logic levels are: 2nd 0; 3rd 1; 4th 1; 5th 0
5. **(b)**

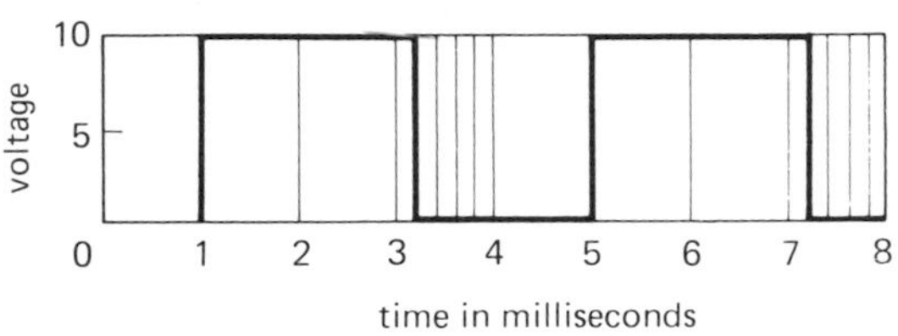

Duration of output pulse = 2.2 ms

81 Communication systems

1. 7000 Hz
2. Lower sideband 790–799 kHz: upper sideband 801–810 kHz

85 Progress questions

4. **(b)** $-1\,M\Omega/1\,M\Omega = -1$
   (c) **(i)** 100 kΩ (more exactly 91 kΩ) **(ii)** 100 kΩ/1 MΩ = 1/10 (more exactly 91/1000)

91 Progress questions

2. **(c)** 1.4 m
6. **(c)** 40 MHz
8. **(c)** **(i)** 312.5 × 50 = 15625 Hz **(ii)** 50 Hz

Index

Index

Index